Engineering Tribology

Engineering Tribology

Edited by
Ross Beckett

Larsen & Keller
www.larsen-keller.com

Engineering Tribology
Edited by Ross Beckett
ISBN: 978-1-63549-279-8 (Hardback)

© 2017 Larsen & Keller

Larsen & Keller

Published by Larsen and Keller Education,
5 Penn Plaza,
19th Floor,
New York, NY 10001, USA

Cataloging-in-Publication Data

Engineering tribology / edited by Ross Beckett.
    p. cm.
Includes bibliographical references and index.
ISBN 978-1-63549-279-8
1. Tribology. 2. Surfaces (Technology). I. Beckett, Ross.
TJ1075 .E54 2017
621.89--dc23

The publisher's policy is to use permanent paper from mills that operate a sustainable forestry policy. Furthermore, the publisher ensures that the text paper and cover boards used have met acceptable environmental accreditation standards.

Printed and bound in the United States of America.

For more information regarding Larsen and Keller Education and its products, please visit the publisher's website www.larsen-keller.com

Table of Contents

Preface

Engineering tribology is a subfield of mechanical engineering and it also has elements of material sciences. It is concerned with the topics like wear, lubrication and friction. It studies the changes and differences which occur in bodies when they interact while being in motion. The aim of this text is to provide students with the basic concepts of engineering tribology. It is complied in such a way that it gives in-depth knowledge of the fundamentals of this subject to the students. Some of the diverse topics covered in this book address the varied branches that fall under this category. This textbook, with its detailed analyses and data, will prove immensely beneficial to students involved in this area at various levels.

To facilitate a deeper understanding of the contents of this book a short introduction of every chapter is written below:

Chapter 1- Tribology is an important branch of mechanical engineering. It studies the science behind interacting surfaces in relative motion. Tribology is an emerging field of study, the following chapter will not only provide an overview, it will also delve deep into the topics related to it.

Chapter 2- The technique that is used to reduce the friction between surfaces by inserting a substance in between them is known as lubrication. The types of lubricants used are ring oilers, motor oils and two-stroke oils. The section on lubrication offers an insightful focus, keeping in mind the complex subject matter.

Chapter 3- The force that causes resistance in two surfaces that come in contact with each other is termed as friction. Friction is the phenomenon behind various devices and concepts found in transportation such as brakes, adhesion railways and road slipperiness. This section is an overview of the topics incorporating all the major aspects of friction.

Chapter 4- Surface tension is a characteristic that is found in a fluid surface; this quality allows the fluid to obtain the least surface area that is possible. The methods and techniques elucidated in this chapter are du Noüy ring method, Wilhelmy plate, spinning drop, maximum bubble pressure method and sessile drop technique.

Chapter 5- Wear is the deformation that results from the exertion of force on a material. Abrasion, galling, adhesion, tribocorrosion and fretting are some of the topics that have been explained in the section. The chapter will provide an integrated understanding of the process of wear.

Chapter 6- The friction and surface tension reducing agents that have been explained in the chapter are bearing, ball bearing, composite bearing, wetting, surface finishing, false brinelling and gear. Bearing is a mechanical device that is used to restrict relative motion to particular desired motion. It also helps in reducing friction between parts that are in motion. The chapter strategically incorporates the main components and key concepts of friction and surface tension reducing agents, providing a complete understanding.

Chapter 7- Stress in quantum mechanics studies the properties of different materials under strain. Stress and strain by themselves also occur in particular ways in particular material. The chapter focuses on topics such as deformation, Cauchy stress tensor and residual stress. The aspects elucidated in the following chapter are of vital importance, and provides a better understanding on stress and deformation in solid materials.

I would like to share the credit of this book with my editorial team who worked tirelessly on this book. I owe the completion of this book to the never-ending support of my family, who supported me throughout the project.

Editor

Introduction to Tribology

Tribology is an important branch of mechanical engineering. It studies the science behind interacting surfaces in relative motion. Tribology is an emerging field of study, the following chapter will not only provide an overview, it will also delve deep into the topics related to it.

Tribology

Tribology is the study of science and engineering of interacting surfaces in relative motion. It includes the study and application of the principles of friction, lubrication and wear. Tribology is a branch of mechanical engineering and materials science.

Etymology

It was coined by the British physicist David Tabor, and also by Peter Jost in 1964, a lubrication expert who noticed the problems with increasing friction on machines, and started the new discipline of tribology.

Fundamentals

The tribological interactions of a solid surface's exposed face with interfacing materials and environment may result in loss of material from the surface. The process leading to loss of material is known as "wear". Major types of wear include abrasion, friction (adhesion and cohesion), erosion, and corrosion. Wear can be minimized by modifying the surface properties of the solids by one or more "surface engineering" processes (also called surface finishing) or by use of lubricants (for frictional or adhesive wear).

Estimated direct and consequential annual loss to industries in the USA due to wear is approximately 1-2% of GDP. (Heinz, 1987). Engineered surfaces extend the working life of both original and recycled and resurfaced equipment, thus saving large sums of money and leading to conservation of material, energy and the environment. Methodologies to minimize wear include systematic approaches to diagnose the wear and to prescribe appropriate solutions. Important methods include:

- Point like contact theory was established by Heinrich Hertz in 1880s.

- Fluid lubrication dynamics was established by Arnold Johannes Sommerfeld in 1900s.

- Terotechnology, where multidisciplinary engineering and management techniques are used to protect equipment and machinery from degradation (Peter Jost, 1972)

- Horst Czichos's systems approach, where appropriate material is selected by checking properties against tribological requirements under operating environment (H. Czichos,1978)

- Asset Management by Material Prognosis - a concept similar to terotechnology which has been introduced by the US Military (DARPA) for upkeep of equipment in good health and start-ready condition for 24 hours. Good health monitoring systems combined with appropriate remedies at maintenance and repair stages have led to improved performance, reliability and extended life cycle of the assets, such as advanced military hardware and civil aircraft.

In recent years,micro- and nanotribology have been gaining ground. Frictional interactions in microscopically small components are becoming increasingly important for the development of new products in electronics, life sciences, chemistry, sensors and by extension for all modern technology.

Friction Regimes

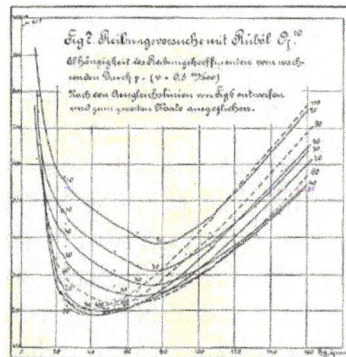

A typical Stribeck curve obtained by Martens

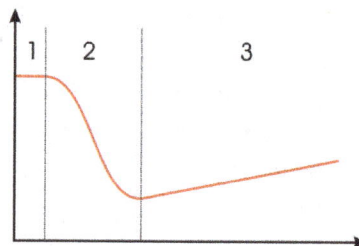

Stribeck curve (Abscissa: Speed, Ordinate: Friction)1. Solid/boundary friction2. Mixed friction3. Fluid friction

Friction regimes for sliding lubricated surfaces have been broadly categorized into:

1. Solid/boundary friction
2. Mixed friction
3. Fluid friction

on the basis of the "Stribeck curve". These curves clearly show the minimum value of friction as the demarcation between full fluid-film lubrication and some solid asperity interactions.

Stribeck and others systematically studied the variation of friction between two liquid lubricated surfaces as a function of a dimensionless lubrication parameter $\eta N/P$, where η is the dynamic viscosity (Ns/m^2), N the sliding speed (m/s) and P the load projected on to the geometrical surface (usually load per unit length of bearing in N/m).

The "Stribeck-curve" has been a classic teaching element in tribology classes.

History

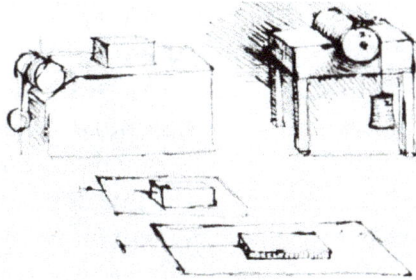

Tribological experiments suggested by Leonardo da Vinci

Duncan Dowson surveyed the history of tribology in his book *History of Tribology* (2nd edition). This comprehensive book covers developments from prehistory, through early civilizations (Mesopotamia, ancient Egypt) and finally the key developments up to the end of the twentieth century.

Historically, Leonardo da Vinci (1452–1519) was the first to enunciate two laws of friction (it was this connection that gave the name to the Leonardo Centre for Tribology, one of the UK's leading research centres on the subject). According to da Vinci, the frictional resistance was the same for two different objects of the same weight but making contacts over different widths and lengths. He also observed that the force needed to overcome friction doubles when the weight doubles. da Vinci's findings remained unpublished in his notebooks. Da Vinci identified the laws of friction in a notebook in 1493 and continued his studies of friction for 20 years.

Guillaume Amontons rediscovered the classic rules (1699), but unlike da Vinci, made his findings public at the Academie Royale des Sciences for verification. They were further developed by Charles-Augustin de Coulomb (1785).

Charles Hatchett (1760–1820) carried out the first reliable test on frictional wear using a simple reciprocating machine to evaluate wear on gold coins. He found that compared to self-mated coins, coins with grits between them wore at a faster rate.

Michael J Neale (1926 - 2012) was a leader the field of Tribology in the mid to late 1900's - For nearly 40 years he specialised in solving problems in machinery design by applying his knowledge of Tribology. Neale was respected as an educator with a gift for integrating theoretical work with his own practical experience to produce easy-to-understand design guides. The Tribology Handbook, which he first edited in 1973 and updated in 1995, is used around the world and forms the basis of numerous training courses for engineering designers.

Stribeck Curve

The "Stribeck curve" or "Stribeck–Hersey curve" (named after Richard Stribeck, who heavily documented and established examples of it, and Mayo D. Hersey), which is used to categorize the friction properties between two surfaces, was developed in the first half of the 20th century. The research of Professor Richard Stribeck (1861–1950) was performed in Berlin at the Royal Prussian Technical Testing Institute (MPA, now BAM). Similar work was previously performed around 1885 by Prof. Adolf Martens (1850–1914) at the same Institute and in the mid-1870s by Dr. Robert H. Thurston at the Stevens Institute of Technology in the U.S. Prof. Dr. Thurston was therefore close to establishing the "Stribeck curve", but he presented no "Stribeck"-like graphs, as he evidently did not fully believe in the relevance of this dependency. Since that time the "Stribeck-curve" has been a classic teaching element in tribology classes.

The graphs of friction force reported by Stribeck stem from a carefully conducted, wide-ranging series of experiments on journal bearings. Stribeck systematically studied the variation of friction between two liquid lubricated surfaces. His results were presented on 5 December 1901 during a public session of the railway society and published on 6 September 1902. They clearly showed the minimum value of friction as the demarcation between full fluid-film lubrication and some solid asperity interactions. Stribeck studied different bearing materials and aspect ratios D/L from 1:1 to 1:2. The maximum sliding speed was 4 m/s and the geometrical contact pressure was limited to 5 MPa.

The reason why the form of the friction curve for liquid lubricated surfaces was later attributed to Stribeck, although both Thurston and Martens achieved their results considerably earlier, (Martens even in the same organization roughly 15 years before), may be because Stribeck published in the most important technical journal in Germany at that time, Zeitschrift des Vereins Deutscher Ingenieure (VDI, Journal of German Mechanical Engineers). Martens published his results "only" in the official journal of the Royal Prussian Technical Testing Institute, which has now become BAM. The VDI journal, as one of the most important journals for engineers, provided wide access to these data and later colleagues rationalized the results into the three classical friction regimes. Thurston however, did not have the experimental means to record a continuous graph of the coefficient of friction but only measured the friction at discrete points; this may be the reason why the minimum in the coefficient of friction was not discovered by him. Instead, Thurston's data did not indicate such a pronounced minimum of friction for a liquid lubricated journal bearing as was demonstrated by the graphs of Martens and Stribeck.

Jost Report

The term *tribology* became widely used following The Jost Report in 1966. The report said that friction, wear and corrosion were costing the UK huge sums of money every year. As a result, the UK set up several national centres for tribology. Since then the term has diffused into the international engineering field, with many specialists now identifying as tribologists.

There are now numerous national and international societies, such as the *Society for Tribologists and Lubrication Engineers* (STLE) in the USA, the *Institution of Mechanical Engineers' Tribology Group* (IMechE Tribology Group) in the UK or the German Society for Tribology (Gesellschaft für Tribologie, www.gft-ev.de) and MYTRIBOS (Malaysian Tribology society).

Most technical universities have researchers working on tribology, often as part of mechanical engineering departments. The limitations in tribological interactions are, however, no longer mainly determined by mechanical designs, but by material limitations. So the discipline of tribology now counts at least as many materials engineers, physicists and chemists as it does mechanical engineers.

New Areas of Tribology

Since the 1990s, new areas of tribology have emerged, including the nanotribology, biotribology, and green tribology. These interdisciplinary areas study the friction, wear and lubrication at the nanoscale (including the Atomic force microscopy and micro/nanoelectromechanical systems, MEMS/NEMS), in biomedical applications (e.g., human joint prosthetics, dental materials), and ecological aspects of friction, lubrication and wear (tribology of clean energy sources, green lubricants, biomimetic tribology).

Recently, intensive studies of superlubricity (phenomenon of vanishing friction) have sparked due to high demand in energy savings. Development of new materials, such as graphene, initiated development of fundamentally new approaches in the lubrication field.

Applications

The study of tribology is commonly applied in bearing design but extends into almost all other aspects of modern technology, even to such unlikely areas as hair conditioners and cosmetics such as lipstick, powders and lip-gloss.

Any product where one material slides or rubs over another is affected by complex tribological interactions, whether lubricated like hip implants and other artificial prostheses, or unlubricated as in high temperature sliding wear in which conventional lubricants cannot be used but in which the formation of compacted oxide layer glazes have been observed to protect against wear.

Tribology plays an important role in manufacturing. In metal-forming operations, friction increases tool wear and the power required to work a piece. This results in increased costs due to more frequent tool replacement, loss of tolerance as tool dimensions shift, and greater forces required to shape a piece. The use of lubricants which minimize direct surface contact reduces tool wear and power requirements.

Nanotribology

Nanotribology is the branch of tribology that studies friction, wear, adhesion and lubrication phenomena at the nanoscale, where atomic interactions and quantum effects are not negligible. The aim of this discipline is characterizing and modifying surfaces for both scientific and technological purposes.

Historically, nanotribological research includes direct investigation with microscopy techniques, such as Scanning Tunneling Microscope (STM), Atomic-Force Microscope (AFM) and Surface Forces Apparatus, (SFA) used to analyze surfaces with extremely high resolution, and thanks to the development of computational methods and power surfaces, we can study these phenomena indirectly as well.

Changing the topology of surfaces at the nanoscale, friction can be either reduced or enhanced more intensively than macroscopic lubrication and adhesion; in this way, superlubrication and superadhesion can be achieved. In micro- and nano-mechanical devices problems of friction and wear, that are critical due to the extremely high surface volume ratio, can be solved covering moving parts with super lubricant coatings. On the other hand, where adhesion is an issue, nanotribological techniques offer a possibility to overcome such difficulties.

History

Friction and wear have been technological issues since ancient periods. On the one hand, the scientific approach of the last centuries towards the comprehension of the underlying mechanisms was focused on macroscopic aspects of tribology. On the other hand, in nanotribology, the systems studied are composed of nanometric structures, where volume forces (such as those related to mass and gravity) can often be considered negligible compared to surface forces. Scientific equipment to study such systems have been developed only in the second half of the 20th century. In 1969 the very first method to study the behavior of a molecularly thin liquid film sandwiched between two smooth surfaces through the SFA was developed. From this starting point, in 1980s researchers would employ other techniques to investigate solid state surfaces at the atomic scale.

Direct observation of friction and wear at the nanoscale started with the first Scanning Tunneling Microscope (STM), which can obtain three-dimensional images of surfaces with atomic resolution; this instrument was developed by Gerd Binnig and Henrich Rohrer in 1981. STM can study only conductive materials, but in 1985 with the invention of the Atomic Force Microscope (AFM) by Binning and his colleagues, also non conductive surfaces can be observed. Afterwards, AFMs were modified to obtain data on normal and frictional forces: these modified microscopes are called Friction Force Microscopes (FFM) or Lateral Force Microscopes (LFM).

From the beginning of the 21st century, computer-based atomic simulation methods have been employed to study the behaviour of single asperities, even those composed by few atoms. Thanks to these techniques, the nature of bonds and interactions in materials can be understood with a high spatial and time resolution.

Surface Analysis

Surface Forces Apparatus

The SFA (*Surface Forces Apparatus*) is an instrument used for measuring physical forces between surfaces, such as adhesion and capillary forces in liquids and vapors, and van der Waals interactions. Since 1969, the year in which the first apparatus of this kind was described, numerous versions of this tool have been developed.

SFA 2000, which has fewer components and is easier to use and clean than previous versions of the apparatus, is one of the currently most advanced equipment utilized for nanotribological purposes on thin films, polymers, nanoparticles and polysaccharides. SFA 2000 has one single cantilever which is able to generate mechanically coarse and electrically fine movements in seven orders of magnitude, respectively with coils and with piezoelectric materials. The extra-fine control enables the user to have a positional accuracy lesser than 1 Å. The sample is trapped by two molecularly smooth surfaces of mica in which it perfectly adheres epitaxially.

Normal forces can be measured by a simple relation:

$$F_{normal}(D) = k(\Delta D_{applied} - \Delta D_{measured})$$

where $\Delta D_{applied}$ is the applied displacement by using one of the control methods mentioned before, k is the spring constant and $\Delta D_{measured}$ is the actual deformation of the sample measured by MBI. Moreover, if $\dfrac{\partial F(D)}{\partial D} > k$

then there is a mechanical instability and therefore the lower surface will jump to a more stable region of the upper surface. And so, the adhesion force is measured with the following formula:

$$F_{adhesion} = k\Delta D_{jump}.$$

Using the DMT model, the interaction energy per unit area can be calculated:

$$W_{flat}(D) = \frac{F_{curved}(D)}{2\pi R}$$

where R is the curvature radius and $F_{curved}(D)$ is the force between cylyndically curved surfaces.

Scanning Probe Microscopy

SPM techniques such as AFM and STM are widely used in nanotribology studies. The Scanning Tunneling Microscope is used mostly for morphological topological investigation of a clean conductive sample, because it is able to give an image of its surface with atomic resolution.

The Atomic Force Microscope is a powerful tool in order to study tribology at a fundamental level. It provides an ultra-fine surface-tip contact with a high refined control over motion and atomic-level precision of measure. The microscope consists, basically, in a high flexible cantilever with a sharp tip, which is the part in contact with the sample and therefore the crossing section must be ideally atomic-size, but actually nanometric (radius of the section varies from 10 to 100 nm). In nanotribology AFM is commonly used for measuring normal and friction forces with a resolution of pico-Newtons.

The tip is brought close to the sample's surface, consequently forces between the last atoms of the tip and the sample's deflect the cantilever proportionally to the intensity of this interactions. Normal forces bend the cantilever vertically up or down of the equilibrium position, depending on the sign of the force. The normal force can be calculated by means of the following equation:

$$F_{normal} = k\Delta V / \sigma$$

where k is the spring constant of the cantilever, ΔV is the output of the photodetector, which is an electric signal, directly with the displacement of the cantilever and σ is the optical-lever sensitivity of the AFM.

On the other hand, lateral forces can be measured with the FFM, which is fundamentally very similar to the AFM. The main difference resides in the tip motion, that slides perpendicularly to its axis. These lateral forces, i.e. friction forces in this case, result in twisting the cantilever, which is controlled to ensure that only the tip touches the surface and not other parts of the probe. At every step the twist is measured and related with the frictional force with this formula:

$$F_{frictional} = \frac{\Delta V k_\phi}{2 h_{eff} \delta}$$

where ΔV is the output voltage, k_ϕ is the torsional constant of the cantilever, h_{eff} is the height of the tip plus the cantilever thickness and δ is the lateral deflection sensitivity.

Since the tip is part of a compliant apparatus, the cantilever, the load can be specified and so the measurement is made in load-control mode; but in this way the cantilever has snap-in and snap-out instabilities and so in some regions measurements cannot be completed stably. These instabilities can be avoided with displacement-controlled techniques, one of this is the interfacial force microscopy.

The tap can be at contact with the sample in the whole measurement process, and this is called contact mode (or static mode), otherwise it can be oscillated and this is called tapping mode (or dynamic mode). Contact mode is commonly applied on hard sample, on which the tip cannot leave any sign of wear, such as scars and debris. For softer materials tapping mode is used to minimize the effects of friction. In this case the tip is vibrated by a piezo and taps the surface at the resonant frequency of the cantilever, i.e. 70-400 kHz, and with an amplitude of 20-100 nm, high enough to allow the tip to not get stuck to the sample because of the adhesion force.

The atomic force microscope can be used as a nanoindenter in order to measure hardness and Young's modulus of the sample. For this application, the tip is made of diamond and it is pressed against the surface for about two seconds, then the procedure is repeated with different loads. The hardness is obtained dividing the maximum load by the residual imprint of the indenter, which can be different from the indenter section because of sink-in or pile-up phenomena. The Young's modulus can be calculated using the Oliver and Pharr method, which allows to obtain a relation between the stiffness of the sample, function of the indentation area, and its Young's and Poisson's moduli.

Atomistic Simulations

Computational methods are particularly useful in nanotribology for studying various phenomena, such as nanoindentation, friction, wear or lubrication. In an atomistic simulation, every single atom's motion and trajectory can be tracked with a very high precision and so this information can be related to experimental results, in order to interpret them, to confirm a theory or to have access to phenomena, that are invisible to a direct study. Moreover, many experimental difficulties do not exist in an atomistic simulation, such as sample preparation and instrument calibration. Theoretically every surface can be created from a flawless one to the most disordered. As well as in the other fields where atomistic simulations are used, the main limitations of these techniques relies on the lack of accurate interatomic potentials and the limited computing power. For this reason, simulation time is very often small (femtoseconds) and the time step is limited to 1 fs for fundamental simulations up to 5 fs for coarse-grained models.

It has been demonstrated with an atomistic simulation that the attraction force between the tip and sample's surface in a SPM measurement produces a jump-to-contact effect. This phenomenon has a completely different origin from the snap-in that occurs in load-controlled AFM, because this latter is originated from the finite compliance of the cantilever. The origin of the atomic resolution of an AFM was discovered and it has been shown that covalent bonds form between the tip and the sample which dominate van der Waals interactions and they are responsible for a such high resolution. Simuling an AFM scansion in contact mode, It has been found that a vacancy or an adatom can be detected only by an atomically sharp tip. Whether in non-contact mode vacancies and adatoms can be distinguished with the so-called frequency modulation technique with a non-atomically sharp tip. In conclusion only in non-contact mode can be achieved atomic resolution with an AFM.

Properties

Friction

Friction, the force opposing to the relative motion, is usually idealized by means of some empirical laws such as Amonton's First and Second laws and Coulomb's law. At the nanoscale, however, such laws may lose their validity. For instance, Amonton's second law states that friction coefficient is independent from the area of contact. Surfaces, in general, have asperities, that reduce the real area of contact and therefore, minimizing such area can minimize friction.

During the scanning process with an AFM or FFM, the tip, sliding on the sample's surface, passes through both low (stable) and high potential energy points, determined, for instance, by atomic positions or, on a larger scale, by surface roughness. Without considering thermal effects, the only force that makes the tip overcome these potential barriers is the spring force given by the support: this causes the stick-slip motion.

At the nanoscale, friction coefficient depends on several conditions. For example, with light loading conditions, tend to be lower than those at the macroscale. With higher loading conditions, such coefficient tends to be similar to the macroscopic one. Temperature and relative motion speed can also affect friction.

Lubricity and Superlubricity at the Atomic Scale

Lubrication is the technique used to reduce friction between two surfaces in mutual contact. Generally, lubricants are fluids introduced between these surfaces in order to reduce friction.

However, in micro- or nano-devices, lubrication is often required and traditional lubricants become too viscous when confined in layers of molecular thickness. A more effective technique is based on thin films, commonly produced by Langmuir-Blodgett deposition, or self-assembled monolayers

Thin films and self-assembled monolayers are also used to increase adhesion phenomena.

Two thin films made of perfluorinated lubricants (PFPE) with different chemical composition were found to have opposite behaviors in humid environment: hydrophobicity increases the adhesive force and decreases lubrication of films with nonpolar end groups; instead, hydrophilicity has the opposite effects with polar end groups.

Superlubricity

"Superlubricity is a frictionless tribological state sometimes occurring in nanoscale material junctions".

At the nanoscale, friction tends to be non isotropic: if two surfaces sliding against each other have incommensurate surface lattice structures, each atom is subject to different amount of force from different directions . Forces, in this situation, can offset each other, resulting in almost zero friction.

The very first proof of this was obtained using a UHV-STM to measure. If lattices are incommensurable, friction was not observed, however, if the surfaces are commensurable, friction force is present. At the atomic level, these tribological properties are directly connected with superlubricity.

An example of this is given by solid lubricants, such as graphite, MoS_2 and Ti_3SiC_2: this can be explained with the low resistance to shear between layers due to the stratified structure of these solids.

Even if at the macroscopic scale friction involves multiple microcontacts with different size and orientation, basing on these experiments one can speculate that a large fraction of contacts will be in superlubric regime. This leads to a great reduction in average friction force, explaining why such solids have a lubricant effect.

Other experiments carried out with the LFM shows that the stick-slip regime is not visible if the applied normal load is negative: the sliding of the tip is smooth and the average friction force seems to be zero

Thermolubricity at the Atomic Scale

With the introduction of AFM and FFM, thermal effects on lubricity at the atomic scale could not be considered negligible any more. Thermal excitation can result in multiple jumps of the tip in the direction of the slide and backward. When the sliding velocity is low, the tip takes a long time to move between low potential energy points and thermal motion can cause it to make a lot of spontaneous forward and reverse jumps: therefore, the required lateral force to make the tip follow the slow support motion is small, so the friction force becomes very low.

For this situation was introduced the term thermolubricity.

Adhesion

Adhesion is the tendency of two surfaces to stay attached together.

The attention in studying adhesion at the micro- and nanoscale increased with the development of AFM: it can be used in nanoindentation experiments, in order to quantify adhesion forces

According to these studies, hardness was found to be constant with film thickness, and it's given by:

$$H = \frac{P_c}{A_c}$$

where A_c is the indentation's area and P_c is the load applied to the indenter.

Stiffness, defined as $S = \frac{dP}{dh}$,

where h is the indentation's depth, can be obtained from r_c, the radius of the indenter-contact line.

$$S = 2 \cdot E' \cdot r_c$$

$$\frac{1}{E'} = \frac{1-v_i^2}{E_i} + \frac{1-v_s^2}{E_s}$$

E' is the reduced Young's modulus, E_i and v_i are the indenter's Young's modulus and Poisson's ratio and E_s, v_s are the same parameters for the sample.

However, r_c can't always be determined from direct observation; it could be deduced from the value of h_c (depth of indentation), but it's possible only if there is no sink-in or pile-up (perfect Sneddon's surface conditions).

If there is sink in, for example, and the indenter is conical the situation is described below.

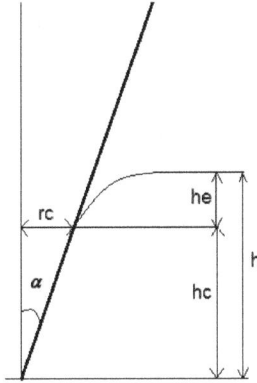

Displacement of the tip (h), elastic displacement of sample surface at the contact line with the indenter (he), contact depth (hc), contact radius (rc) and cone angle (α) of the indenter are shown.

From the image, we can see that:

$$h = h_c + h_e \text{ and } r_c = h_c \cdot \tan \alpha$$

From Oliver and Pharr's study

$$h_e = \epsilon \cdot h$$

where ε depends on the geometry of the indenter; $\epsilon = 1 - \dfrac{2}{\pi}$ if it's conical, $\epsilon = \dfrac{1}{2}$ if it's spherical and $\epsilon = 1$ if it's a flat cylinder.

Oliver and Pharr, therefore, did not consider adhesive force, but only elastic force, so they concluded:

$$F_e = - \cdot E' \cdot \tan \quad \cdot (h - h_f)$$

Considering adhesive force

$$P = F_e + F_a$$

Introducing W_a as the adhesion energy and γ_a as the work of adhesion:

$$W_a = \frac{-\gamma_a \cdot 4 \cdot \tan \alpha}{\pi \cdot \cos \alpha} \cdot h_c^2$$

obtaining

$$F_a = -\frac{\gamma_a \cdot 8 \tan \alpha}{\pi \cdot \cos \alpha} \cdot (h - h_f)$$

In conclusion:

$$P(h) = \frac{2E' \cdot \tan \alpha}{\pi} \cdot (h - h_f)^2 - \frac{\gamma_a \cdot 8 \tan \alpha}{\pi \cdot \cos \alpha} \cdot (h - h_f)$$

The consequences of the additional term of adhesion is visible in the following graph:

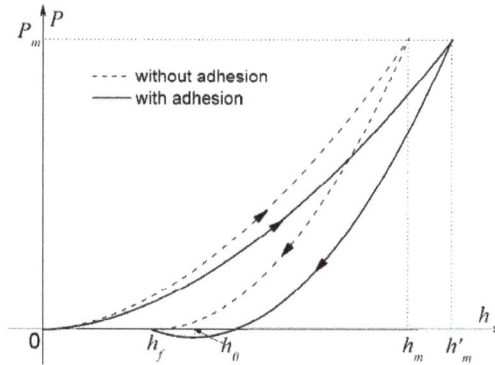

Load-displacement curves that shows the effect of adhesion force

During loading, indentation depth is higher when adhesion is not neglibible: adhesion forces contributes to the work of indentation; on the other hand, during unloading process, adhesion forces opposes indentation process.

Applications of Adhesion Studies

This phenomenon is very important in thin films, because a mismatch between the film and the surface can cause internal stresses and, consequently interface debonding.

When a normal load is applied with an indenter, the film deforms plastically, until the load reaches a critical value: an interfacial fracture starts to develop. The crack propagates radially, until the film is buckled.

On the other hand, adhesion was also investigated for its biomimetic applications: several creatures including insects, spiders, lizards and geckos have developed a unique climbing ability that are trying to be replicated in synthetic materials .

It was shown that a multi-level hierarchical structure produces adhesion enhancement: a synthetic adhesive replicating gecko feet organization was created using nanofabrication techniques and self-assembly.

Wear

Wear is related to the removal and the deformation of a material caused by the mechanical actions. At the nanoscale, wear is not uniform. The mechanism of wear generally begins on the surface of material. The relative motion of two surfaces can cause indentations obtained by the removal and deformation of surface material. Continued motion can eventually grow in both width and depth these indentations.

At the macro scale wear is measured by quantifying the volume (or mass) of material loss or by measuring the ratio of wear volume per energy dissipated. At the nanoscale, however, measuring such volume can be difficult and therefore, it is possible to use evaluate wear by analyzing modifications in surface topology, generally by means of AFM scanning.

References

- H. Czichos, K.-H. Habig, Tribologie-Handbuch (Tribology handbook), Vieweg Verlag, Wiesbaden, 2nd edition, 2003, ISBN 3-528-16354-2

- Duncan Dowson, History of Tribology, Second Edition, Professional Engineering Publishing, 1997, ISBN 1-86058-070-X

- Bhushan, Bharat (2013). Principles and applications of tribology, 2nd edition. New York: John Wiley & Sons, Ltd., Publication. ISBN 978-1-119-94454-6.

- Bhushan, Bharat (2013). Principles and applications of tribology, 2nd edition. New York: John Wiley & Sons, Ltd., Publication. pp. 711–713. ISBN 978-1-119-94454-6.

- Hutchings, Ian M. (2016-08-15). "Leonardo da Vinci's studies of friction". Wear. 360–361: 51–66. doi:10.1016/j.wear.2016.04.019.

- "Study reveals Leonardo da Vinci's 'irrelevant' scribbles mark the spot where he first recorded the laws of friction". Retrieved 2016-07-26.

- Hod, Oded (2012-08-20). "Superlubricity - a new perspective on an established paradigm". Physical Review B. 86 (7). arXiv:1204.3749 . doi:10.1103/PhysRevB.86.075444. ISSN 1098-0121.

- Mitchell, Luke (November 2012). Ward, Jacob, ed. "The Fiction of Nonfriction". Popular Science. No. 5. 281 (November 2012): 40.

Lubrication: A Comprehensive Study

The technique that is used to reduce the friction between surfaces by inserting a substance in between them is known as lubrication. The types of lubricants used are ring oilers, motor oils and two-stroke oils. The section on lubrication offers an insightful focus, keeping in mind the complex subject matter.

Lubrication

Lubrication of the ship steam engine crankshaft. The two bottles of lubricant are attached to the piston and move while the engine is operating.

Lubrication is the process or technique employed to reduce friction between, and wear of one or both, surfaces in proximity and moving relative to each other, by interposing a substance called a lubricant in between them. The lubricant can be a solid, (e.g. Molybdenum disulfide MoS_2) a solid/liquid dispersion, a liquid such as oil or water, a liquid-liquid dispersion (a grease) or a gas.

With fluid lubricants the applied load is either carried by pressure generated within the liquid due to the frictional viscous resistance to motion of the lubricating fluid between the surfaces, or by the liquid being pumped under pressure between the surfaces.

Lubrication can also describe the phenomenon where reduction of friction occurs unintentionally, which can be hazardous such as hydroplaning on a road.

The science of friction, lubrication and wear is called tribology.

Adequate lubrication allows smooth continuous operation of equipment, reduces the rate of wear, and prevents excessive stresses or seizures at bearings. When lubrication breaks down, components can rub destructively against each other, causing heat, local welding, destructive damage and failure.

The Regimes of Lubrication

As the load increases on the contacting surfaces three distinct situations can be observed with respect to the mode of lubrication, which are called regimes of lubrication:

- Fluid film lubrication is the lubrication regime in which, through viscous forces, the load is fully supported by the lubricant within the space or gap between the parts in motion relative to one another object (the lubricated conjunction) and solid–solid contact is avoided.

  - Hydrostatic lubrication is when an external pressure is applied to the lubricant in the bearing, to maintain the fluid lubricant film where it would otherwise be squeezed out.

  - Hydrodynamic lubrication is where the motion of the contacting surfaces, and the exact design of the bearing is used to pump lubricant around the bearing to maintain the lubricating film. This design of bearing may wear when started, stopped or reversed, as the lubricant film breaks down.

- Elastohydrodynamic lubrication: Mostly for nonconforming surfaces or higher load conditions, the bodies suffer elastic strains at the contact. Such strain creates a load-bearing area, which provides an almost parallel gap for the fluid to flow through. Much as in hydrodynamic lubrication, the motion of the contacting bodies generates a flow induced pressure, which acts as the bearing force over the contact area. In such high pressure regimes, the viscosity of the fluid may rise considerably. At full film elastohydrodynamic lubrication the generated lubricant film completely separates the surfaces. Contact between raised solid features, or *asperities*, can occur, leading to a mixed-lubrication or boundary lubrication regime.

- Boundary lubrication (also called boundary film lubrication): The hydrodynamic effects are negligible. The bodies come into closer contact at their asperities; the heat developed by the local pressures causes a condition which is called stick-slip and some asperities break off. At the elevated temperature and pressure conditions chemically reactive constituents of the lubricant react with the contact surface forming a highly resistant tenacious layer, or film on the moving solid surfaces (boundary film) which is capable of supporting the load and major wear or breakdown is avoided. Boundary lubrication is also defined as that regime in which the load is carried by the surface asperities rather than by the lubricant.

- Mixed lubrication: This regime is in between the full film elastohydrodynamic and boundary lubrication regimes. The generated lubricant film is not enough to separate the bodies completely, but hydrodynamic effects are considerable.

Besides supporting the load the lubricant may have to perform other functions as well, for instance it may cool the contact areas and remove wear products. While carrying out these functions the lubricant is constantly replaced from the contact areas either by the relative movement (hydrodynamics) or by externally induced forces.

Lubrication is required for correct operation of mechanical systems pistons, pumps, cams, bearings, turbines, cutting tools etc. where without lubrication the pressure between the surfaces in close proximity would generate enough heat for rapid surface damage which in a coarsened condition may literally weld the surfaces together, causing seizure.

In some applications, such as piston engines, the film between the piston and the cylinder wall also seals the combustion chamber, preventing combustion gases from escaping into the crankcase.

Lubricant

A lubricant is a substance introduced to reduce friction between surfaces in mutual contact, which ultimately reduces the heat generated when the surfaces move. It may also have the function of transmitting forces, transporting foreign particles, or heating or cooling the surfaces. The property of reducing friction is known as lubricity.

In addition to industrial applications, lubricants are used for many other purposes. Other uses include cooking (oils and fats in use in frying pans, in baking to prevent food sticking), bio-medical applications on humans (e.g. lubricants for artificial joints), ultrasound examination, medical examinations, and the use of personal lubricant for sexual purposes.

Properties

A good lubricant generally possesses the following characteristics:

- high boiling point and low freezing point (in order to stay liquid within a wide range of temperature)

- high viscosity index

- thermal stability

- hydraulic stability

- demulsibility

- corrosion prevention

- high resistance to oxidation.

Formulation

Typically lubricants contain 90% base oil (most often petroleum fractions, called mineral oils) and less than 10% additives. Vegetable oils or synthetic liquids such as hydrogenated polyolefins, esters, silicones, fluorocarbons and many others are sometimes used as base oils. Additives deliver reduced friction and wear, increased viscosity, improved viscosity index, resistance to corrosion and oxidation, aging or contamination, etc.

Non-liquid lubricants include grease, powders (dry graphite, PTFE, molybdenum disulphide, tungsten disulphide, etc.), PTFE tape used in plumbing, air cushion and others. Dry lubricants such as graphite, molybdenum disulphide and tungsten disulphide also offer lubrication at temperatures (up to 350 °C) higher than liquid and oil-based lubricants are able to operate. Limited interest has been shown in low friction properties of compacted oxide glaze layers formed at several hundred degrees Celsius in metallic sliding systems, however, practical use is still many years away due to their physically unstable nature.

Additives

A large number of additives are used to impart performance characteristics to the lubricants. The main families of additives are:

- Antioxidants

- Detergents

- Anti-wear

- Metal deactivators

- Corrosion inhibitors, Rust inhibitors

- Friction modifiers

- Extreme Pressure

- Anti-foaming agents

- Viscosity index improvers

- Demulsifying/Emulsifying

- Stickiness improver, provide adhesive property towards tool surface (in metal-working)

- Complexing agent (in case of greases)

Note that many of the basic chemical compounds used as detergents (example: calcium sulfonate) serve the purpose of the first seven items in the list as well. Usually it is not economically or technically feasible to use a single do-it-all additive compound. Oils for hypoid gear lubrication will contain high content of EP additives. Grease lubricants may contain large amount of solid particle friction modifiers, such as graphite, molybdenum sulfide.

Types of Lubricants

In 1999, an estimated 37,300,000 tons of lubricants were consumed worldwide. Automotive applications dominate, but other industrial, marine, and metal working applications are also big consumers of lubricants. Although air and other gas-based lubricants are known (e.g., in fluid bearings), liquid and solid lubricants dominate the market, especially the former.

Lubricants are generally composed of a majority of base oil plus a variety of additives to impart desirable characteristics. Although generally lubricants are based on one type of base oil, mixtures of the base oils also are used to meet performance requirements.

Base Oil Groups

Mineral oil term is used to encompass lubricating base oil derived from crude oil. The American Petroleum Institute (API) designates several types of lubricant base oil:

- Group I – Saturates < 90% and/or sulfur > 0.03%, and Society of Automotive Engineers (SAE) viscosity index (VI) of 80 to 120

  Manufactured by solvent extraction, solvent or catalytic dewaxing, and hydro-finishing processes. Common Group I base oil are 150SN (solvent neutral), 500SN, and 150BS (brightstock)

- Group II – Saturates > 90% and sulfur < 0.03%, and SAE viscosity index of 80 to 120

  Manufactured by hydrocracking and solvent or catalytic dewaxing processes. Group II base oil has superior anti-oxidation properties since virtually all hydrocarbon molecules are saturated. It has water-white color.

- Group III – Saturates > 90%, sulfur < 0.03%, and SAE viscosity index over 120

Manufactured by special processes such as isohydromerization. Can be manu-factured from base oil or slax wax from dewaxing process.

- Group IV 3 b– Polyalphaolefins (PAO)

- Group V – All others not included above such as naphthenics, PAG, l 8oesters.

The lubricant industry commonly extends this group terminology to include:

- Group I+ with a Viscosity Index of 103–108

- Group II+ with a Viscosity Index of 113–119

- Group III+ with a Viscosity Index of at least 140

Can also be classified into three categories depending on the prevailing compositions:

- Paraffinic

- Naphthenic

- Aromatic9

  Lubricants for internal combustion engines contain additives to reduce oxida-tion and improve lubrication. The main constituent of such lubricant product is called the base oil, base stock. While it is advantageous to have a high-grade base oil in a lubricant, proper selection of the lubricant additives is equally as important. Thus some poorly selected formulation of PAO lubricant may not last as long as more expensive formulation of Group III+ lubricant.

Biolubricants Made from Vegetable Oils and Other Renewable Sources

These are primarily triglyceride esters derived from plants and animals. For lubricant base oil use the vegetable derived materials are preferred. Common ones include high oleic canola oil, castor oil, palm oil, sunflower seed oil and rapeseed oil from vegetable, and Tall oil from tree sources. Many vegetable oils are often hydrolyzed to yield the acids which are subsequently combined selectively to form specialist synthetic esters. Other naturally derived lubricants include lanolin (wool grease, a natural water repellent).

Whale oil was a historically important lubricant, with some uses up to the latter part of the 20th century as a friction modifier additive for automatic transmission fluid.

In 2008, the biolubricant market was around 1% of UK lubricant sales in a total lubri-cant market of 840,000 tonnes/year.

Lanolin is a natural water repellent, derived from sheep wool grease, and is an alter-native to the more common petro-chemical based lubricants. This lubricant is also a corrosion inhibitor, protecting against rust, salts, and acids.

Water can also be used on its own, or as a major component in combination with one of the other base oils. Commonly used in engineering processes, such as milling and lathe turning.

Synthetic Oils

- Polyalpha-olefin (PAO)

- Synthetic esters

- Polyalkylene glycols (PAG)

- Phosphate esters

- Alkylated naphthalenes (AN)

- Silicate esters

- Ionic fluids

- Multiply alkylated cyclopentanes (MAC)

Solid Lubricants

PTFE: polytetrafluoroethylene (PTFE) is typically used as a coating layer on, for example, cooking utensils to provide a non-stick surface. Its usable temperature range up to 350 °C and chemical inertness make it a useful additive in special greases. Under extreme pressures, PTFE powder or solids is of little value as it is soft and flows away from the area of contact. Ceramic or metal or alloy lubricants must be used then. "Teflon®" is a brand of PTFE owned by DuPont Co.

Inorganic solids: Graphite, hexagonal boron nitride, molybdenum disulfide and tungsten disulfide are examples of materials that can be used as solid lubricants, often to very high temperature. The use of some such materials is sometimes restricted by their poor resistance to oxidation (e.g., molybdenum disulfide can only be used up to 350 °C in air, but 1100 °C in reducing environments).

Metal/alloy: Metal alloys, composites and pure metals can be used as grease additives or the sole constituents of sliding surfaces and bearings. Cadmium and Gold are used for plating surfaces which gives them good corrosion resistance and sliding properties, Lead, Tin, Zinc alloys and various Bronze alloys are used as sliding bearings, or their powder can be used to lubricate sliding surfaces alone.

Aqueous Lubrication

Aqueous lubrication is of interest in a number of technological applications. Strongly hydrated brush polymers such as PEG can act as lubricants at liquid solid interfaces. By continuous rapid exchange of bound water with other free water molecules, these polymer films keep the surfaces separated while maintaining a high fluidity at the brush–brush interface at high compressions, thus leading to a very low coefficient of friction.

Applications

Lubricants perform the following key functions:

- Keep moving parts apart
- Reduce friction
- Transfer heat
- Carry away contaminants & debris
- Transmit power
- Protect against wear
- Prevent corrosion
- Seal for gases
- Stop the risk of smoke and fire of objects
- Prevent rust.

One of the single largest applications for lubricants, in the form of motor oil, is protecting the internal combustion engines in motor vehicles and powered equipment.

Lubricants such as 2-cycle oil are added to fuels like gasoline which has low lubricity. Sulfur impurities in fuels also provide some lubrication properties, which has to be taken in account when switching to a low-sulfur diesel; biodiesel is a popular diesel fuel additive providing additional lubricity.

Another approach to reducing friction and wear is to use bearings such as ball bearings, roller bearings or air bearings, which in turn require internal lubrication themselves, or to use sound, in the case of acoustic lubrication.

Keep Moving Parts Apart

Lubricants are typically used to separate moving parts in a system. This has the benefit of reducing friction and surface fatigue, together with reduced heat generation, operating noise and vibrations. Lubricants achieve this in several ways. The most common is by forming a physical barrier i.e., a thin layer of lubricant separates the moving parts. This is analogous to hydroplaning, the loss of friction observed when a car tire is separated from the road surface by moving through standing water. This is termed hydrodynamic lubrication. In cases of high surface pressures or temperatures, the fluid film is much thinner and some of the forces are transmitted between the surfaces through the lubricant..

Reduce Friction

Typically the lubricant-to-surface friction is much less than surface-to-surface friction in a system without any lubrication. Thus use of a lubricant reduces the overall system friction. Reduced friction has the benefit of reducing heat generation and reduced formation of wear particles as well as improved efficiency. Lubricants may contain additives known as friction modifiers that chemically bind to metal surfaces to reduce surface friction even when there is insufficient bulk lubricant present for hydrodynamic lubrication, e.g. protecting the valve train in a car engine at startup.

Transfer Heat

Both gas and liquid lubricants can transfer heat. However, liquid lubricants are much more effective on account of their high specific heat capacity. Typically the liquid lubricant is constantly circulated to and from a cooler part of the system, although lubricants may be used to warm as well as to cool when a regulated temperature is required. This circulating flow also determines the amount of heat that is carried away in any given unit of time. High flow systems can carry away a lot of heat and have the additional benefit of reducing the thermal stress on the lubricant. Thus lower cost liquid lubricants may be used. The primary drawback is that high flows typically require larger sumps and bigger cooling units. A secondary drawback is that a high flow system that relies on the flow rate to protect the lubricant from thermal stress is susceptible to catastrophic failure during sudden system shut downs. An automotive oil-cooled turbocharger is a typical example. Turbochargers get red hot during operation and the oil that is cooling them only survives as its residence time in the system is very short (i.e. high flow rate). If the system is shut down suddenly (pulling into a service area after a high-speed drive and stopping the engine) the oil that is in the turbo charger immediately oxidizes and will clog the oil ways with deposits. Over time these deposits can completely block the oil ways, reducing the cooling with the result that the turbo charger experiences total failure, typically with seized bearings. Non-flowing lubricants such as greases and pastes are not effective at heat transfer although they do contribute by reducing the generation of heat in the first place.

Carry Away Contaminants and Debris

Lubricant circulation systems have the benefit of carrying away internally generated debris and external contaminants that get introduced into the system to a filter where they can be removed. Lubricants for machines that regularly generate debris or contaminants such as automotive engines typically contain detergent and dispersant additives to assist in debris and contaminant transport to the filter and removal. Over time the filter will get clogged and require cleaning or replacement, hence the recommendation to change a car's oil filter at the same time as changing the oil. In closed systems such as gear boxes the filter may be supplemented by a magnet to attract any iron fines that get created.

It is apparent that in a circulatory system the oil will only be as clean as the filter can make it, thus it is unfortunate that there are no industry standards by which consumers can readily assess the filtering ability of various automotive filters. Poor filtration significantly reduces the life of the machine (engine) as well as making the system inefficient.

Transmit Power

Lubricants known as hydraulic fluid are used as the working fluid in hydrostatic power transmission. Hydraulic fluids comprise a large portion of all lubricants produced in the world. The automatic transmission's torque converter is another important application for power transmission with lubricants.

Protect Against Wear

Lubricants prevent wear by keeping the moving parts apart. Lubricants may also contain anti-wear or extreme pressure additives to boost their performance against wear and fatigue.

Prevent Corrosion

Good quality lubricants are typically formulated with additives that form chemical bonds with surfaces, or exclude moisture, to prevent corrosion and rust. It reduces corrosion between two metallic surface and avoids contact between these surfaces to avoid immersed corrosion.

Seal for Gases

Lubricants will occupy the clearance between moving parts through the capillary force, thus sealing the clearance. This effect can be used to seal pistons and shafts.

Application by Fluid Types

- Automotive
    - Engine oils
        - Petrol (Gasolines) engine oils
        - Diesel engine oils
    - Automatic transmission fluid
    - Gearbox fluids
    - Brake fluids
    - Hydraulic fluids

- Tractor (one lubricant for all systems)
    - o Universal Tractor Transmission Oil – UTTO
    - o Super Tractor Oil Universal – STOU – includes engine
- Other motors
    - o 2-stroke engine oils
    - o Personal lubricant
- Industrial
    - o Hydraulic oils
    - o Air compressor oils
    - o Food Grade lubricants
    - o Gas Compressor oils
    - o Gear oils
    - o Bearing and circulating system oils
    - o Refrigerator compressor oils
    - o Steam and gas turbine oils
- Aviation
    - o Gas turbine engine oils
    - o Piston engine oils
- Marine
    - o Crosshead cylinder oils
    - o Crosshead Crankcase oils
    - o Trunk piston engine oils
    - o Stern tube lubricants
- Horological

Other Relevant Phenomena

"Glaze" Formation (High Temperature Wear)

A further phenomenon that has undergone investigation in relation to high temperature wear prevention and lubrication, is that of a compacted oxide layer glaze formation. This is the generation of a compacted oxide layer which sinters together to form a crystalline 'glaze' (not the amorphous layer seen in pottery) generally at high temperatures, from metallic surfaces sliding against each other (or a metallic surface against a ceramic surface). Due to the elimination of metallic contact and adhesion by the generation of oxide, friction and wear is reduced. Effectively, such a surface is self-lubricating.

As the "glaze" is already an oxide, it can survive to very high temperatures in air or oxidising environments. However, it is disadvantaged by it being necessary for the base metal (or ceramic) having to undergo some wear first to generate sufficient oxide debris.

Marketing

The global lubricant market is generally competitive with numerous manufacturers and marketers. Overall the western market may be considered mature with a flat to declining overall volumes with growth in the emerging economies. The lubricant marketers generally pursue one or more of the following strategies when pursuing business.

- Specification:

Lubricants are descirbed by specifications, which is often supported by a logo, symbol or words that inform the consumer that the lubricant marketer has obtained independent verification of conformance to the specification. Examples include the API's donut logo or the NSF tick mark. A widely perceived is viscosity specification SAE (SAE) 10W-40. Lubricity specifications are institute- and manufacturer-based. In the U.S. institute: API S applies to petrol engines, API C applies to diesel engines. Higher second letter marks better oil properties, such lower engine wear supported by tests. In EU the ACEA specifications are used. Classes A, B, C, E are followed by a number. Japan introduced the JASO specification for motorbike engines. In the industrial market place the specification may take the form of a legal contract to supply a conforming fluid or purchasers may choose to buy on the basis of a manufacturers own published specification.

- Original equipment manufacturer (OEM) approval:

Specifications often denote a minimum acceptable performance levels. Thus many equipment manufacturers add on their own particular requirements or tighten the tolerance on a general specification to meet their particular needs (or doing a different set of tests or using different/own testbed engine). This gives the lubricant marketer an avenue to differentiate their product by designing it to meet an OEM specification. Often, the OEM carries out extensive testing and maintains an active list of approved products. This is a powerful marketing tool in the lubricant marketplace. Text on the back of the motor oil label usually has a list of conformity to some OEM specifications, such as MB, MAN, Volvo, Cummins, VW, BMW or others. Manufactures may have vastly different specifications for the range of engines they make; one may not be completely suitable for some other.

- Performance:

The lubricant marketer claims benefits for the customer based on the superior performance of the lubricant. Such marketing is supported by glamorous advertising, sponsorships of typically sporting events and endorsements. Unfortunately broad performance claims are common in the consumer marketplace, which are difficult or impossible for a typical consumer to verify. In the B2B market place the marketer is normally expected to show data that supports the claims, hence reducing the use of broad claims. Increasing performance, reducing wear and fuel consumption is also aim of the later API, ACEA and car manufacturer oil specifications, so lubricant marketers can back their claims by doing extensive (and expensive) testing.

- Longevity:

The marketer claims that their lubricant maintains its performance over a longer period of time. For example, in the consumer market, a typical motor oil change interval is around the 3,000–6,000 miles (5,000–10,000 km). The lubricant marketer may offer a lubricant that lasts for 12,000 miles (19,000 km) or more to convince a user to pay a premium. Typically, the consumer would need to check or balance the longer life and any warranties offered by the lubricant manufacturer with the possible loss of equipment manufacturer warranties by not following its schedule. Many car and engine manufacturers support extended drain intervals, but request extended drain interval certified oil used in that case; and sometimes a special oil filter. Example: In older Mercedes-Benz engines and in truck engines one can use engine oil MB 228.1 for basic drain interval. Engine oils conforming with higher specification MB 228.3 may be used twice as long, oil of MB 228.5 specification 3x longer. Note that the oil drain interval is valid for new engine with fuel conforming car manufacturer specification. When using lower grade fuel, or worn engine the oil change interval has to shorten accordingly. In general oils approved for extended use are of higher specification and reduce wear. In the industrial market place the longevity is generally measured in time units and the lubricant marketer can suffer large financial penalties if their claims are not substantiated.

- Efficiency:

The lubricant marketer claims improved equipment efficiency when compared to rival products or technologies, the claim is usually valid when comparing lubricant of higher specification with previous grade. Typically the efficiency is proved by showing a reduction in energy costs to operate the system. Guaranteeing improved efficiency is the goal of some oil test specifications such as API CI-4 Plus for diesel engines. Some car/engine manufacturers also specifically request certain higher efficiency level for lubricants for extended drain intervals.

- Operational tolerance:

The lubricant is claimed to cope with specific operational environment needs. Some common environments include dry, wet, cold, hot, fire risk, high load, high or low speed, chemical compatibility, atmospheric compatibility, pressure or vacuum and various combinations. The usual thermal characteristics is outlined with SAE viscosity given for 100 °C, like SAE 30, SAE 40. For low temperature viscosity the SAE xxW mark is used. Both markings can be combined together to form a SAE 0W-60 for example. Viscosity index (VI) marks viscosity change with temperature, with higher VI numbers being more temperature stable.

- Economy:

The marketer offers a lubricant at a lower cost than rivals either in the same grade or a similar one that will fill the purpose for lesser price. (Stationary installations with short drain intervals.) Alternative may be offering a more expensive lubricant and promise return in lower wear, specific fuel consumption or longer drain intervals. (Expensive machinery, un-affordable downtimes.)

- Environment friendly:

The lubricant is said to be environmentally friendly. Typically this is supported by qualifying statements or conformance to generally accepted approvals. Several organizations, typically government sponsored, exist globally to qualify and approve such lubricants by evaluating their potential for environmental harm. Typically, the lubricant manufacturer is allowed to indicate such approval by showing some special mark. Examples include the German "Blue Angel", European "Daisy" Eco label, Global Eco-Label "GEN mark", Nordic, "White Swan", Japanese "Earth friendly mark"; USA "Green Seal", Canadian "Environmental Choice", Chinese "Huan", Singapore "Green Label" and the French "NF Environment mark".

- Composition:

The marketer claims novel composition of the lubricant which improves some tangible performance over its rivals. Typically the technology is protected via formal patents or other intellectual property protection mechanism to prevent rivals from copying. Lot of claims in this area are simple marketing buzzwords, since most of them are related to a manufacturer specific process naming (which achieves similar results than other ones) but the competition is prohibited from using a trademark.

- Quality:

The marketer claims broad superior quality of its lubricant with no factual evidence. The quality is "proven" by references to famous brand, sporting figure, racing team, some professional endorsement or some similar subjective claim. All motor oil labels wear mark similar to "of outstanding quality" or "quality additives," the actual comparative evidence is always lacking.

Disposal and Environmental

It is estimated that 40% of all lubricants are released into the environment. Common

Disposal methods include Recycling, burning, landfill and discharge into water, though typically disposal in landfill and discharge into water are strictly regulated in most countries, as even small amount of lubricant can contaminate a large amount of water. Most regulations permit a threshold level of lubricant that may be present in waste streams and companies spend hundreds of millions of dollars annually in treating their waste waters to get to acceptable levels.

Burning the lubricant as fuel, typically to generate electricity, is also governed by regulations mainly on account of the relatively high level of additives present. Burning generates both airborne pollutants and ash rich in toxic materials, mainly heavy metal compounds. Thus lubricant burning takes place in specialized facilities that have incorporated special scrubbers to remove airborne pollutants and have access to landfill sites with permits to handle the toxic ash.

Unfortunately, most lubricant that ends up directly in the environment is due to general public discharging it onto the ground, into drains and directly into landfills as trash. Other direct contamination sources include runoff from roadways, accidental spillages, natural or man-made disasters and pipeline leakages.

Improvement in filtration technologies and processes has now made recycling a viable option (with rising price of base stock and crude oil). Typically various filtration systems remove particulates, additives and oxidation products and recover the base oil. The oil may get refined during the process. This base oil is then treated much the same as virgin base oil however there is considerable reluctance to use recycled oils as they are generally considered inferior. Basestock fractionally vacuum distilled from used lubricants has superior properties to all natural oils, but cost effectiveness depends on many factors. Used lubricant may also be used as refinery feedstock to become part of crude oil. Again there is considerable reluctance to this use as the additives, soot and wear metals will seriously poison/deactivate the critical catalysts in the process. Cost prohibits carrying out both filtration (soot, additives removal) and re-refining (distilling, isomerisation, hydrocrack, etc.) however the primary hindrance to recycling still remains the collection of fluids as refineries need continuous supply in amounts measured in cisterns, rail tanks.

Occasionally, unused lubricant requires disposal. The best course of action in such situations is to return it to the manufacturer where it can be processed as a part of fresh batches.

Environment: Lubricants both fresh and used can cause considerable damage to the environment mainly due to their high potential of serious water pollution. Further the additives typically contained in lubricant can be toxic to flora and fauna. In used fluids the oxidation products can be toxic as well. Lubricant persistence in the environment largely depends upon the base fluid, however if very toxic additives are used they may negatively affect the persistence. Lanolin lubricants are non-toxic making them the environmental alternative which is safe for both users and the environment.

Societies and Industry Bodies

- American Petroleum Institute (API)

- Society of Tribologists and Lubrication Engineers (STLE)

- National Lubricating Grease Institute (NLGI)

- Society of Automotive Engineers (SAE)

- Independent Lubricant Manufacturer Association (ILMA)

- European Automobile Manufacturers Association (ACEA)

- Japanese Automotive Standards Organization (JASO)

- Petroleum Packaging Council (PPC)

Types of Lubricants

Ring Oiler

A ring oiler or oil ring is a form of oil-lubrication system for bearings.

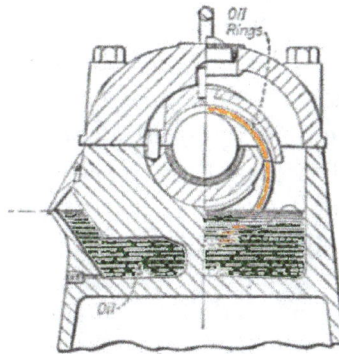

Section through a bearing, showing the oil sump beneath *(green)* and the ring oiler *(orange)* in place around the shaft

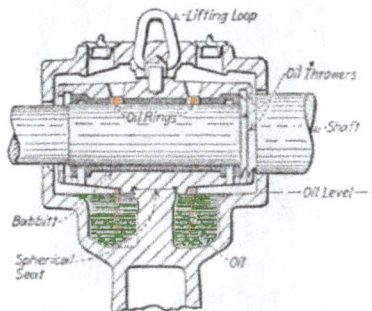

Section though a long Babbitt metal sleeve bearing, with two ring oilers fitted through grooves in the upper part of the bearing.

Ring oilers were used for medium-speed applications with moderate loads, during the first half of the 20th century. These represented the later years of the stationary steam engine, and the beginnings of the high-speed steam engine, the internal combustion oil engine and electrical generating equipment. Before this time plain bearings were lubricated by drip-feed oil cups or manually by an engine tender with an oil can. As speeds or bearing loads later increased, forced pressure lubrication became more prevalent and the ring oiler fell from use.

A ring oiler is a simple device, consisting of a large metal ring placed around a horizontal shaft, adjacent to a bearing. An oil sump is underneath this shaft and the ring is large enough to dip into the oil. As the shaft rotates, the ring is carried round with it. The rotating ring in turn picks up some oil and deposits it onto the shaft, from where it flows sideways and lubricates the bearings. The oil ring is effectively a simple lubrication pump, with only one moving part and no complex or high-precision components. The device is crude, but automatic, effective and reliable. Unlike a drip oiler, there is also no need to close off the oiler or remove oil wicks when the machine is stopped.

Ring oilers were used for speeds up to around 1,000 rpm. Above this, the oil tended to be thrown centrifugally from the ring, rather than carried by it (although it is still currently applied on steam turbines with speeds around 3200 rpm). The bearing must also remain horizontal and stable, so although suitable for crankshaft main bearings, they could not be used on connecting rod big end bearings. They were not used on vehicles for similar reasons, although the engines concerned at this time were anyway too large and heavy for practical mobile use. Automatic ring oilers were particularly useful for large engines with multiple horizontally opposed cylinders, where it was otherwise difficult to access the central main bearings. Ring oilers were most suited where bearing side-loads were relatively light, but the bearing capacity required more lubrication than could be supplied by a drip feed oiler. For this reason they were widely used on larger electric motors and generators.

Motor Oil

Motor oil sample

Motor oil, engine oil, or engine lubricant is any of various substances (comprising oil enhanced with additives, for example, in many cases, extreme pressure additives) that are used for lubrication of internal combustion engines. The main function of motor oil is to reduce wear on moving parts; it also cleans moving parts from the sludge, inhibits corrosion, improves sealing, and cools the engine by carrying heat away from moving parts.

Motor oils are derived from petroleum-based and non-petroleum-synthesized chemical compounds. Motor oils today are mainly blended by using base oils composed of hydrocarbons, polyalphaolefins (PAO), and polyinternal olefins (PIO), organic compounds consisting entirely of carbon and hydrogen. The base oils of some high-performance motor oils contain up to 20% by weight of esters.

History

On September 6, 1866 American John Ellis founded the Continuous Oil Refining Company (Later to become Valvoline). While studying the possible healing powers of crude oil, Dr. Ellis was disappointed to find no real medicinal value, but was intrigued by its potential lubricating properties. He eventually abandoned the medical practice to devote his time to the development of an all-petroleum, high viscosity lubricant for steam engines – then using inefficient combinations of petroleum and animal and vegetable fats. He made his breakthrough when he developed an oil that worked effectively in high temperatures. This meant no more gummed valves, corroded cylinders or leaking seals. In 1873 Ellis officially renamed the company to Valvoline after the steam engine valves the product lubricated.

Use

Motor oil is a lubricant used in internal combustion engines, which power cars, motorcycles, lawnmowers, engine-generators, and many other machines. In engines, there are parts which move against each other, and the friction wastes otherwise useful power by converting the kinetic energy to heat. It also wears away those parts, which could lead to lower efficiency and degradation of the engine. This increases fuel consumption, decreases power output, and can lead to engine failure.

Lubricating oil creates a separating film between surfaces of adjacent moving parts to minimize direct contact between them, decreasing heat caused by friction and reducing wear, thus protecting the engine. In use, motor oil transfers heat through convection as it flows through the engine by means of air flow over the surface of the oil pan, an oil cooler and through the buildup of oil gases evacuated by the Positive Crankcase Ventilation (PCV) system.

In petrol (gasoline) engines, the top piston ring can expose the motor oil to temperatures of 160 °C (320 °F). In diesel engines the top ring can expose the oil to temperatures over 315 °C (600 °F). Motor oils with higher viscosity indices thin less at these higher temperatures.

Coating metal parts with oil also keeps them from being exposed to oxygen, inhibiting oxidation at elevated operating temperatures preventing rust or corrosion. Corrosion inhibitors may also be added to the motor oil. Many motor oils also have detergents and dispersants added to help keep the engine clean and minimize oil sludge build-up. The oil is able to trap soot from combustion in itself, rather than leaving it deposited on the internal surfaces. It is a combination of this, and some singeing that turns used oil black after some running.

Rubbing of metal engine parts inevitably produces some microscopic metallic particles from the wearing of the surfaces. Such particles could circulate in the oil and grind against moving parts, causing wear. Because particles accumulate in the oil, it is typically circulated through an oil filter to remove harmful particles. An oil pump, a vane or gear pump powered by the engine, pumps the oil throughout the engine, including the oil filter. Oil filters can be a *full flow* or *bypass* type.

In the crankcase of a vehicle engine, motor oil lubricates rotating or sliding surfaces between the crankshaft journal bearings (main bearings and big-end bearings), and rods connecting the pistons to the crankshaft. The oil collects in an oil pan, or sump, at the bottom of the crankcase. In some small engines such as lawn mower engines, dippers on the bottoms of connecting rods dip into the oil at the bottom and splash it around the crankcase as needed to lubricate parts inside. In modern vehicle engines, the oil pump takes oil from the oil pan and sends it through the oil filter into oil galleries, from which the oil lubricates the main bearings holding the crankshaft up at the main journals and camshaft bearings operating the valves. In typical modern vehicles, oil pressure-fed from the oil galleries to the main bearings enters holes in the main journals of the crankshaft.

From these holes in the main journals, the oil moves through passageways inside the crankshaft to exit holes in the rod journals to lubricate the rod bearings and connecting rods. Some simpler designs relied on these rapidly moving parts to splash and lubricate the contacting surfaces between the piston rings and interior surfaces of the cylinders. However, in modern designs, there are also passageways through the rods which carry oil from the rod bearings to the rod-piston connections and lubricate the contacting surfaces between the piston rings and interior surfaces of the cylinders. This oil film also serves as a seal between the piston rings and cylinder walls to separate the combustion chamber in the cylinder head from the crankcase. The oil then drips back down into the oil pan.

Motor oil may also serve as a cooling agent. In some constructions oil is sprayed through a nozzle inside the crankcase onto the piston to provide cooling of specific parts that undergo high temperature strain. On the other hand, the thermal capacity of the oil pool has to be filled, i.e. the oil has to reach its designed temperature range before it can protect the engine under high load. This typically takes longer than heating the main cooling agent — water or mixtures thereof — up to its operating temperature. In order to inform the driver about the oil temperature, some older and most high-performance or racing engines feature an oil thermometer.

Due to its high viscosity, motor oil is not always the preferred oil for certain applications. Some applications make use of lighter products such as WD-40, when a lighter oil is desired, or honing oil if the desired viscosity needs to be mid-range.

Non-vehicle Motor Oils

An example is lubricating oil for four-stroke or four-cycle internal combustion engines such as those used in portable electricity generators and "walk behind" lawn mowers. Another example is two-stroke oil for lubrication of two-stroke or two-cycle internal combustion engines found in snow blowers, chain saws, model air planes, gasoline powered gardening equipment like hedge trimmers, leaf blowers and soil cultivators. Often, these motors are not exposed to as wide service temperature ranges as in vehicles, so these oils may be single viscosity oils.

In small two-stroke engines, the oil may be pre-mixed with the gasoline or fuel, often in a rich gasoline:oil ratio of 25:1, 40:1 or 50:1, and burned in use along with the gasoline. Larger two-stroke engines used in boats and motorcycles may have a more economical oil injection system rather than oil pre-mixed into the gasoline. The oil injection system is not used on small engines used in applications like snowblowers and trolling motors as the oil injection system is too expensive for small engines and would take up too much room on the equipment. The oil properties will vary according to the individual needs of these devices. Non-smoking two-stroke oils are composed of esters or polyglycols. Environmental legislation for leisure marine applications, especially in Europe, encouraged the use of ester-based two cycle oil.

Properties

Most motor oils are made from a heavier, thicker petroleum hydrocarbon base stock derived from crude oil, with additives to improve certain properties. The bulk of a typical motor oil consists of hydrocarbons with between 18 and 34 carbon atoms per molecule. One of the most important properties of motor oil in maintaining a lubricating film between moving parts is its viscosity. The viscosity of a liquid can be thought of as its "thickness" or a measure of its resistance to flow. The viscosity must be high enough to maintain a lubricating film, but low enough that the oil can flow around the engine parts under all conditions. The viscosity index is a measure of how much the oil's viscosity changes as temperature changes. A higher viscosity index indicates the viscosity changes less with temperature than a lower viscosity index.

Motor oil must be able to flow adequately at the lowest temperature it is expected to experience in order to minimize metal to metal contact between moving parts upon starting up the engine. The *pour point* defined first this property of motor oil, as defined by ASTM D97 as "... an index of the lowest temperature of its utility ..." for a given application, but the "cold cranking simulator" and "Mini-Rotary Viscometer", ASTM D4684-08) are today the properties required in motor oil specs and define the SAE classifications.

Oil is largely composed of hydrocarbons which can burn if ignited. Still another important property of motor oil is its flash point, the lowest temperature at which the oil gives off vapors which can ignite. It is dangerous for the oil in a motor to ignite and burn, so a high flash point is desirable. At a petroleum refinery, fractional distillation separates a motor oil fraction from other crude oil fractions, removing the more volatile components, and therefore increasing the oil's flash point (reducing its tendency to burn).

Another manipulated property of motor oil is its Total base number (TBN), which is a measurement of the reserve alkalinity of an oil, meaning its ability to neutralize acids. The resulting quantity is determined as mg KOH/ (gram of lubricant). Analogously, Total acid number (TAN) is the measure of a lubricant's acidity. Other tests include zinc, phosphorus, or sulfur content, and testing for excessive foaming.

The NOACK volatility (ASTM D-5800) Test determines the physical evaporation loss of lubricants in high temperature service. A maximum of 14% evaporation loss is allowable to meet API SL and ILSAC GF-3 specifications. Some automotive OEM oil specifications require lower than 10%.

Grades

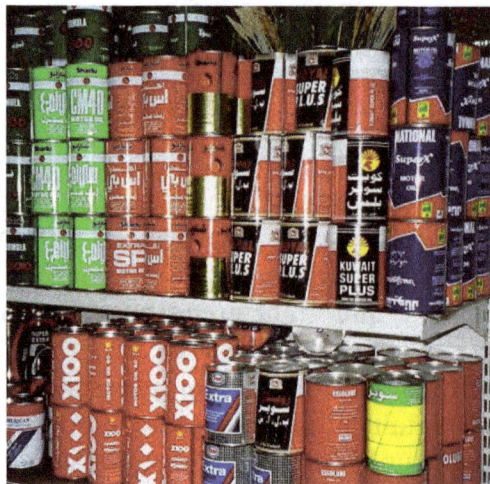

Range of motor oils on display in Kuwait in now-obsolete cardboard cans with steel lids.

The Society of Automotive Engineers (SAE) has established a numerical code system for grading motor oils according to their viscosity characteristics. SAE viscosity gradings include the following, from low to high viscosity: 0, 5, 10, 15, 20, 25, 30, 40, 50 or 60. The numbers 0, 5, 10, 15 and 25 are suffixed with the letter W, designating they are "winter" (not "weight") or cold-start viscosity, at lower temperature. The number 20 comes with or without a W, depending on whether it is being used to denote a cold or hot viscosity grade. The document SAE J300 defines the viscometrics related to these grades.

A common grade is 10W-30.

Kinematic viscosity is graded by measuring the time it takes for a standard amount of oil to flow through a standard orifice, at standard temperatures. The longer it takes, the higher the viscosity and thus higher SAE code. Bigger numbers are thicker.

The SAE has a separate viscosity rating system for gear, axle, and manual transmission oils, SAE J306, which should not be confused with engine oil viscosity. The higher numbers of a gear oil (e.g., 75W-140) do not mean that it has higher viscosity than an engine oil. In anticipation of new lower engine oil viscosity grades, to avoid confusion with the "winter" grades of oil the SAE adopted SAE 16 as a standard to follow SAE 20 instead of SAE 15. Regarding the change Michael Covitch of Lubrizol, Chair of the SAE International Engine Oil Viscosity Classification (EOVC) task force was quoted stating "If we continued to count down from SAE 20 to 15 to 10, etc., we would be facing continuing customer confusion problems with popular low-temperature viscosity grades such as SAE 10W, SAE 5W, and SAE 0W," he noted. "By choosing to call the new viscosity grade SAE 16, we established a precedent for future grades, counting down by fours instead of fives: SAE 12, SAE 8, SAE 4."

Single-grade

A single-grade engine oil, as defined by SAE J300, cannot use a polymeric Viscosity Index Improver (also referred to as Viscosity Modifier) additive. SAE J300 has established eleven viscosity grades, of which six are considered Winter-grades and given a W designation. The 11 viscosity grades are 0W, 5W, 10W, 15W, 20W, 25W, 20, 30, 40, 50, and 60. These numbers are often referred to as the "weight" of a motor oil, and single-grade motor oils are often called "straight-weight" oils.

For single winter grade oils, the dynamic viscosity is measured at different cold temperatures, specified in J300 depending on the viscosity grade, in units of mPa·s, or the equivalent older non-SI units, centipoise (abbreviated cP), using two different test methods. They are the Cold Cranking Simulator (ASTMD5293) and the Mini-Rotary Viscometer (ASTM D4684). Based on the coldest temperature the oil passes at, that oil is graded as SAE viscosity grade 0W, 5W, 10W, 15W, 20W, or 25W. The lower the viscosity grade, the lower the temperature the oil can pass. For example, if an oil passes at the specifications for 10W and 5W, but fails for 0W, then that oil must be labeled as an SAE 5W. That oil cannot be labeled as either 0W or 10W.

For single non-winter grade oils, the kinematic viscosity is measured at a temperature of 100 °C (212 °F) in units of mm²/s (millimeter squared per second) or the equivalent older non-SI units, centistokes (abbreviated cSt). Based on the range of viscosity the oil falls in at that temperature, the oil is graded as SAE viscosity grade 20, 30, 40, 50, or 60. In addition, for SAE grades 20, 30, and 1000, a minimum viscosity measured at 150 °C (302 °F) and at a high-shear rate is also required. The higher the viscosity, the higher the SAE viscosity grade is.

Multi-grade

The temperature range the oil is exposed to in most vehicles can be wide, ranging from cold temperatures in the winter before the vehicle is started up, to hot operating temperatures when the vehicle is fully warmed up in hot summer weather. A specific oil will have high viscosity when cold and a lower viscosity at the engine's operating temperature. The difference in viscosities for most single-grade oil is too large between the extremes of temperature. To bring the difference in viscosities closer together, special polymer additives called viscosity index improvers, or VIIs are added to the oil. These additives are used to make the oil a *multi-grade* motor oil, though it is possible to have a multi-grade oil without the use of VIIs. The idea is to cause the multi-grade oil to have the viscosity of the base grade when cold and the viscosity of the second grade when hot. This enables one type of oil to be used all year. In fact, when multi-grades were initially developed, they were frequently described as *all-season oil*. The viscosity of a multi-grade oil still varies logarithmically with temperature, but the slope representing the change is lessened. This slope representing the change with temperature depends on the nature and amount of the additives to the base oil.

The SAE designation for multi-grade oils includes two viscosity grades; for example, *10W-30* designates a common multi-grade oil. The first number '10W' is the viscosity of the oil at cold temperature and the second number is the viscosity at 100 °C (212 °F). The two numbers used are individually defined by SAE J300 for single-grade oils. Therefore, an oil labeled as 10W-30 must pass the SAE J300 viscosity grade requirement for both 10W and 30, and all limitations placed on the viscosity grades (for example, a 10W-30 oil must fail the J300 requirements at 5W). Also, if an oil does not contain any VIIs, and can pass as a multi-grade, that oil can be labelled with either of the two SAE viscosity grades. For example, a very simple multi-grade oil that can be easily made with modern base oils without any VII is a 20W-20. This oil can be labeled as 20W-20, 20W, or 20. Note, if any VIIs are used however, then that oil cannot be labeled as a single grade.

Breakdown of VIIs under shear is a concern in motorcycle applications, where the transmission may share lubricating oil with the motor. For this reason, synthetic oil or motorcycle-specific oil is sometimes recommended. The necessity of higher-priced motorcycle-specific oil has also been challenged by at least one consumer organization.

Standards

American Petroleum Institute (API)

Engine lubricants are evaluated against the American Petroleum Institute (API), SJ, SL, SM, SN, CH-4, CI-4, CI-4 PLUS and CJ-4 as well as International Lubricant Standardization and Approval Committee (ILSAC) GF-3, GF-4 and GF-5, and Cummins, Mack and John Deere requirements. These evaluations include chemical and physical properties using bench test methods as well as actual running engine tests to quantify engine sludge, oxidation, component wear, oil consumption, piston deposits and fuel economy.

The API sets minimum for performance standards for lubricants. Motor oil is used for the lubrication, cooling, and cleaning of internal combustion engines. Motor oil may be composed of a lubricant base stock only in the case of non-detergent oil, or a lubricant base stock plus additives to improve the oil's detergency, extreme pressure performance, and ability to inhibit corrosion of engine parts.

Groups: Lubricant base stocks are categorized into five groups by the API. Group I base stocks are composed of fractionally distilled petroleum which is further refined with solvent extraction processes to improve certain properties such as oxidation resistance and to remove wax. Group II base stocks are composed of fractionally distilled petroleum that has been hydrocracked to further refine and purify it. Group III base stocks have similar characteristics to Group II base stocks, except that Group III base stocks have higher viscosity indexes. Group III base stocks are produced by further hydrocracking of either Group II base stocks or hydroisomerized slack wax (a Group I and II dewaxing process by-product). Group IV base stock are polyalphaolefins (PAOs). Group V is a catch-all group for any base stock not described by Groups I to IV. Examples of group V base stocks include polyolesters (POE), polyalkylene glycols (PAG), and perfluoropolyalkylethers (PFPAEs). Groups I and II are commonly referred to as mineral oils, group III is typically referred to as synthetic (except in Germany and Japan, where they must not be called synthetic) and group IV is a synthetic oil. Group V base oils are so diverse that there is no catch-all description.

The API service classes have two general classifications: *S* for "service/spark ignition" (typical passenger cars and light trucks using gasoline engines), and *C* for "commercial/compression ignition" (typical diesel equipment). Engine oil which has been tested and meets the API standards may display the API Service Symbol (also known as the "Donut") with the service designation on containers sold to oil users.

The latest API service standard designation is SN for gasoline automobile and light-truck engines. The SN standard refers to a group of laboratory and engine tests, including the latest series for control of high-temperature deposits. Current API service categories include SN, SM, SL and SJ for gasoline engines. All previous service designations are obsolete, although motorcycle oils commonly still use the SF/SG standard.

All the current gasoline categories (including the obsolete SH) have placed limitations on the phosphorus content for certain SAE viscosity grades (the xW-20, xW-30) due to the chemical poisoning that phosphorus has on catalytic converters. Phosphorus is a key anti-wear component in motor oil and is usually found in motor oil in the form of zinc dithiophosphate (ZDDP). Each new API category has placed successively lower phosphorus and zinc limits, and thus has created a controversial issue of obsolescent oils needed for older engines, especially engines with sliding (flat/cleave) tappets. API and ILSAC, which represents most of the world's major automobile/engine manufacturers, state API SM/ILSAC GF-4 is fully backwards compatible, and it is noted that one of the engine tests required for API SM, the Sequence IVA, is a sliding tappet design to test specifically for cam wear protection. Not everyone is in agreement with backwards compatibility, and in addition, there are special situations, such as "performance" engines or fully race built engines, where the engine protection requirements are above and beyond API/ILSAC requirements. Because of this, there are specialty oils out in the market place with higher than API allowed phosphorus levels. Most engines built before 1985 have the flat/cleave bearing style systems of construction, which is sensitive to reducing zinc and phosphorus. For example, in API SG rated oils, this was at the 1200-1300 ppm level for zinc and phosphorus, where the current SM is under 600 ppm. This reduction in anti-wear chemicals in oil has caused premature failures of camshafts and other high pressure bearings in many older automobiles and has been blamed for premature failure of the oil pump drive/cam position sensor gear that is meshed with camshaft gear in some modern engines.

There are three diesel engine service designations which are current: CJ-4, CI-4, and CH-4. Some manufacturers continue to use obsolete designations such as CC for small or stationary diesel engines. In addition, API created a separated CI-4 PLUS designation in conjunction with CJ-4 and CI-4 for oils that meet certain extra requirements, and this marking is located in the lower portion of the API Service Symbol "Donut".

It is possible for an oil to conform to both the gasoline and diesel standards. In fact, it is the norm for all diesel rated engine oils to carry the "corresponding" gasoline specification. For example, API CJ-4 will almost always list either SL or SM, API CI-4 with SL, API CH-4 with SJ, and so on.

Motorcycle Oil

The API oil classification structure has eliminated specific support for wet-clutch motorcycle applications in their descriptors, and API SJ and newer oils are referred to be specific to automobile and light truck use. Accordingly, motorcycle oils are subject to their own unique standards. As discussed above, motorcycle oils commonly still use the obsolescent SF/SG standard.

ILSAC

The International Lubricant Standardization and Approval Committee (ILSAC) also has standards for motor oil. Introduced in 2004, GF-4 applies to SAE 0W-20, 5W-20, 0W-30, 5W-30, and 10W-30 viscosity grade oils. In general, ILSAC works with API in creating the newest gasoline oil specification, with ILSAC adding an extra requirement of fuel economy testing to their specification. For GF-4, a Sequence VIB Fuel Economy Test (ASTM D6837) is required that is not required in API service category SM.

A key new test for GF-4, which is also required for API SM, is the Sequence IIIG, which involves running a 3.8 L (232 c.i.d.), GM 3.8 L V-6 at 125 hp (93 kW), 3,600 rpm, and 150 °C (300 °F) oil temperature for 100 hours. These are much more severe conditions than any API-specified oil was designed for: cars which typically push their oil temperature consistently above 100 °C (212 °F) are most turbocharged engines, along with most engines of European or Japanese origin, particularly small capacity, high power output.

The IIIG test is about 50% more difficult than the previous IIIF test, used in GF-3 and API SL oils. Engine oils bearing the API starburst symbol since 2005 are ILSAC GF-4 compliant.

To help consumers recognize that an oil meets the ILSAC requirements, API developed a "starburst" certification mark.

A new set of specifications, GF-5, took effect in October 2010. The industry has one year to convert their oils to GF-5 and in September 2011, ILSAC will no longer offer licensing for GF-4.

ACEA

The ACEA (*Association des Constructeurs Européens d'Automobiles*) performance/ quality classifications A3/A5 tests used in Europe are arguably more stringent than the API and ILSAC standards. CEC (The Co-ordinating European Council) is the development body for fuel and lubricant testing in Europe and beyond, setting the standards via their European Industry groups; ACEA, ATIEL, ATC and CONCAWE.

Lubrizol, a supplier of additives to nearly all motor oil companies, hosts a Relative Performance Tool which directly compares the manufacturer and industry specs. Differences in their performance is apparent in the form of interactive spider graphs, which both expert and novice can appreciate.

JASO

The Japanese Automotive Standards Organization (JASO) has created their own set of performance and quality standards for petrol engines of Japanese origin.

For four-stroke gasoline engines, the JASO T904 standard is used, and is particularly relevant to motorcycle engines. The JASO T904-MA and MA2 standards are designed to distinguish oils that are approved for wet clutch use, with MA2 lubricants delivering higher friction performance. The JASO T904-MB standard denotes oils not suitable for wet clutch use, and are therefore used in scooters equipped with continuously variable transmissions. The addition of friction modifiers to JASO MB oils can contribute to greater fuel economy in these applications.

For two-stroke gasoline engines, the JASO M345 (FA, FB, FC, FD) standard is used, and this refers particularly to low ash, lubricity, detergency, low smoke and exhaust blocking.

These standards, especially JASO-MA (for motorcycles) and JASO-FC, are designed to address oil-requirement issues not addressed by the API service categories. One element of the JASO-MA standard is a friction test designed to determine suitability for wet clutch usage. An oil that meets JASO-MA is considered appropriate for wet clutch operations. Oils marketed as motorcycle-specific will carry the JASO-MA label.

ASTM

A 1989 American Society for Testing and Materials (ASTM) report stated that its 12-year effort to come up with a new high-temperature, high-shear (HTHS) standard was not successful. Referring to SAE J300, the basis for current grading standards, the report stated:

The rapid growth of non-Newtonian multigraded oils has rendered kinematic viscosity as a nearly useless parameter for characterising "real" viscosity in critical zones of an engine... There are those who are disappointed that the twelve-year effort has not resulted in a redefinition of the SAE J300 Engine Oil Viscosity Classification document so as to express high-temperature viscosity of the various grades ... In the view of this writer, this redefinition did not occur because the automotive lubricant market knows of no field failures unambiguously attributable to insufficient HTHS oil viscosity.

Other Additives

In addition to the viscosity index improvers, motor oil manufacturers often include other additives such as detergents and dispersants to help keep the engine clean by minimizing sludge buildup, corrosion inhibitors, and alkaline additives to neutralize acidic oxidation products of the oil. Most commercial oils have a minimal amount of zinc dialkyldithiophosphate as an anti-wear additive to protect contacting metal surfaces with zinc and other compounds in case of metal to metal contact. The quantity of zinc dialkyldithiophosphate is limited to minimize adverse effect on catalytic converters. Another aspect for after-treatment devices is the deposition of oil ash, which increases the exhaust back pressure and reduces fuel economy over time. The so-called "chemical box" limits today the concentrations of sulfur, ash and phosphorus (SAP).

There are other additives available commercially which can be added to the oil by the user for purported additional benefit. Some of these additives include:

- EP additives, like zinc dialkyldithiophosphate (ZDDP) additives and sulfonates, preferably calcium sulfonates, are available to consumers for additional protection under extreme-pressure conditions or in heavy duty performance situations. Calcium sulfonates additives are also added to protect motor oil from oxidative breakdown and to prevent the formation of sludge and varnish deposits. Both were the main basis of additive packages used by lubricant manufacturers up until the 1990s when the need for ashless additives arose. Main advantage was very low price and wide availability (sulfonates were originally waste byproducts). Currently there are ashless oil lubricants without these additives, which can only fulfill the qualities of the previous generation with more expensive basestock and more expensive organic or organometallic additive compounds. Some new oils are not formulated to provide the level of protection of previous generations to save manufacturing costs. Lately API specifications reflect that

- Some molybdenum disulfide containing additives to lubricating oils are claimed to reduce friction, bond to metal, or have anti-wear properties. MoS_2 particles can be shear-welded on steel surface and some engine components were even treated with MoS_2 layer during manufacture, namely liners in engines. (Trabant for example). They were used in World War II in flight engines and became commercial after World War II until the 1990s. They were commercialized in the 1970s (ELF ANTAR Molygraphite) and are today still available (Liqui Moly MoS_2 10 W-40, www.liqui-moly.de). Main disadvantage of molybdenum disulfide is anthracite black color, so oil treated with it is hard to distinguish from a soot filled engine oil with metal shavings from spun crankshaft bearing.

- In the 1980s and 1990s, additives with suspended PTFE particles were available, e.g., "Slick50", to consumers to increase motor oil's ability to coat and protect metal surfaces. There is controversy as to the actual effectiveness of these products, as they can coagulate and clog the oil filter and tiny oil passages in the engine. It is supposed to work under boundary lubricating conditions, which good engine designs tend to avoid anyway. Also, Teflon alone has little to no ability to firmly stick on a sheared surface, unlike molybdenum disulfide, for example.

- Various extreme-pressure (EP) additives and antiwear additives.

- Many patents proposed use perfluoropolymers to reduce friction between metal parts, such as PTFE (Teflon), or micronized PTFE. However, the application obstacle of PTFE is insolubility in lubricant oils. Their application is questionable and depends mainly on the engine design — one that can not maintain reasonable lubricating conditions might benefit, while properly designed engine with oil film thick enough would not see any difference. PTFE is a very soft material, thus its friction coefficient becomes worse than that of hardened steel-to-steel mating surfaces under common loads. PTFE is used in composition of sliding bearings where it improves lubrication under relatively light load until the oil pressure builds up to full hydrodynamic lubricating conditions.

EP additives may be incompatible with some motorcycles which share wet clutch lubrication with the engine.

Environmental Effects

Blue drain and yellow fish symbol used by the UK Environment Agency to raise awareness of the ecological impacts of contaminating surface drainage

Due to its chemical composition, world-wide dispersion and effects on the environment, used motor oil is considered a serious environmental problem. Most current motor oil lubricants contain petroleum base stocks, which are toxic to the environment and difficult to dispose of after use. Over 40% of the pollution in America's waterways is from used motor oil. Used oil is considered the largest source of oil pollution in the U.S. harbor and waterways, at 385 million gallons per year, mostly from improper disposal. By far, the greatest cause of motor oil pollution in our oceans comes from drains and urban street runoff, much of which is from improper disposal of engine oil. One gallon of used oil can create an eight-acre slick on surface water, threatening fish, waterfowl and other aquatic life. According to the U.S. EPA, films of oil on the surface of water prevent the replenishment of dissolved oxygen, impair photosynthetic processes, and block sunlight. Toxic effects of used oil on freshwater and marine organisms vary, but significant long-term effects have been found at concentrations of 310 ppm in several freshwater fish species and as low as 1 ppm in marine life forms. Motor oil can have an incredibly detrimental effect on the environment, particularly to plants that depend on healthy soil to grow. There are three main ways that motor oil affects plants: contaminating water supplies, contaminating soil, and poisoning plants. Used motor oil dumped on land reduces soil productivity. Improperly disposed used oil ends up in landfills, sewers, backyards, or storm drains where soil, groundwater and drinking water may be contaminated.

Synthetic Oils

Synthetic lubricants were first synthesized, or man-made, in significant quantities as replacements for mineral lubricants (and fuels) by German scientists in the late 1930s and early 1940s because of their lack of sufficient quantities of crude for their (primarily military) needs. A significant factor in its gain in popularity was the ability of synthetic-based lubricants to remain fluid in the sub-zero temperatures of the Eastern front in wintertime, temperatures which caused petroleum-based lubricants to solidify owing to their higher wax content. The use of synthetic lubricants widened through the 1950s and 1960s owing to a property at the other end of the temperature spectrum, the ability to lubricate aviation engines at temperatures that caused mineral-based lubricants to break down. In the mid-1970s, synthetic motor oils were formulated and commercially applied for the first time in automotive applications. The same SAE system for designating motor oil viscosity also applies to synthetic oils. Synthetic oils are derived from either Group III, Group IV, or some Group V bases. Synthetics include classes of lubricants like synthetic esters as well as "others" like GTL (Methane Gas-to-Liquid) (Group V) and polyalpha-olefins (Group IV). Higher purity and therefore better property control theoretically means synthetic oil has better mechanical properties at extremes of high and low temperatures. The molecules are made large and "soft" enough to retain good viscosity at higher temperatures, yet branched molecular structures interfere with solidification and therefore allow flow at lower temperatures. Thus, although the viscosity still decreases as temperature increases, these synthetic motor oils have a higher viscosity index over the traditional petroleum base. Their specially designed properties allow a wider temperature range at higher and lower temperatures and often include a lower pour point. With their improved viscosity index, synthetic oils need lower levels of viscosity index improvers, which are the oil components most vulnerable to thermal and mechanical degradation as the oil ages, and thus they do not degrade as quickly as traditional motor oils. However, they still fill up with particulate matter, although the matter better suspends within the oil, and the oil filter still fills and clogs up over time. So, periodic oil and filter changes should still be done with synthetic oil; but some synthetic oil suppliers suggest that the intervals between oil changes can be longer, sometimes as long as 16,000-24,000 km (10,000–15,000 mi) primarily due to reduced degradation by oxidation.

Tests show that fully synthetic oil is superior in extreme service conditions to conventional oil, and may perform better for longer under standard conditions. But in the vast majority of vehicle applications, mineral oil based lubricants, fortified with additives and with the benefit of over a century of development, continue to be the predominant lubricant for most internal combustion engine applications.

Bio-based Oils

Bio-based oils existed prior to the development of petroleum-based oils in the 19th century. They have become the subject of renewed interest with the advent of bio-fuels and the push for green products. The development of canola-based motor oils began in 1996 in order to pursue environmentally friendly products. Purdue University has funded a project to develop and test such oils. Test results indicate satisfactory performance from the oils tested. A review on the status of bio-based motor oils and base oils globally, as well as in the U.S, shows how bio-based lubricants show promise in augmenting the current petroleum-based supply of lubricating materials, as well as replacing it in many cases.

The USDA National Center for Agricultural Utilization Research developed an Estolide lubricant technology made from vegetable and animal oils. Estolides have shown great promise in a wide range of applications, including engine lubricants. Working with the USDA, a California-based company Biosynthetic Technologies has developed a high-performance "drop-in" biosynthetic oil using Estolide technology for use in motor oils and industrial lubricants. This biosynthetic oil American Petroleum Institute (API) has the potential to greatly reduce environmental challenges associated with petroleum. Independent testing not only shows biosynthetic oils to be among the highest-rated products for protecting engines and machinery; they are also bio-based, biodegradable, non-toxic and do not bioaccumulate in marine organisms. Also, motor oils and lubricants formulated with biosynthetic base oils can be recycled and re-refined with petroleum-based oils. The U.S.-based company Green Earth Technologies manufactures a bio-based motor oil, called G-Oil, made from animal oils.

Maintenance

The oil and the oil filter need to be periodically replaced. While there is a full industry surrounding regular oil changes and maintenance, an oil change is a fairly simple operation that most car owners can do themselves.

In engines, there is some exposure of the oil to products of internal combustion, and microscopic coke particles from black soot accumulate in the oil during operation. Also the rubbing of metal engine parts produces some microscopic metallic particles from the wearing of the surfaces. Such particles could circulate in the oil and grind against the part surfaces causing wear. The oil filter removes many of the particles and sludge, but eventually the oil filter can become clogged, if used for extremely long periods.

The motor oil and especially the additives also undergo thermal and mechanical degradation, which reduce the viscosity and reserve alkalinity of the oil. At reduced viscosity, the oil is not as capable of lubricating the engine, thus increasing wear and the chance of overheating. Reserve alkalinity is the ability of the oil to resist formation of acids. Should the reserve alkalinity decline to zero, those acids form and corrode the engine.

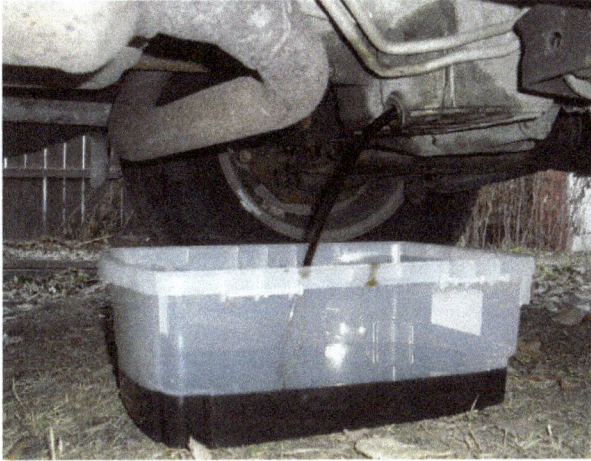

Oil being drained from a car

Some engine manufacturers specify which SAE viscosity grade of oil should be used, but different viscosity motor oil may perform better based on the operating environment. Many manufacturers have varying requirements and have designations for motor oil they require to be used. In general, unless specified by the manufacturer, heavier weight oils are not necessarily better than lighter weight oils; heavy oils tend to stick longer to parts between two moving surfaces, and this degrades the oil faster than a lighter weight oil that flows better, allowing fresh oil in its place sooner. Cold weather has a thickening effect on conventional oil, and this is one reason lighter weight oils are manufacturer recommended in places with cold winters.

Motor oil changes are usually scheduled based on the time in service or the distance that the vehicle has traveled.These are rough indications of the real factors that control when an oil change is appropriate, which include how long the oil has been run at elevated temperatures, how many heating cycles the engine has been through, and how hard the engine has worked. The vehicle distance is intended to estimate the time at high temperature, while the time in service is supposed to correlate with the number of vehicle trips and capture the number of heating cycles. Oil does not degrade significantly just sitting in a cold engine. On the other hand, if a car is driven just for very short distances, the oil is not allowed to fully heat-up, and contaminants such as water accumulates in the oil, due to lack of sufficient heat to boil off the water. Oil of this nature, just sitting in an engine, can cause problems.

Also important is the quality of the oil used, especially with synthetics (synthetics are more stable than conventional oils). Some manufacturers address this (for example, BMW and VW with their respective long-life standards), while others do not.

Time-based intervals account for the short-trip drivers who drive short distances, which build up more contaminants. Manufacturers advise to not exceed their time or distance-driven interval for a motor oil change. Many modern cars now list somewhat higher intervals for changing oil and filter, with the constraint of "severe" service requiring more frequent changes with less-than ideal driving. This applies to short trips of under 15 km (10 mi), where the oil does not get to full operating temperature long enough to burn off condensation, excess fuel, and other contamination that leads to "sludge", "varnish", "acids", or other deposits. Many manufacturers have engine computer calculations to estimate the oil's condition based on the factors which degrade it, such as RPM, temperatures, and trip length; one system adds an optical sensor for determining the clarity of the oil in the engine. These systems are commonly known as Oil Life Monitors or OLMs.

Some quick oil change shops recommended intervals of 5,000 km (3,000 mi) or every three months, which is not necessary, according to many automobile manufacturers. This has led to a campaign by the California EPA against the 3,000 mile myth, promoting vehicle manufacturer's recommendations for oil change intervals over those of the oil change industry. This is still an active debate within the industry however and service technicians still recommend 3000 or 5000 miles service intervals in the conservative North American market, as it suits the customer to have their vehicle inspected regularly in order to prevent larger problems from developing (for example slight coolant leaks left unnoticed could lead to an overheat condition). Also, in many vehicles engine "sludge" from longer oil change intervals has become a problem and led to very costly repairs sometimes including complete engine overhauls. Oil consumption is also a problem when using longer intervals and vehicle owners need to be aware of this and check their oil levels regularly. the average percentage of loss in the oil is around 12% to 14% according to many automobile manufactures .Severe engine damage will result from running the oil level too low. On top of that many manufactures are now using turbochargers and lack of proper lubrication is the primary cause of premature turbo failure. This lack of lubrication can be caused by sludge build up in the oil lines causing restriction of flow and/or simply low oil levels.

The engine user can, in replacing the oil, adjust the viscosity for the ambient temperature change, thicker for summer heat and thinner for the winter cold. Lower viscosity oils are common in newer vehicles.

By the mid-1980s, recommended viscosities had moved down to 10W-30, primarily to improve fuel efficiency. A modern typical application would be Honda motor's use of 5W-20 (and in their newest vehicles, 0W-20) viscosity oil for 12,000 km (7,500 mi). Engine designs are evolving to allow the use of even lower-viscosity oils without the risk of excessive metal-to-metal abrasion, principally in the cam and valve mechanism areas. In line with car manufacturers push towards these lower viscosities in search of better fuel economy, on April 2, 2013 the Society of Automotive Engineers (SAE) introduced a SAE 16 viscosity rating, a break from its traditional "divisible by 10" numbering system for its high-temperature viscosity ratings that spanned from low-viscosity SAE 20 to high-viscosity SAE 60.

Future

A new process to break down polyethylene, a common plastic product found in many consumer containers, is used to make a paraffin-like wax with the correct molecular properties for conversion into a lubricant, bypassing the expensive Fischer-Tropsch process. The plastic is melted and then pumped into a furnace. The heat of the furnace breaks down the molecular chains of polyethylene into wax. Finally, the wax is subjected to a catalytic process that alters the wax's molecular structure, leaving a clear oil.

Biodegradable Motor Oils based on esters or hydrocarbon-ester blends appeared in the 1990s followed by formulations beginning in 2000 which respond to the bio-no-tox-criteria of the European preparations directive (EC/1999/45). This means, that they not only are biodegradable according to OECD 301x test methods, but also the aquatic toxicities (fish, algae, daphnie) are each above 100 mg/L.

Another class of base oils suited for engine oil are the polyalkylene glycols. They offer zero-ash, bio-no-tox properties and lean burn characteristics.

Re-refined Motor Oil

The oil in a motor oil product does break down and burns as it is used in an engine — it also gets contaminated with particles and chemicals that make it a less effective lubricant. Re-refining cleans the contaminants and used additives out of the dirty oil. From there, this clean "base stock" is blended with some virgin base stock and a new additives package to make a finished lubricant product that can be just as effective as lubricants made with all-virgin oil. The United States Environmental Protection Agency (EPA) defines re-refined products as containing at least 25% re-refined base stock, but other standards are significantly higher. The California State public contract code defines a re-refined motor oil as one that contains at least 70% re-refined base stock.

Packaging

Motor oils were sold at retail in glass bottles, metal cans and metal/cardboard cans, before the advent of the current polyethylene plastic bottle, which began to appear in the early 1980s. Reusable spouts were made separately from the cans; with a piercing point like that of a can opener, these spouts could be used to puncture the top of the can and to provide an easy way to pour the oil.

Today, motor oil is generally sold in bottles of either 1 U.S. quart (946mL) or 1L as well as in larger plastic containers ranging from approximately 4.4 to 5 liters (4.6 to

5.3 U.S. qt) due to most small to mid-size engines requiring around 3.6 to 5.2 liters (3.8 to 5.5 U.S. qt) of engine oil.

There is a growing trend to sell motor oil in flexible packaging, for instance stand-up pouches.

Distribution to larger users (such as drive-through oil change shops) is often in bulk, by tanker truck or in 1 barrel (160 l) drums.

Two-stroke Oil

An example of two-stroke oil bottle with measurement cap. Oil is dyed blue to make it easier to recognize it in the gasoline. Because it's not diluted, it appears black in this bottle.

Two-stroke oil (also referred to as two-cycle oil, 2-cycle oil, 2T oil, 2-stroke oil or petroil) is a special type of motor oil intended for use in crankcase compression two-stroke engines.

Unlike a four-stroke engine, whose crankcase is closed except for its ventilation system, a two-stroke engine uses the crankcase as part of the induction tract, and therefore, oil must be mixed with gasoline to be distributed throughout the engine for lubrication. The resultant mix is referred to as petroil. This oil is ultimately burned along with the fuel as a total-loss oiling system. This results in increased exhaust emissions, sometimes with excess smoke and/or a distinctive odor.

The oil-base stock can be petroleum, castor oil, semi-synthetic or synthetic oil and is mixed (or metered by injection) with petrol/gasoline at a fuel-to-oil ratio ranging from 16:1 to as low as 100:1. To avoid the high emissions and oily deposits on spark plugs, modern two-strokes, especially for small engines such as garden equipment and chainsaws, may now demand a synthetic oil and can suffer from oiling problems otherwise.

Engine original equipment manufacturers (OEMs) introduced pre-injection systems (sometimes known as "auto-lube") to engines to operate from a 32:1 to 100:1 ratio. Oils must meet or exceed the following typical specifications: TC-W3TM, NMMA, [API] TC, JASO FC, ISO-L-EGC.

Comparing regular lubricating oil with two-stroke oil, the relevant difference is that two-stroke oil must have a much lower ash content. This is required to minimize deposits that tend to form if ash is present in the oil which is burned in the engine's combustion chamber. Additionally a non-2T-specific oil can turn to gum in a matter of days if mixed with gasoline and not immediately consumed. Another important factor is that 4-stroke engines have a different requirement for 'stickiness' than 2-strokes do. Since the 1980s different types of two-stroke oil have been developed for specialized uses such as outboard motor two-strokes, premix two-stroke oil, as well as the more standard auto lube (motorcycle) two-stroke oil. As a rule of thumb, most containers of oil commercially offered will have somewhere on the label printed that it is compatible with 'Autolube' or injector pumps. Those bottles tend to have the consistency of liquid dish soap if shaken. A more viscous oil cannot reliably be passed through an injection system, although a premix machine can be run on either type.

"Racing" oil or castor-based does offer excellent lubricity - at the expense of premature coking. For the average moped/scooter/trail rider it will not garner an appreciable increase in performance and will require very frequent teardowns.

Additive Ingredients

Additives for two-stroke oils fall into several general categories: Detergent/Dispersants, Antiwear agents, Biodegradability components and antioxidants (Zinc compounds). Some of the higher quality include a fuel stabilizer as well.

Automatic Lubrication

Automatic lubrication (also called autolube or auto-lube) refers to a lubrication system on a two-stroke engine, in which the oil is automatically mixed with fuel and manual oil-fuel pre-mixing is not necessary. The oil is contained in a reservoir that connects to a small oil pump in the engine, which needs to be periodically refilled.

This system is commonly used for motorcycles as it eliminates the need of pre-mixing fuel and two-stroke oil. Vespa Sprint is an example where pre-mixing of two-stroke oil is required. Automatic lubrication was introduced for motorcycles by Velocette in 1913.

An example of application of automatic lubrication system is Suzuki AX100 motorcycle. The motorcycle has a separate oil reservoir on its right side which supplies the cylinder with two-stroke oil proportional to engine speed.

Oil injection pump on a Yamaha DX100- just behind the carburettor (visible on the left) It is the primary component of two-stroke Automatic Lubrication System. Amount of two-stroke oil injected by the pump depends on the throttle position. A cable from the throttle is connected to the oil pump indicating throttle's position. A tube ensures flow of oil from the reservoir to the oil pump.

Advantages

1. Consistent lubrication and oil consumption is reduced greatly

2. More effective lubrication results because the oil enters the engine in larger size droplets

3. There is much less unwanted carbon deposited on the spark plugs, cylinder heads, pistons and exhaust system.

4. There is much less exhaust smoke

5. Refueling is simplified

Disdvantages

1. The system is more complicated compared to manual pre-mixing, although it is easier for the end user.

2. For any reason, if the two-stroke oil pump fails to operate properly, chance of damaging the engine is very high.

3. The two-stroke oil tank in scooters and motorcycles is usually hidden from direct view of the rider and needs filling up occasionally. Without any indicator to indicate oil level, it is possible for a novice rider to forget to fill up the oil tank. This can end up starving the engine of oil and cause damage.

Total-loss Oiling System

A total-loss oiling system is an engine lubrication system whereby oil is introduced into the engine, and then either burned or ejected overboard. Now rare in four-stroke engines, total loss oiling is still used in many two-stroke engines.

Sight-glass lubricator.A needle valve adjusts the rate of flow, which may be seen as drops passing through the window beneath the glass reservoir.

Steam Engines

Steam engines used many separate oil boxes, dotted around the engine. Each one was filled before starting and often refilled during running. Where access was difficult, usually because the oil box was on a moving component, the oil box had to be large enough to contain enough oil for a long working shift. To control the flow rate of oil from the reservoir to the bearing, the oil would flow through an oil wick by capillary action, rather than downwards under gravity.

On steamships that ran their engines for days at a time, some crew members would be "oilers" whose prime duty was to continuously monitor and maintain oil boxes.

Displacement lubricator for adding oil to a steam supply

On steam locomotives, access would be impossible during running, so in some cases centralised mechanical lubricators were used. These devices comprised a large oil tank with a multiple-outlet pump which fed the engine's bearings through a pipe system. Lubrication of the engine's internal valves was done by adding oil to the steam supply, using a displacement lubricator.

Oil Recirculation

The first recirculating systems used a collection sump, but no pumped circulation, merely 'splash' lubrication where the connecting rod dipped into the oil surface and splashed it around. These first appeared on high-speed steam engines. Later, splash lubrication engines added a 'dipper', a metal rod whose only function was to dip into the oil and spread it around.

As engines became faster and more powerful, the amount of oil required became so great that a total loss system would have been impractical, both technically and for cost.

Splash lubrication was also used on the first internal combustion engines. It persisted for some time, even in the first high-performance cars. One of Ettore Bugatti's first technical innovations was a minor improvement to the splash lubrication of crankshafts, helping to establish his reputation as an innovative engineer.

A more sophisticated form of splash lubrication, long-used for rotating motor shafts rather than reciprocating engines, was the ring oiler.

Pumped Oil

Later systems collect oil in a sump, from where it can be collected and pumped around the engine again, usually after rudimentary filtering. This system has long been the norm for larger internal combustion engines.

A pumped oil system can use higher oil pressures and so makes the use of hydrostatic bearings easier. These gave a greater load capacity and soon became essential for small, lightweight engines such as in cars. It was this bearing design that saw the end of splash lubrication and total loss oiling. It disappeared from nearly all cars in the 1920s, although total loss continued in small low power stationary engines into the 1950s. Chevrolet used splash lubrication for their rod bearings until 1953, where it was phased out for the 235 'Six,' and then in 1954 when the 216 was eliminated from their line, and both the solid lifter and hydraulic lifter versions of the 235 had full-pressure lubrication.

Two Stroke Engines and Petroil Mixtures

Two-stroke engines have a total-loss lubrication system. Lubricating oil is mixed with the fuel, either manually beforehand (the petroil method), or automatically via an oil pump. Prior to being burned in the combustion chamber, this air/fuel/oil mixture passes through the engine's crankcase, lubricating the moving parts as it does so. In order to reduce exhaust smoke, the Kawasaki H2 750 cc (46 cu in) 2-stroke triple motorcycle had a scavenge pump with a spring-loaded ball-valve under each crankcase to return surplus oil to the tank for reuse.

Wankel Engines

Wankel engines are internal combustion engines using an eccentric rotary design to convert pressure into rotating motion. These engines exhibit some features of both four stroke and two stroke engines. Lubrication is total loss, but there may be some variations. For instance, the MidWest AE series of wankel aero-engines were not only both water-cooled and air-cooled, but also the engine had a lubrication system is a semi-total-loss system. Silkolene 2-stroke oil was directly injected into the inlet tracts and onto the main roller bearings. The oil that entered the combustion chamber lubricated the rotor tips and was then total-loss, but the oil that fed the bearings became a mist within the rotor-cooling air, and around 30% of that oil was recovered and returned to the remote oil tank.

Dry Sump

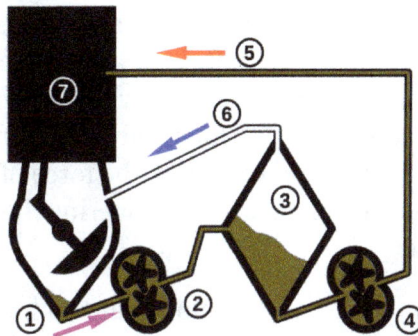

Schematic diagram of a basic dry sump engine lubrication system.Engine lubricating oil collects in sump (1), is withdrawn continuously by scavenge pump (2) and travels to the reservoir (3), where gases entrained in the oil separate and the oil cools. Gases (6) are returned to the engine sump. Pressure pump (4) forces the de-gassed and cooled oil (5) back to the engine cylinder head (7).

A dry sump is a lubricating motor oil management method for four-stroke and large two-stroke piston internal combustion engines that uses two or more oil pumps and a secondary reservoir for oil, as compared to a conventional wet sump system, which uses only the main sump (U.S.: oil pan) below the engine and a single pump. A dry sump can also be implemented with a single vacuum pump that creates positive and negative pressures to pull and push the oil out and into the engine. A dry sump engine requires a pressure relief valve to regulate negative pressure inside the engine, so internal seals are not inverted.

Engines are both lubricated and cooled by oil that circulates throughout the engine, feeding various bearings and other moving parts and then draining, via gravity, into the sump at the base of the engine. In the wet sump system of most production automobile engines, a pump collects this oil from the sump and circulates it back through the engine.

In a dry sump, the oil still falls to the base of the engine, but rather than collecting in a reservoir-style oil sump, it falls into a much shallower sump, where one or more scavenge pumps draw it off and transfer it to a (usually external) reservoir, where it is both cooled and deaerated. A pressure pump then draws this oil and circulates it through the engine. Often, dry sump designs mount the pressure pump and scavenge pumps on a common shaft, so that one pulley at the front of the system can run as many pumps as the engine design requires. It is common practice to have one scavenge pump per crankcase section and in the case of a V-type engine an additional scavenge pump to remove oil fed to the valve gear. Therefore, a V8 engine would have four scavenge pumps and a pressure pump in the pump *stack*.

Advantages

A dry sump offers many advantages. The most obvious are increased oil capacity afforded by the remote reservoir, and the capability to mount the engine lower in the vehicle because of the lower sump profile—lowering the overall center of gravity. The external reservoir can also be relocated to another part of the car to improve weight distribution. Increased oil capacity by using a larger external reservoir than would be practical in a wet-sump system cools the oil more and releases entrained gasses from ring blow-by and the action of the crankshaft. Increased oil capacity also aids with cooling because of the longer time it takes to heat saturate the oil. Furthermore, dry-sump designs are not susceptible to the oil movement problems from high cornering forces that wet sump systems can suffer. In a wet sump, the force of the vehicle cornering can force the oil to one side of the oil pan, possibly uncovering the oil pump pickup tube and causing a loss of oil pressure.

Because scavenge pumps are typically mounted at the lowest point on the engine, the oil flows into the pump intake by gravity rather than having to be lifted up into the intake of the pump as in a wet sump. Also, the scavenge pumps can be of a design that is more tolerant of entrained gasses than the typical pressure pump, which can lose suction if too much air mixes into the oil. Since the pressure pump is typically lower than the external oil tank, it always has a positive pressure on its suction regardless of cornering forces. Another phenomenon that occurs in high-performance car engines is oil frothing up inside the crank-case due to the very high revs agitating the oil. Lastly, having the pumps external to the engine makes them easier to maintain or replace.

Dry sumps are common on larger diesel engines such as those used for ship propulsion. Many racing cars, high performance sports cars, and aerobatic aircraft also use dry-sump equipped engines because they prevent oil-starvation at high *g* loads, and because their lower center of gravity positively affects performance.

The main purpose of the dry sump system is to contain all the stored oil in a separate tank, or reservoir. This reservoir is usually tall and round or narrow and specially designed with internal baffles, and an oil outlet (supply) at the very bottom for uninhibited oil supply. The dry sump oil pump is a minimum of 2 stages, with as many as 5 or 6. One stage is for pressure and is supplied the oil from the bottom of the reservoir, and along with an adjustable pressure regulator, supplies the oil under pressure through the filter and into the engine. The remaining stages "scavenge" the oil out of the dry sump pan and return the oil (and air) to the top of the tank or reservoir. If an oil cooler is used usually it is mounted inline between the scavenge outlets and the tank. The dry sump pump is usually driven by a Gilmer or HTD timing belt and pulleys, off the front of the crankshaft, at approximately one half crank speed. The dry sump pump is designed with multiple stages, to insure that all the oil is scavenged from the pan. This also results in removing excess air from the crankcase, and is the reason they are called "dry sump" meaning the oil pan is essentially dry. Increased engine reliability from the consistent oil pressure provided by the dry sump system is the reason dry sumps were invented. Other benefits mentioned earlier are: shallower oil pan allowing engine to be lowered in chassis, horsepower increase due to less viscous drag (oil resistance due to sloshing into rotating assembly) and cooler oil.

Disadvantages

Dry sump systems add cost, complexity, and weight. The extra pumps and lines require additional oil and maintenance. Also, the performance-enhancing features of dry sump lubrication can hurt a car's day-to-day driveability. A good example is the classic Mercedes-Benz 300SL, a car that was designed for racing but sold to the general public and used on-road. The car had high oil capacity and a dry sump system to cope with continuous high-speed running while racing. Owners found in general use, however, that the oil never achieved the correct operating temperature because the system was so efficient at cooling the oil. A makeshift solution was devised to deliberately block the oil cooler airflow to boost the oil temperature.

Dry Sump Motorcycle Engines

The advantages of dry sump lubrication are particularly beneficial to motorcycles, which tend to be ridden (driven) more vigorously than other road vehicles. The classic British parallel twin motorcycles such as BSA, Triumph and Norton all used dry sump lubrication. Traditionally, the oil tank was a remote item, but some late-model BSAs and the Meriden Triumphs used "oil-in-the-frame" designs. Although motorcycles such as the Honda CB750 (1969) features a dry-sump engine, modern motorcycles tend to use a wet-sump design. This is understandable with across-the-frame inline four-cylinder engines, since these wide engines must be mounted fairly high in the frame (for ground clearance), so the space below may as well be used for a wet sump. However, narrower engines can be mounted lower and ideally should use dry sump lubrication.

The Yamaha TRX850 270-degree parallel twin motorcycle has a dry sump engine with its oil reservoir not remote, but integral to the engine, sitting atop the gearbox. This design eliminates external oil lines, allowing simpler engine removal and providing faster oil warm up.

The Yamaha SR400/500 uses a dry sump design where the bike's frame tubing doubles as the oil reservoir and cooling system.

Harley-Davidson has used dry sump type lubricating oil systems in their engines since the 1930s.

The Rotax engined Aprilia RSV Mille, and the Aprilia RST1000 Futura both incorporate a dry sump, along with sister bikes, the SL1000 Falco and ETV1000 Caponord.

The Honda XR500R, XR600R, XR650R and XR650L four stroke dirt bikes utilize a dry sump with the oil in the frame tubing.

Grease (Lubricant)

Grease is a semisolid lubricant. Grease generally consists of a soap emulsified with mineral or vegetable oil. The characteristic feature of greases is that they possess a high initial viscosity, which upon the application of shear, drops to give the effect of an oil-lubricated bearing of approximately the same viscosity as the base oil used in the grease. This change in viscosity is called shear thinning. Grease is sometimes used to describe lubricating materials that are simply soft solids or high viscosity liquids, but these materials do not exhibit the shear-thinning properties characteristic of the classical grease. For example, petroleum jellies such as Vaseline are not generally classified as greases.

Greases are applied to mechanisms that can only be lubricated infrequently and where a lubricating oil would not stay in position. They also act as sealants to prevent ingress of water and incompressible materials. Grease-lubricated bearings have greater frictional characteristics due to their high viscosity.

Properties

A *true* grease consists of an oil and/or other fluid lubricant that is mixed with a thickener, typically a *soap*, to form a solid or semisolid. Greases are a type of *shear-thinning* or pseudo-plastic fluid, which means that the viscosity of the fluid is reduced under shear. After sufficient force to shear the grease has been applied, the viscosity drops and approaches that of the base lubricant, such as the mineral oil. This sudden drop in shear force means that grease is considered a plastic fluid, and the reduction of shear force with time makes it thixotropic. It is often applied using a grease gun, which applies the grease to the part being lubricated under pressure, forcing the solid grease into the spaces in the part.

Thickeners

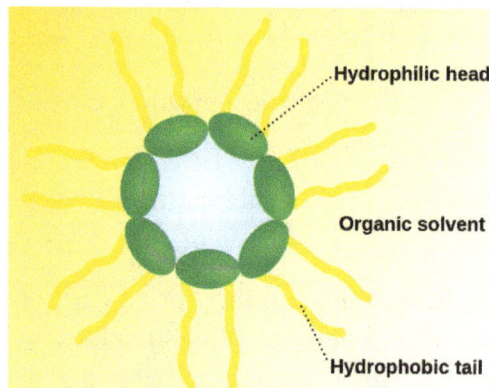

An inverse micelle formed when a soap is dispersed in an oil. This structure is broken reversibly upon shearing the grease.

Soaps are the most common emulsifying agent used, and the selection of the type of soap is determined by the application. Soaps include calcium stearate, sodium stearate, lithium stearate, as well as mixtures of these components. Fatty acids derivatives other than stearates are also used, especially lithium 12-hydroxystearate. The nature of the soaps influences the temperature resistance (relating to the viscosity), water resistance, and chemical stability of the resulting grease.

Powdered Solid Greases

Powdered solids may also be used as thickeners, especially as clays, which are used in some inexpensive, low performance greases. Fatty oil-based greases have also been prepared with other thickeners, such as tar, graphite, or mica, which also increase the durability of the grease.

Engineering Assessment and Analysis of Greases

Lithium-based greases are the most commonly used; sodium and lithium-based greases have higher melting point (dropping point) than calcium-based greases but are not resistant to the action of water. Lithium-based grease has a dropping point at 190 to 220 °C (350 to 400 °F). However the maximum usable temperature for lithium-based grease is 120 °C.

The amount of grease in a sample can be determined in a laboratory by extraction with a solvent followed by e.g. gravimetric determination.

Additives

Solid lubricating additives including PTFE, graphite, and molybdenum disulphide are added to some greases to improve their lubricating properties. Gear greases consist of rosin oil, condensed with lime and stirred with mineral oil, with some percentage of water. Special-purpose greases contain glycerol and sorbitan esters. They are used, for example, in low-temperature conditions. Some greases are labeled "EP", which indicates "extreme pressure". Under high pressure or shock loading, normal grease can be compressed to the extent that the greased parts come into physical contact, causing friction and wear. EP grease contains solid lubricants, usually graphite and/or molybdenum disulfide, to provide protection under heavy loadings. The solid lubricants bond to the surface of the metal, and prevent metal-to-metal contact and the resulting friction and wear when the lubricant film gets too thin.

Copper is added to some greases for high pressure applications, or where corrosion could prevent dis-assembly of components later in their service life. Copaslip is the registered trademark of one such grease produced by Molyslip Atlantic Ltd, and has become a generic term (often misspelled as "copperslip" or "coppaslip") for anti-seize lubricants which contain copper.

History

Grease from the early Egyptian or Roman eras is thought to have been prepared by combining lime with olive oil. The lime saponifies some of the triglyceride that comprises oil to give a calcium grease. In the middle of the 19th century, soaps were intentionally added as thickeners to oils. Over the centuries, all manner of materials have been employed as greases. For example, black slugs *Arion ater* were used as axle-grease to lubricate wooden axle-trees or carts in Sweden.

Classification and Standards

Red wheel bearing grease for automotive applications.

Jointly developed by ASTM International, the National Lubricating Grease Institute (NLGI) and SAE International, standard ASTM D4950 *"standard classification and specification for automotive service greases"* was first published in 1989 by ASTM International. It categorizes greases suitable for the lubrication of chassis components and wheel bearings of vehicles, based on performance requirements, using codes adopted from the NLGI's *"chassis and wheel bearing service classification system"*:

- LA and LB: chassis lubricants (suitability up to mild and severe duty respectively)

- GA, GB and GC: wheel-bearings (suitability up to mild, moderate and severe duty respectively)

A given performance category may include greases of different consistencies.

The measure of the consistency of grease is commonly expressed by its NLGI consistency number.

The main elements of standard ATSM D4950 and NLGI's consistency classification are reproduced and described in standard SAE J310 *"automotive lubricating greases"* published by SAE International.

Standard ISO 6743-9 *"lubricants, industrial oils and related products (class L) — classification — part 9: family X (greases)"*, first released in 1987 by the International Organization for Standardization, establishes a detailed classification of greases used for the lubrication of equipment, components of machines, vehicles, etc. It assigns a single multi-part code to each grease based on its operational properties (including temperature range, effects of water, load, etc.) and its NLGI consistency number.

Other Greases

Silicone Grease

Silicone grease is an amorphous fumed-silica thickened, polysiloxane-based compound, which can be used to provide lubrication and corrosion resistance. Since it is not oil-based, it is often used where oil-based lubricants would attack rubber seals. Silicone greases also maintain stability under high temperatures. They are often used, in pure form or mixed with zinc oxide, to join heat sinks to computer CPUs.

Fluoroether-based Grease

Fluoropolymers containing C-O-C (ether) bonds for flexibility are soft, and often used as greases in demanding environments due to their inertness. Fomblin by Solvay Solexis and Krytox by duPont are prominent examples.

Laboratory Grease

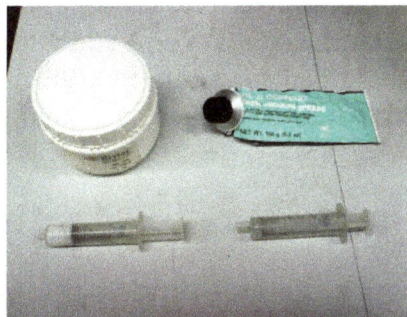

Grease is used to lubricate glass stopcocks and joints. Some laboratories fill them into syringes for easy application. Two typical examples: Left - Krytox, a fluoroether-based grease; Right - a silicone-based high vacuum grease by Dow Corning.

Apiezon, silicone-based, and fluoroether-based greases are all used commonly in laboratories for lubricating stopcocks and ground glass joints. The grease helps to prevent joints from "freezing", as well as ensuring high vacuum systems are properly sealed. Apiezon or similar hydrocarbon based greases are the cheapest, and most suitable for high vacuum applications. However, they dissolve in many organic solvents. This quality makes clean-up with pentane or hexanes trivial, but also easily leads to contamination of reaction mixtures.

Silicone-based greases are cheaper than fluoroether-based greases. They are relatively inert and generally do not affect reactions, though reaction mixtures often get contaminated (detected through NMR near δ 0). Silicone-based greases are not easily removed with solvent, but they are removed efficiently by soaking in a base bath.

Fluoroether-based greases are inert to many substances including solvents, acids, bases, and oxidizers. They are, however, expensive, and are not easily cleaned away.

Food-grade Grease

Food-grade greases are those greases that come in contact with food. Food-grade lubricant base oil are generally low sulfur petrochemical, less easily oxidized and emulsified. Another commonly used poly-α olefin base oil as well The United States Department of Agriculture (USDA) has three food-grade designations: H1, H2 and H3. H1 lubricants are food-grade lubricants used in food-processing environments where there is the possibility of incidental food contact. H2 lubricants are food-grade lubricants used on equipment and machine parts in locations with no possibility of contact. H3 lubricants are food-grade lubricants, typically edible oils, used to prevent rust on hooks, trolleys and similar equipment.

Water-soluble Grease Analogs

In some cases, the lubrication and high viscosity of a grease are desired in situations where non-toxic, non-oil based materials are required. Carboxymethyl cellulose, or CMC, is one popular material used to create a water-based analog of greases. CMC serves to both thicken the solution and add a lubricating effect, and often silicone-based lubricants are added for additional lubrication. The most familiar example of this type of lubricant, used as a surgical and personal lubricant, is K-Y Jelly.

Cork Grease

Cork Grease is a lubricant used to lubricate cork, for example in musical wind instruments. It is usually applied using small lip-balm/lip-stick like applicators.

Dropping Point

The dropping point of a soap-thickened lubricating grease is the temperature at which it passes from a semi-solid to a liquid state under specific test conditions. It is an indication of the type of thickener used, and a measure of the cohesiveness of the oil and thickener of a grease.

Dropping point is used in combination with other testable properties to determine the suitability of greases for specific applications. It is applicable only to greases that contain soap thickeners. Greases with other thickeners, such as many synthetic greases, do not change state. Instead, they separate oil, and the dropping point as a phase transition does not apply.

ASTM Test Procedure

The dropping point test procedures are given in ASTM standards D-566 and D-2265. The test apparatus consists of a grease cup with a small hole in the bottom, test tube, two thermometers, a container, stirring device if required and an electric heater. The inside surfaces of the grease cup are coated with the grease to be tested. A thermometer is inserted into the cup and held in place so that the thermometer does not touch the grease. This assembly is placed inside a test tube. The test tube is lowered into the container which is filled with oil in D-566 and has an aluminum block in D-2265. Another thermometer is inserted into the oil/block.

To execute a test, the oil/block is heated, while being stirred, at a rate of 8 °F (4.4 °C) to 12 °F (6.7 °C) per minute until the temperature is approximately 30 °F (17 °C) below the expected dropping point. The heat is reduced until the test tube temperature is at most 4 °F (2.2 °C) less than the oil/block temperature. Once the temperature has stabilized the sample is inserted. The dropping point is the temperature recorded on the test tube thermometer, plus a correction factor for the oil/block temperature, when a drop of grease falls through the hole in the grease cup. If the drop trails a thread, the dropping temperature is the temperature at which the thread breaks. D-2265 explains that the dropping point is useful to assist in identifying the type of grease, and for establishing and maintaining benchmarks for quality control. It adds that the results are not sufficient to assess service performance because dropping point is a static test.

Other Test Procedures

Equivalent to D566 and D2265:

- IP 132

- ISO 2176:1995 Petroleum products—Lubricating grease—Determination of dropping point

- DIN 51806

Other:

- National Standard of People's Republic of China GB/T 4929 "Test Methods for Dropping Point of Grease"

- S 1448(P-52)

- GOST 7134-73, Method B

- JIS K2220:2003 Lubricating grease

- DIN 51801 Dropping Point of Lubricating Grease

Silicone Grease

Silicone grease is a waterproof grease made by combining a silicone oil with a thickener. Most commonly, the silicone oil is polydimethylsiloxane (PDMS) and the thickener is amorphous fumed silica. Using this formulation, silicone grease is a translucent white viscous paste, with exact properties dependent on the type and proportion of the components. More specialized silicone greases are made from fluorinated silicones or, for low temperature applications, PDMS containing some phenyl substituents in place of methyl groups. For food applications, the thickener is calcium stearate. For applications involving highly reactive substances, powdered Teflon is the thickener.

Use in Industry

Silicone grease is commonly used for lubricating and preserving rubber parts, such as O-rings. Additionally, silicone grease does not swell or soften the rubber, which can be a problem with hydrocarbon based greases. It functions well as a corrosion-inhibitor and lubricant for purposes that require a thicker lubricant.

Thermal grease often consists of a silicone grease base, along with added thermally conductive fillers. It is used for heat transfer abilities, rather than friction reduction.

Special versions of silicone grease are also used widely by the plumbing industry in faucets and seals, as well as dental equipment. These special versions are formulated using components not known to be an ingestion hazard. Electrical utilities use silicone grease to lubricate separable elbows on lines which must endure high temperatures. Silicone greases generally have an operating temperature range of approximately −40 to 200 °C (−40 to 392 °F) with some high-temperature versions extending that range slightly.

Use in the Chemical Laboratory

Silicone grease is widely used as a temporary sealant and a lubricant for interconnecting ground glass joints, as is typically used in laboratory glassware. Although silicones are normally assumed to be chemically inert, several historically significant compounds have resulted from unintended reactions with silicones. The first salts of crown ethers $(OSi(CH_3)_2)_n$ (n = 6, 7) were produced by reactions of organolithium and organopotassium compounds with silicone greases or the serendipitous reaction of stannanetriol with silicone grease to afford a cage-like compound having three Sn-O-Si-O-Sn linkages in the molecule.

Silicone grease is soluble in organic solvents, and lubrication of an apparatus with silicone grease may result in the reaction mixture being contaminated with the grease. The impurity may be carried through purification by chromatography in undesirable amounts. In NMR spectroscopy, the methyl groups in polydimethylsiloxane display ${}^{1}H$ and ${}^{13}C$ chemical shifts similar to trimethylsilane (TMS), the reference compound for those forms of NMR spectroscopy. As with TMS, the signal is a singlet. In ${}^{1}H$ NMR, silicone grease appears at a singlet at δ = 0.07 ppm in $CDCl_3$, 0.09 in CD_3CN, 0.29 in C_6D_6, and -0.06 ppm in $(CD_3)_2SO$. In ^{13}C NMR, it appears at δ = 1.19 ppm in $CDCl_3$ and 1.38 ppm in C_6D_6. Tables of impurities commonly found in NMR spectroscopy have been prepared, and such tables include silicone grease.

Consumer Uses

Silicone-based lubricants are often used by consumers in applications where other common consumer lubricants, such as petroleum jelly, would damage certain products, such as latex rubber condoms and gaskets on dry-suits. It can be used to lubricate fountain pen filling mechanisms and threads. It is used to seal and preserve O-rings in flashlights, plumbing, waterproof watches, and air rifles. Silicone grease is widely used to lubricate threads of water submersible flashlights used for diving and spearfishing. This grease improves water resistance of the flashlights and protects threads from wearing out. Silicone grease is used with waterproof devices as it has a very thick body and doesn't dissolve in water, as most spirits and other liquids would. Silicone-based lubricants are also commonly used for remote control hobbies, or for the insulation of CPU sockets when Overclocking at very low temperatures.

Various household uses include lubricating door hinges, shower heads, threads on bolts, garden hose threads or any thread or mechanism that can be lubricated.

As a Sealant Around Electrical Contacts

Dielectric Grease

Dielectric grease is electrically insulating and does not break down when high voltage is applied. It is often applied to electrical connectors, particularly those containing rubber gaskets, as a means of lubricating and sealing rubber portions of the connector without arcing.

A common use of dielectric grease is in high-voltage connections associated with gasoline engine spark plugs. The grease is applied to the rubber boot of the plug wire. This helps the rubber boot slide onto the ceramic insulator of the plug. The grease also acts to seal the rubber boot, while at the same time preventing the rubber from becoming stuck to the ceramic. Generally, spark plugs are located in areas of high temperature and the grease is formulated to withstand the temperature range expected. It can be applied to the actual contact as well, because the contact pressure is sufficient to penetrate the grease film. Doing so on such high pressure contact surfaces between different metals has the advantage of sealing the contact area against electrolytes that might cause rapid deterioration from galvanic corrosion.

Another common use of dielectric grease is on the rubber mating surfaces or gaskets of multi-pin electrical connectors used in automotive and marine engines. The grease again acts as a lubricant and a sealant on the nonconductive mating surfaces of the connector. It is not recommended to be applied to the actual electrical conductive contacts of the connector because it could interfere with the electrical signals passing through the connector in cases where the contact pressure is very low. Products designed as electronic connector lubricants, on the other hand, should be applied to such connector contacts and can dramatically extend their useful life. Polyphenyl Ether, rather than silicone grease, is the active ingredient in some such connector lubricants.

Silicone grease should not be applied to (or next to) any switch contact that might experience arcing, as silicone can convert to silicon-carbide under arcing conditions, and accumulation of the silicon-carbide can cause the contacts to prematurely fail. (British Telecom had this problem in the 1970s when silicone Symel® sleeving was used in telephone exchanges. Vapor from the sleeving migrated to relay contacts and the resultant silicon-carbide caused intermittent connection.)

Ramsay Grease

Ramsay grease (Ramsay-Fett in German) is a vacuum grease, used as a lubrication and a sealant of ground glass joints and cocks on laboratory glassware, e.g. burettes. It is usable to about 10^{-2} mbar (about 1 Pa) and about 30 °C. Its vapor pressure at 20 °C is about 10^{-4} mbar (0.01 Pa). It is named after Sir William Ramsay.

Different grades exist (e.g. thick or viscous, soft). The viscous one is used for standard stopcocks and ground joins. The soft grade is for large stopcocks and ground joints, desiccators, and for lower temperature use. Ramsay grease consists of paraffin wax, vaseline, and crude natural rubber, in ratio 1:3:7 to 1:8:16. Due to the rubber content it has less tendency to flow.

One recipe for a grease usable up to 25 °C consists of 6 parts of vaseline, 1 part of paraffin wax, and 6 parts of Pará rubber.

The dropping point of Leybold-brand Ramsay grease is 56 °C; its maximum service temperature is 25-30 °C. Its vapor pressure at 25 °C is 10^{-7} torr (0.013 mPa), at 38 °C it is 10^{-4} torr (13 mPa).

An equivalent of Ramsay grease can be made by cooking lanolin with natural rubber extracted from golf balls.

References

- G. Corsico, L. Mattei, A. Roselli and C. Gommellini, Poly(internal olefins)- Synthetic Lubricants and high-performance functional fluids,, Marcel Dekker, 1999,Chapter 2, p. 53-62, ISBN 0-8247-0194-1

- Chris Collins (2007), "Implementing Phytoremediation of Petroleum Hydrocarbons, Methods in Biotechnology" 23:99-108. Humana Press. ISBN 1-58829-541-9.

- Company, DIANE Publishing (1994-04-01). How to Set Up a Local Program to Recycle Used Oil. DIANE Publishing. ISBN 9780788106576.

- Nunney, Malcom J. (2007). Light and Heavy Vehicle Technology (4th ed.). Elsevier Butterworth-Heinemann. p. 7. ISBN 978-0-7506-8037-0.

- Smith, Philip H. (1965). The High-Speed Two-Stroke Petrol Engine. London: Foulis. pp. 236–237. ISBN 085429-049-4.

- Rand, Salvatore J., ed. (2003). Significance of tests for petroleum products. "ASTM manual" series, volume 1 (7th ed.). ASTM International. p. 166. ISBN 978-0-8031-2097-6.

- Totten, G.E., Handbook of Lubrication and Tribology Volume 1: Application and Maintenance, CRC Press, 2006, ISBN 0-8493-2095-X

- "Technical Description - The Dry Sump System". Armstrong Race Engineering, Gary Armstrong, DrySump.com, 08-03-2016.

- Lucian C. Pop and M. Saito (2015). "Serendipitous Reactions Involving a Silicone Grease". Coordination Chemistry Reviews. doi:10.1016/j.ccr.2015.07.005.

- "Article 4. Recycled Materials, Goods, and Supplies - California Public Contract Code Section 12209". Retrieved 25 September 2015.

- "FAQs - Used Motor Oil Collection and Recycling - American Petroleum Institute". www.recycleoil.org. Retrieved 2015-10-28.

- "Understanding JASO MA and MB: Specific Performance for the Right Applications". www.mceo.com. The Lubrizol Corporation. Retrieved 6 November 2015.

- "Two-stroke motorcycle oils - JASO M345 standard". www.mceo.com. The Lubrizol Corporation. Retrieved 23 January 2014.

- Vijayendran, Bhima (2014-03-24). "Biobased Motor Oils Are Ready for Primetime". Industrial Biotechnology. 10 (2): 64–68. doi:10.1089/ind.2014.1505. ISSN 1550-9087.

- Lucian C. Pop; et al. (2014). "Synthesis and structures of monomeric group 14 triols and their reactivity". Canadian Journal of Chemistry. 92 (6): 542–548. doi:10.1139/cjc-2013-0496.

- What is Honing Oil?. Complete Multi-tool Sharpening Kit. Swiss Army Supplies Website. 2011. Retrieved 13 December 2012.

Friction: An Overview

The force that causes resistance in two surfaces that come in contact with each other is termed as friction. Friction is the phenomenon behind various devices and concepts found in transportation such as brakes, adhesion railways and road slipperiness. This section is an overview of the topics incorporating all the major aspects of friction.

Friction

Friction is the force resisting the relative motion of solid surfaces, fluid layers, and material elements sliding against each other. There are several types of friction:

- Dry friction resists relative lateral motion of two solid surfaces in contact. Dry friction is subdivided into *static friction* ("stiction") between non-moving surfaces, and *kinetic friction* between moving surfaces.

- Fluid friction describes the friction between layers of a viscous fluid that are moving relative to each other.

- Lubricated friction is a case of fluid friction where a lubricant fluid separates two solid surfaces.

- Skin friction is a component of drag, the force resisting the motion of a fluid across the surface of a body.

- Internal friction is the force resisting motion between the elements making up a solid material while it undergoes deformation.

When surfaces in contact move relative to each other, the friction between the two surfaces converts kinetic energy into thermal energy (that is, it converts work to heat). This property can have dramatic consequences, as illustrated by the use of friction created by rubbing pieces of wood together to start a fire. Kinetic energy is converted to thermal energy whenever motion with friction occurs, for example when a viscous fluid is stirred. Another important consequence of many types of friction can be wear, which may lead to performance degradation and/or damage to components. Friction is a component of the science of tribology.

Friction is not itself a fundamental force. Dry friction arises from a combination of inter-surface adhesion, surface roughness, surface deformation, and surface contamination. The complexity of these interactions makes the calculation of friction from first principles impractical and necessitates the use of empirical methods for analysis and the development of theory.

Friction is a non-conservative force - work done against friction is path dependent. In the presence of friction, some energy is always lost in the form of heat. Thus mechanical energy is not conserved.

History

The Greeks, including Aristotle, Vitruvius, and Pliny the Elder, were interested in the cause and mitigation of friction. They were aware of differences between static and kinetic friction with Themistius stating in 350 A.D. that "it is easier to further the motion of a moving body than to move a body at rest".

The classic laws of sliding friction were discovered by Leonardo da Vinci in 1493, a pioneer in tribology, but the laws documented in his notebooks, were not published and remained unknown. These laws were rediscovered by Guillaume Amontons in 1699. Amontons presented the nature of friction in terms of surface irregularities and the force required to raise the weight pressing the surfaces together. This view was further elaborated by Bernard Forest de Bélidor and Leonhard Euler (1750), who derived the angle of repose of a weight on an inclined plane and first distinguished between static and kinetic friction. A different explanation was provided by John Theophilus Desaguliers (1725), who demonstrated the strong cohesion forces between lead spheres of which a small cap is cut off and which were then brought into contact with each other.

The understanding of friction was further developed by Charles-Augustin de Coulomb (1785). Coulomb investigated the influence of four main factors on friction: the nature of the materials in contact and their surface coatings; the extent of the surface area; the normal pressure (or load); and the length of time that the surfaces remained in contact (time of repose). Coulomb further considered the influence of sliding velocity, temperature and humidity, in order to decide between the different explanations on the nature of friction that had been proposed. The distinction between static and dynamic friction is made in Coulomb's friction law, although this distinction was already drawn by Johann Andreas von Segner in 1758. The effect of the time of repose was explained by Pieter van Musschenbroek (1762) by considering the surfaces of fibrous materials, with fibers meshing together, which takes a finite time in which the friction increases.

John Leslie (1766–1832) noted a weakness in the views of Amontons and Coulomb: If friction arises from a weight being drawn up the inclined plane of successive asperities, why then isn't it balanced through descending the opposite slope? Leslie was equally skeptical about the role of adhesion proposed by Desaguliers, which should on the whole have the same tendency to accelerate as to retard the motion. In Leslie's view, friction should be seen as a time-dependent process of flattening, pressing down asperities, which creates new obstacles in what were cavities before.

Arthur Jules Morin (1833) developed the concept of sliding versus rolling friction. Osborne Reynolds (1866) derived the equation of viscous flow. This completed the classic empirical model of friction (static, kinetic, and fluid) commonly used today in engineering. In 1877, Fleeming Jenkin and J. A. Ewing investigated the continuity between static and kinetic friction.

The focus of research during the 20th century has been to understand the physical mechanisms behind friction. Frank Philip Bowden and David Tabor (1950) showed that, at a microscopic level, the actual area of contact between surfaces is a very small fraction of the apparent area. This actual area of contact, caused by "asperities" (roughness) increases with pressure. The development of the atomic force microscope (ca. 1986) enabled scientists to study friction at the atomic scale, showing that, on that scale, dry friction is the product of the inter-surface shear stress and the contact area. These two discoveries explain the macroscopic proportionality between normal force and static frictional force between dry surfaces.

Laws of Dry Friction

The elementary property of sliding (kinetic) friction were discovered by experiment in the 15th to 18th centuries and were expressed as three empirical laws:

- Amontons' First Law: The force of friction is directly proportional to the applied load.

- Amontons' Second Law: The force of friction is independent of the apparent area of contact.

- Coulomb's Law of Friction: Kinetic friction is independent of the sliding velocity.

Dry Friction

Dry friction resists relative lateral motion of two solid surfaces in contact. The two regimes of dry friction are 'static friction' ("stiction") between non-moving surfaces, and *kinetic friction* (sometimes called sliding friction or dynamic friction) between moving surfaces.

Coulomb friction, named after Charles-Augustin de Coulomb, is an approximate model used to calculate the force of dry friction. It is governed by the model:

$$F_f \le \mu F_n,$$

where

- F_f is the force of friction exerted by each surface on the other. It is parallel to the surface, in a direction opposite to the net applied force.

- μ is the coefficient of friction, which is an empirical property of the contacting materials,

- F_n is the normal force exerted by each surface on the other, directed perpendicular (normal) to the surface.

The Coulomb friction F_f may take any value from zero up to μF_n , and the direction of the frictional force against a surface is opposite to the motion that surface would experience in the absence of friction. Thus, in the static case, the frictional force is exactly what it must be in order to prevent motion between the surfaces; it balances the net force tending to cause such motion. In this case, rather than providing an estimate of the actual frictional force, the Coulomb approximation provides a threshold value for this force, above which motion would commence. This maximum force is known as traction.

The force of friction is always exerted in a direction that opposes movement (for kinetic friction) or potential movement (for static friction) between the two surfaces. For example, a curling stone sliding along the ice experiences a kinetic force slowing it down. For an example of potential movement, the drive wheels of an accelerating car experience a frictional force pointing forward; if they did not, the wheels would spin, and the rubber would slide backwards along the pavement. Note that it is not the direction of movement of the vehicle they oppose, it is the direction of (potential) sliding between tire and road.

Normal Force

A block on a ramp

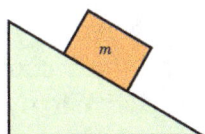

Free body diagram
of just the block

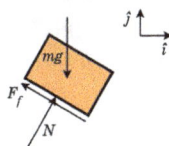

Free-body diagram for a block on a ramp. Arrows are vectors indicating directions and magnitudes of forces. N is the normal force, mg is the force of gravity, and F_f is the force of friction.

The normal force is defined as the net force compressing two parallel surfaces together; and its direction is perpendicular to the surfaces. In the simple case of a mass resting on a horizontal surface, the only component of the normal force is the force due to gravity, where $N = mg$. In this case, the magnitude of the friction force is the product of the mass of the object, the acceleration due to gravity, and the coefficient of friction. However, the coefficient of friction is not a function of mass or volume; it depends only on the material. For instance, a large aluminum block has the same coefficient of friction as a small aluminum block. However, the magnitude of the friction force itself depends on the normal force, and hence on the mass of the block.

If an object is on a level surface and the force tending to cause it to slide is horizontal, the normal force N between the object and the surface is just its weight, which is equal to its mass multiplied by the acceleration due to earth's gravity, g. If the object is on a tilted surface such as an inclined plane, the normal force is less, because less of the force of gravity is perpendicular to the face of the plane. Therefore, the normal force, and ultimately the frictional force, is determined using vector analysis, usually via a free body diagram. Depending on the situation, the calculation of the normal force may include forces other than gravity.

Coefficient of Friction

The coefficient of friction (COF), often symbolized by the Greek letter μ, is a dimensionless scalar value which describes the ratio of the force of friction between two bodies and the force pressing them together. The coefficient of friction depends on the materials used; for example, ice on steel has a low coefficient of friction, while rubber on pavement has a high coefficient of friction. Coefficients of friction range from near zero to greater than one.

For surfaces at rest relative to each other $\mu = \mu_s$, where μ_s is the *coefficient of static friction*. This is usually larger than its kinetic counterpart.

For surfaces in relative motion $\mu = \mu_k$, where μ_k is the *coefficient of kinetic friction*. The Coulomb friction is equal to F_f, and the frictional force on each surface is exerted in the direction opposite to its motion relative to the other surface.

Arthur Morin introduced the term and demonstrated the utility of the coefficient of friction. The coefficient of friction is an empirical measurement – it has to be measured experimentally, and cannot be found through calculations. Rougher surfaces tend to have higher effective values. Both static and kinetic coefficients of friction depend on the pair of surfaces in contact; for a given pair of surfaces, the coefficient of static friction is *usually* larger than that of kinetic friction; in some sets the two coefficients are equal, such as teflon-on-teflon.

Most dry materials in combination have friction coefficient values between 0.3 and 0.6. Values outside this range are rarer, but teflon, for example, can have a coefficient as low as 0.04. A value of zero would mean no friction at all, an elusive property. Rubber in contact with other surfaces can yield friction coefficients from 1 to 2. Occasionally it is maintained that μ is always < 1, but this is not true. While in most relevant applications μ < 1, a value above 1 merely implies that the force required to slide an object along the surface is greater than the normal force of the surface on the object. For example, silicone rubber or acrylic rubber-coated surfaces have a coefficient of friction that can be substantially larger than 1.

While it is often stated that the COF is a "material property," it is better categorized as a "system property." Unlike true material properties (such as conductivity, dielectric constant, yield strength), the COF for any two materials depends on system variables like temperature, velocity, atmosphere and also what are now popularly described as aging and deaging times; as well as on geometric properties of the interface between the materials. For example, a copper pin sliding against a thick copper plate can have a COF that varies from 0.6 at low speeds (metal sliding against metal) to below 0.2 at high speeds when the copper surface begins to melt due to frictional heating. The latter speed, of course, does not determine the COF uniquely; if the pin diameter is increased so that the frictional heating is removed rapidly, the temperature drops, the pin remains solid and the COF rises to that of a 'low speed' test.

Approximate Coefficients of Friction

Materials		Static Friction,		Kinetic/Sliding Friction,	
Dry and clean		Lubricated	Dry and clean	Lubricated	
Aluminium	Steel	0.61		0.47	
Alumina ceramic	Silicon Nitride ceramic				0.004 (wet)
BAM (Ceramic alloy AlMgB$_{14}$)	Titanium boride (TiB$_2$)	0.04–0.05	0.02		
Brass	Steel	0.35-0.51	0.19	0.44	
Cast iron	Copper	1.05		0.29	
Cast iron	Zinc	0.85		0.21	
Concrete	Rubber	1.0	0.30 (wet)	0.6-0.85	0.45-0.75 (wet)
Concrete	Wood	0.62			
Copper	Glass	0.68			
Copper	Steel	0.53		0.36	
Glass	Glass	0.9-1.0		0.4	
Human synovial fluid	Cartilage		0.01		0.003
Ice	Ice	0.02-0.09			
Polyethene	Steel	0.2	0.2		
PTFE (Teflon)	PTFE (Teflon)	0.04	0.04		0.04

Steel	Ice	0.03			
Steel	PTFE (Teflon)	0.04-0.2	0.04		0.04
Steel	Steel	0.74-0.80	0.16	0.42-0.62	
Wood	Metal	0.2-0.6	0.2 (wet)		
Wood	Wood	0.25-0.5	0.2 (wet)		

Under certain conditions some materials have very low friction coefficients. An example is (highly ordered pyrolytic) graphite which can have a friction coefficient below 0.01. This ultralow-friction regime is called superlubricity.

Static Friction

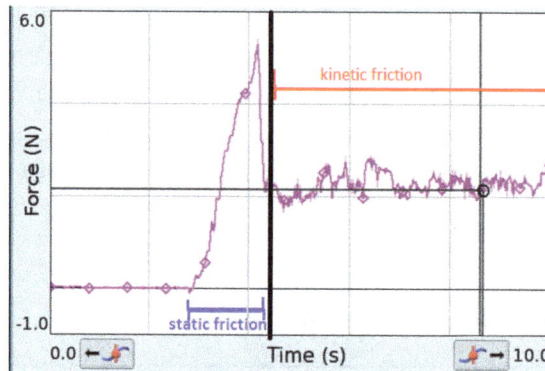

When the mass is not moving, the object experiences static friction. The friction increases as the applied force increases until the block moves. After the block moves, it experiences kinetic friction, which is less than the maximum static friction.

Static friction is friction between two or more solid objects that are not moving relative to each other. For example, static friction can prevent an object from sliding down a sloped surface. The coefficient of static friction, typically denoted as μ_s, is usually higher than the coefficient of kinetic friction.

The static friction force must be overcome by an applied force before an object can move. The maximum possible friction force between two surfaces before sliding begins is the product of the coefficient of static friction and the normal force: $F_{max} = \mu_s F_n$. . When there is no sliding occurring, the friction force can have any value from zero up to F_{max}. Any force smaller than F_{max} attempting to slide one surface over the other is opposed by a frictional force of equal magnitude and opposite direction. Any force larger than F_{max} overcomes the force of static friction and causes sliding to occur. The instant sliding occurs, static friction is no longer applicable—the friction between the two surfaces is then called kinetic friction.

An example of static friction is the force that prevents a car wheel from slipping as it rolls on the ground. Even though the wheel is in motion, the patch of the tire in contact with the ground is stationary relative to the ground, so it is static rather than kinetic friction.

The maximum value of static friction, when motion is impending, is sometimes referred to as limiting friction, although this term is not used universally.

Kinetic Friction

Kinetic friction, also known as dynamic friction or sliding friction, occurs when two objects are moving relative to each other and rub together (like a sled on the ground). The coefficient of kinetic friction is typically denoted as μ_k, and is usually less than the coefficient of static friction for the same materials. However, Richard Feynman comments that "with dry metals it is very hard to show any difference." The friction force between two surfaces after sliding begins is the product of the coefficient of kinetic friction and the normal force: $F_k = \mu_k F_n$.

New models are beginning to show how kinetic friction can be greater than static friction. Kinetic friction is now understood, in many cases, to be primarily caused by chemical bonding between the surfaces, rather than interlocking asperities; however, in many other cases roughness effects are dominant, for example in rubber to road friction. Surface roughness and contact area affect kinetic friction for micro- and nano-scale objects where surface area forces dominate inertial forces.

The origin of kinetic friction at nanoscale can be explained by thermodynamics. Upon sliding, new surface forms at the back of a sliding true contact, and existing surface disappears at the front of it. Since all surfaces involve the thermodynamic surface energy, work must be spent in creating the new surface, and energy is released as heat in removing the surface. Thus, a force is required to move the back of the contact, and frictional heat is released at the front.

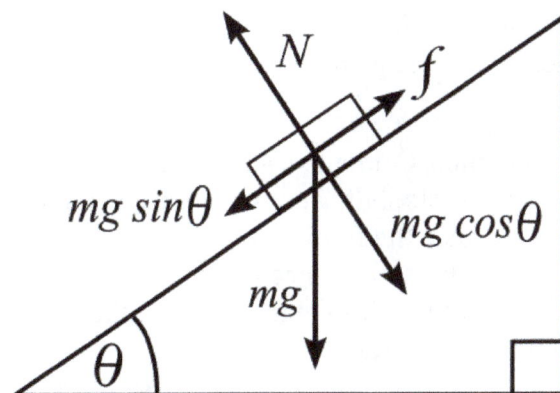

Angle of friction, θ, when block just starts to slide.

Angle of Friction

For certain applications it is more useful to define static friction in terms of the maximum angle before which one of the items will begin sliding. This is called the *angle of friction* or *friction angle*. It is defined as:

$$\tan\theta = \mu_s$$

where θ is the angle from horizontal and μ_s is the static coefficient of friction between the objects. This formula can also be used to calculate μ_s from empirical measurements of the friction angle.

Friction at the Atomic Level

Determining the forces required to move atoms past each other is a challenge in designing nanomachines. In 2008 scientists for the first time were able to move a single atom across a surface, and measure the forces required. Using ultrahigh vacuum and nearly zero temperature (5 K), a modified atomic force microscope was used to drag a cobalt atom, and a carbon monoxide molecule, across surfaces of copper and platinum.

Limitations of the Coulomb Model

The Coulomb approximation mathematically follows from the assumptions that surfaces are in atomically close contact only over a small fraction of their overall area, that this contact area is proportional to the normal force (until saturation, which takes place when all area is in atomic contact), and that the frictional force is proportional to the applied normal force, independently of the contact area. Such reasoning aside, however, the approximation is fundamentally an empirical construct. It is a rule of thumb describing the approximate outcome of an extremely complicated physical interaction. The strength of the approximation is its simplicity and versatility. Though in general the relationship between normal force and frictional force is not exactly linear (and so the frictional force is not entirely independent of the contact area of the surfaces), the Coulomb approximation is an adequate representation of friction for the analysis of many physical systems.

When the surfaces are conjoined, Coulomb friction becomes a very poor approximation (for example, adhesive tape resists sliding even when there is no normal force, or a negative normal force). In this case, the frictional force may depend strongly on the area of contact. Some drag racing tires are adhesive for this reason. However, despite the complexity of the fundamental physics behind friction, the relationships are accurate enough to be useful in many applications.

"Negative" Coefficient of Friction

As of 2012, a single study has demonstrated the potential for an *effectively negative coefficient of friction in the low-load regime,* meaning that a decrease in normal force leads to an increase in friction. This contradicts everyday experience in which an increase in normal force leads to an increase in friction. This was reported in the journal *Nature* in October 2012 and involved the friction encountered by an atomic force microscope stylus when dragged across a graphene sheet in the presence of graphene-adsorbed oxygen.

Numerical Simulation of the Coulomb Model

Despite being a simplified model of friction, the Coulomb model is useful in many numerical simulation applications such as multibody systems and granular material. Even its most simple expression encapsulates the fundamental effects of sticking and sliding which are required in many applied cases, although specific algorithms have to be designed in order to efficiently numerically integrate mechanical systems with Coulomb friction and bilateral and/or unilateral contact. Some quite nonlinear effects, such as the so-called Painlevé paradoxes, may be encountered with Coulomb friction.

Dry Friction and Instabilities

Dry friction can induce several types of instabilities in mechanical systems which display a stable behaviour in the absence of friction. These instabilities may be caused by the decrease of the friction force with an increasing velocity of sliding, by material expansion due to heat generation during friction (the thermo-elastic instabilities), or by pure dynamic effects of sliding of two elastic materials (the Adams-Martins instabilities). The latter were originally discovered in 1995 by George G. Adams and João Armé-nio Correia Martins for smooth surfaces and were later found in periodic rough surfaces. In particular, friction-related dynamical instabilities are thought to be responsible for brake squeal and the 'song' of a glass harp, phenomena which involve stick and slip, modelled as a drop of friction coefficient with velocity.

A practically important case is the self-oscillation of the strings of bowed instruments such as the violin, cello, hurdy-gurdy, erhu etc.

A connection between dry friction and flutter instability in a simple mechanical system has been discovered, watch the movie for more details.

Frictional instabilities can lead to the formation of new self-organized patterns (or "secondary structures") at the sliding interface, such as in-situ formed tribofilms which are utilized for the reduction of friction and wear in so-called self-lubricating materials.

Fluid Friction

Fluid friction occurs between fluid layers that are moving relative to each other. This internal resistance to flow is named *viscosity*. In everyday terms, the viscosity of a fluid is described as its "thickness". Thus, water is "thin", having a lower viscosity, while honey is "thick", having a higher viscosity. The less viscous the fluid, the greater its ease of deformation or movement.

All real fluids (except superfluids) offer some resistance to shearing and therefore are viscous. For teaching and explanatory purposes it is helpful to use the concept of an inviscid fluid or an ideal fluid which offers no resistance to shearing and so is not viscous.

Lubricated Friction

Lubricated friction is a case of fluid friction where a fluid separates two solid surfaces. Lubrication is a technique employed to reduce wear of one or both surfaces in close proximity moving relative to each another by interposing a substance called a lubricant between the surfaces.

In most cases the applied load is carried by pressure generated within the fluid due to the frictional viscous resistance to motion of the lubricating fluid between the surfaces. Adequate lubrication allows smooth continuous operation of equipment, with only mild wear, and without excessive stresses or seizures at bearings. When lubrication breaks down, metal or other components can rub destructively over each other, causing heat and possibly damage or failure.

Skin Friction

Skin friction arises from the interaction between the fluid and the skin of the body, and is directly related to the area of the surface of the body that is in contact with the fluid. Skin friction follows the drag equation and rises with the square of the velocity.

Skin friction is caused by viscous drag in the boundary layer around the object. There are two ways to decrease skin friction: the first is to shape the moving body so that smooth flow is possible, like an airfoil. The second method is to decrease the length and cross-section of the moving object as much as is practicable.

Internal Friction

Internal friction is the force resisting motion between the elements making up a solid material while it undergoes deformation.

Plastic deformation in solids is an irreversible change in the internal molecular structure of an object. This change may be due to either (or both) an applied force or a change in temperature. The change of an object's shape is called strain. The force causing it is called stress.

Elastic deformation in solids is reversible change in the internal molecular structure of an object. Stress does not necessarily cause permanent change. As deformation occurs, internal forces oppose the applied force. If the applied stress is not too large these opposing forces may completely resist the applied force, allowing the object to assume a new equilibrium state and to return to its original shape when the force is removed. This is known as elastic deformation or elasticity.

Radiation Friction

As a consequence of light pressure, Einstein in 1909 predicted the existence of "radiation friction" which would oppose the movement of matter. He wrote, "radiation will exert pressure on both sides of the plate. The forces of pressure exerted on the two sides are equal if the plate is at rest. However, if it is in motion, more radiation will be reflected on the surface that is ahead during the motion (front surface) than on the back surface. The backwardacting force of pressure exerted on the front surface is thus larger than the force of pressure acting on the back. Hence, as the resultant of the two forces, there remains a force that counteracts the motion of the plate and that increases with the velocity of the plate. We will call this resultant 'radiation friction' in brief."

Other Types of Friction

Rolling Resistance

Rolling resistance is the force that resists the rolling of a wheel or other circular object along a surface caused by deformations in the object and/or surface. Generally the force of rolling resistance is less than that associated with kinetic friction. Typical values for the coefficient of rolling resistance are 0.001. One of the most common examples of rolling resistance is the movement of motor vehicle tires on a road, a process which generates heat and sound as by-products.

Braking Friction

Any wheel equipped with a brake is capable of generating a large retarding force, usually for the purpose of slowing and stopping a vehicle or piece of rotating machinery. Braking friction differs from rolling friction because the coefficient of friction for rolling friction is small whereas the coefficient of friction for braking friction is designed to be large by choice of materials for brake pads.

Triboelectric Effect

Rubbing dissimilar materials against one another can cause a build-up of electrostatic charge, which can be hazardous if flammable gases or vapours are present. When the static build-up discharges, explosions can be caused by ignition of the flammable mixture.

Belt Friction

Belt friction is a physical property observed from the forces acting on a belt wrapped around a pulley, when one end is being pulled. The resulting tension, which acts on both ends of the belt, can be modeled by the belt friction equation.

In practice, the theoretical tension acting on the belt or rope calculated by the belt friction equation can be compared to the maximum tension the belt can support. This helps a designer of such a rig to know how many times the belt or rope must be wrapped around the pulley to prevent it from slipping. Mountain climbers and sailing crews demonstrate a standard knowledge of belt friction when accomplishing basic tasks.

Reducing Friction

Devices

Devices such as wheels, ball bearings, roller bearings, and air cushion or other types of fluid bearings can change sliding friction into a much smaller type of rolling friction.

Many thermoplastic materials such as nylon, HDPE and PTFE are commonly used in low friction bearings. They are especially useful because the coefficient of friction falls with increasing imposed load. For improved wear resistance, very high molecular weight grades are usually specified for heavy duty or critical bearings.

Lubricants

A common way to reduce friction is by using a lubricant, such as oil, water, or grease, which is placed between the two surfaces, often dramatically lessening the coefficient of friction. The science of friction and lubrication is called tribology. Lubricant technology is when lubricants are mixed with the application of science, especially to industrial or commercial objectives.

Superlubricity, a recently discovered effect, has been observed in graphite: it is the substantial decrease of friction between two sliding objects, approaching zero levels. A very small amount of frictional energy would still be dissipated.

Lubricants to overcome friction need not always be thin, turbulent fluids or powdery solids such as graphite and talc; acoustic lubrication actually uses sound as a lubricant.

Another way to reduce friction between two parts is to superimpose micro-scale vibration to one of the parts. This can be sinusoidal vibration as used in ultrasound-assisted cutting or vibration noise, known as dither.

Energy of Friction

According to the law of conservation of energy, no energy is destroyed due to friction, though it may be lost to the system of concern. Energy is transformed from other forms into thermal energy. A sliding hockey puck comes to rest because friction converts its kinetic energy into heat which raises the thermal energy of the puck and the ice surface. Since heat quickly dissipates, many early philosophers, including Aristotle, wrongly concluded that moving objects lose energy without a driving force.

When an object is pushed along a surface along a path C, the energy converted to heat is given by a line integral, in accordance with the definition of work

$$E_{th} = \int_C \mathbf{F}_{\text{fric}}(\mathbf{x}) \cdot d\mathbf{x} = \int_C \mu_k \, \mathbf{F}_n(\mathbf{x}) \cdot d\mathbf{x},$$

where

$\mathbf{F}_{\text{fric}}$ is the friction force,

$\mathbf{F}_n$ is the vector obtained by multiplying the magnitude of the normal force by a unit vector pointing *against* the object's motion,

μ_k is the coefficient of kinetic friction, which is inside the integral because it may vary from location to location (e.g. if the material changes along the path),

$\mathbf{x}$ is the position of the object.

Energy lost to a system as a result of friction is a classic example of thermodynamic irreversibility.

Work of Friction

In the reference frame of the interface between two surfaces, static friction does *no* work, because there is never displacement between the surfaces. In the same reference frame, kinetic friction is always in the direction opposite the motion, and does *negative* work. However, friction can do *positive* work in certain frames of reference. One can see this by placing a heavy box on a rug, then pulling on the rug quickly. In this case, the box slides backwards relative to the rug, but moves forward relative to the frame of reference in which the floor is stationary. Thus, the kinetic friction between the box and rug accelerates the box in the same direction that the box moves, doing *positive* work.

The work done by friction can translate into deformation, wear, and heat that can affect the contact surface properties (even the coefficient of friction between the surfaces). This can be beneficial as in polishing. The work of friction is used to mix and join materials such as in the process of friction welding. Excessive erosion or wear of mating sliding surfaces occurs when work due to frictional forces rise to unacceptable levels. Harder corrosion particles caught between mating surfaces in relative motion (fretting) exacerbates wear of frictional forces. Bearing seizure or failure may result from excessive wear due to work of friction. As surfaces are worn by work due to friction, fit and surface finish of an object may degrade until it no longer functions properly.

Applications

Friction is an important factor in many engineering disciplines.

Transportation

- Automobile brakes inherently rely on friction, slowing a vehicle by converting its kinetic energy into heat. Incidentally, dispersing this large amount of heat safely is one technical challenge in designing brake systems.

- Rail adhesion refers to the grip wheels of a train have on the rails.

- Road slipperiness is an important design and safety factor for automobiles

    o Split friction is a particularly dangerous condition arising due to varying friction on either side of a car.

    o Road texture affects the interaction of tires and the driving surface.

Measurement

- A tribometer is an instrument that measures friction on a surface.

- A profilograph is a device used to measure pavement surface roughness.

Household Usage

- Friction is used to heat and ignite matchsticks (friction between the head of a matchstick and the rubbing surface of the match box).

Application of Friction in Transportation

Brake

Disc brake on a motorcycle

A brake is a mechanical device that inhibits motion by absorbing energy from a moving system. It is used for slowing or stopping a moving vehicle, wheel, axle, or to prevent its motion, most often accomplished by means of friction.

Background

Most brakes commonly use friction between two surfaces pressed together to convert the kinetic energy of the moving object into heat, though other methods of energy conversion may be employed. For example, regenerative braking converts much of the energy to electrical energy, which may be stored for later use. Other methods convert kinetic energy into potential energy in such stored forms as pressurized air or pressurized oil. Eddy current brakes use magnetic fields to convert kinetic energy into electric current in the brake disc, fin, or rail, which is converted into heat. Still other braking methods even transform kinetic energy into different forms, for example by transferring the energy to a rotating flywheel.

Brakes are generally applied to rotating axles or wheels, but may also take other forms such as the surface of a moving fluid (flaps deployed into water or air). Some vehicles use a combination of braking mechanisms, such as drag racing cars with both wheel brakes and a parachute, or airplanes with both wheel brakes and drag flaps raised into the air during landing.

Since kinetic energy increases quadratically with velocity , an object moving at 10 m/s has 100 times as much energy as one of the same mass moving at 1 m/s, and consequently the theoretical braking distance, when braking at the traction limit, is 100 times as long. In practice, fast vehicles usually have significant air drag, and energy lost to air drag rises quickly with speed.

Almost all wheeled vehicles have a brake of some sort. Even baggage carts and shopping carts may have them for use on a moving ramp. Most fixed-wing aircraft are fitted with wheel brakes on the undercarriage. Some aircraft also feature air brakes designed to reduce their speed in flight. Notable examples include gliders and some World War II-era aircraft, primarily some fighter aircraft and many dive bombers of the era. These allow the aircraft to maintain a safe speed in a steep descent. The Saab B 17 dive bomber and Vought F4U Corsair fighter used the deployed undercarriage as an air brake.

Friction brakes on automobiles store braking heat in the drum brake or disc brake while braking then conduct it to the air gradually. When traveling downhill some vehicles can use their engines to brake.

When the brake pedal of a modern vehicle with hydraulic brakes is pushed against the master cylinder, ultimately a piston pushes the brake pad against the brake disc which slows the wheel down. On the brake drum it is similar as the cylinder pushes the brake shoes against the drum which also slows the wheel down.

Types

Rendering of a drum brake

Single pivot side-pull bicycle caliper brake.

Brakes may be broadly described as using friction, pumping, or electromagnetics. One brake may use several principles: for example, a pump may pass fluid through an orifice to create friction:

Frictional

Frictional brakes are most common and can be divided broadly into "shoe" or "pad" brakes, using an explicit wear surface, and hydrodynamic brakes, such as parachutes, which use friction in a working fluid and do not explicitly wear. Typically the term "friction brake" is used to mean pad/shoe brakes and excludes hydrodynamic brakes, even though hydrodynamic brakes use friction. Friction (pad/shoe) brakes are often rotating devices with a stationary pad and a rotating wear surface. Common configurations include shoes that contract to rub on the outside of a rotating drum, such as a band brake; a rotating drum with shoes that expand to rub the inside of a drum, commonly called a "drum brake", although other drum configurations are possible; and pads that pinch a rotating disc, commonly called a "disc brake". Other brake configurations are used, but less often. For example, PCC trolley brakes include a flat shoe which is clamped to the rail with an electromagnet; the Murphy brake pinches a rotating drum, and the Ausco Lambert disc brake uses a hollow disc (two parallel discs with a structural bridge) with shoes that sit between the disc surfaces and expand laterally.

A drum brake is a vehicle brake in which the friction is caused by a set of brake shoes that press against the inner surface of a rotating drum. The drum is connected to the rotating roadwheel hub.

Drum brakes generally can be found on older car and truck models. However, because of their low production cost, drum brake setups are also installed on the rear of some low-cost newer vehicles. Compared to modern disc brakes, drum brakes wear out faster due to their tendency to overheat.

The disc brake is a device for slowing or stopping the rotation of a road wheel. A brake disc (or rotor in U.S. English), usually made of cast iron or ceramic, is connected to the wheel or the axle. To stop the wheel, friction material in the form of brake pads (mounted in a device called a brake caliper) is forced mechanically, hydraulically, pneumatically or electromagnetically against both sides of the disc. Friction causes the disc and attached wheel to slow or stop.

Ceramic brakes, also called "carbon ceramic", are high-end type of frictional brakes with brake pads and rotors made from porcelain compound blends, that feature better stopping capability and greater resistance to overheat. Due to their high production cost, ceramic brakes aren't widely used as factory equipment, and their availability on the automotive aftermarket is low compared to traditional metallic brakes. However, being performance-oriented equipment, ceramic brakes are popular among racers.

Pumping

Pumping brakes are often used where a pump is already part of the machinery. For example, an internal-combustion piston motor can have the fuel supply stopped, and then internal pumping losses of the engine create some braking. Some engines use a valve override called a Jake brake to greatly increase pumping losses. Pumping brakes can dump energy as heat, or can be regenerative brakes that recharge a pressure reservoir called a hydraulic accumulator.

Electromagnetic

Electromagnetic brakes are likewise often used where an electric motor is already part of the machinery. For example, many hybrid gasoline/electric vehicles use the electric motor as a generator to charge electric batteries and also as a regenerative brake. Some diesel/electric railroad locomotives use the electric motors to generate electricity which is then sent to a resistor bank and dumped as heat. Some vehicles, such as some transit buses, do not already have an electric motor but use a secondary "retarder" brake that is effectively a generator with an internal short-circuit. Related types of such a brake are eddy current brakes, and electro-mechanical brakes (which actually are magnetically driven friction brakes, but nowadays are often just called "electromagnetic brakes" as well).

Electromagnetic brakes slow an object through electromagnetic induction, which creates resistance and in turn either heat or electricity. Friction brakes apply pressure on two separate objects to slow the vehicle in a controlled manner.

Characteristics

Brakes are often described according to several characteristics including:

- Peak force – The peak force is the maximum decelerating effect that can be obtained. The peak force is often greater than the traction limit of the tires, in which case the brake can cause a wheel skid.

- Continuous power dissipation – Brakes typically get hot in use, and fail when the temperature gets too high. The greatest amount of power (energy per unit time) that can be dissipated through the brake without failure is the continuous power dissipation. Continuous power dissipation often depends on e.g., the temperature and speed of ambient cooling air.

- Fade – As a brake heats, it may become less effective, called brake fade. Some designs are inherently prone to fade, while other designs are relatively immune. Further, use considerations, such as cooling, often have a big effect on fade.

- Smoothness – A brake that is grabby, pulses, has chatter, or otherwise exerts varying brake force may lead to skids. For example, railroad wheels have little traction, and friction brakes without an anti-skid mechanism often lead to skids, which increases maintenance costs and leads to a "thump thump" feeling for riders inside.

- Power – Brakes are often described as "powerful" when a small human application force leads to a braking force that is higher than typical for other brakes in the same class. This notion of "powerful" does not relate to continuous power dissipation, and may be confusing in that a brake may be "powerful" and brake strongly with a gentle brake application, yet have lower (worse) peak force than a less "powerful" brake.

- Pedal feel – Brake pedal feel encompasses subjective perception of brake power output as a function of pedal travel. Pedal travel is influenced by the fluid displacement of the brake and other factors.

- Drag – Brakes have varied amount of drag in the off-brake condition depending on design of the system to accommodate total system compliance and deformation that exists under braking with ability to retract friction material from the rubbing surface in the off-brake condition.

- Durability – Friction brakes have wear surfaces that must be renewed periodically. Wear surfaces include the brake shoes or pads, and also the brake disc or drum. There may be tradeoffs, for example a wear surface that generates high peak force may also wear quickly.

- Weight – Brakes are often "added weight" in that they serve no other function. Further, brakes are often mounted on wheels, and unsprung weight can significantly hurt traction in some circumstances. "Weight" may mean the brake itself, or may include additional support structure.

- Noise – Brakes usually create some minor noise when applied, but often create squeal or grinding noises that are quite loud.

Brake Boost

Brake booster from a Geo Storm.

Most modern vehicles use a vacuum assisted brake system that greatly increases the force applied to the vehicle's brakes by its operator. This additional force is supplied by the manifold vacuum generated by air flow being obstructed by the throttle on a running engine. This force is greatly reduced when the engine is running at fully open throttle, as the difference between ambient air pressure and manifold (absolute) air pressure is reduced, and therefore available vacuum is diminished. However, brakes are rarely applied at full throttle; the driver takes the right foot off the gas pedal and moves it to the brake pedal - unless left-foot braking is used.

Because of low vacuum at high RPM, reports of unintended acceleration are often accompanied by complaints of failed or weakened brakes, as the high-revving engine, having an open throttle, is unable to provide enough vacuum to power the brake booster. This problem is exacerbated in vehicles equipped with automatic transmissions as the vehicle will automatically downshift upon application of the brakes, thereby increasing the torque delivered to the driven-wheels in contact with the road surface.

Noise

Brake lever on a horse-drawn hearse

Although ideally a brake would convert all the kinetic energy into heat, in practice a significant amount may be converted into acoustic energy instead, contributing to noise pollution.

For road vehicles, the noise produced varies significantly with tire construction, road surface, and the magnitude of the deceleration. Noise can be caused by different things. These are signs that there may be issues with brakes wearing out over time.

Fires

Railway braking produces sparks and is an important cause of forest fires.

Inefficiency

A significant amount of energy is always lost while braking, even with regenerative braking which is not perfectly efficient. Therefore, a good metric of efficient energy use while driving is to note how much one is braking. If the majority of deceleration is from unavoidable friction instead of braking, one is squeezing out most of the service from the vehicle. Minimizing brake use is one of the fuel economy-maximizing behaviors.

While energy is always lost during a brake event, a secondary factor that influences efficiency is "off-brake drag", or drag that occurs when the brake is not intentionally actuated. After a braking event, hydraulic pressure drops in the system, allowing the brake caliper pistons to retract. However, this retraction must accommodate all compliance in the system (under pressure) as well as thermal distortion of components like the brake disc or the brake system will drag until the contact with the disc, for example, knocks the pads and pistons back from the rubbing surface. During this time, there can be significant brake drag. This brake drag can lead to significant parasitic power loss, thus impact fuel economy and overall vehicle performance.

Adhesion Railway

Driving wheel of steam locomotive

Adhesion railway or adhesion traction is the most common type of railway, where power is applied by driving some or all of the wheels of the locomotive. Rail adhesion relies on the friction between a steel wheel and a steel rail. The term is particularly used when discussing conventional railways to distinguish from other forms of traction such as funicular or cog railway.

Traction or friction can be reduced when the rails are greasy, because of rain, oil or de-composing leaves which compact into a hard slippery lignin coating. Measures against reduced adhesion due to leaves include application of "Sandite" (a gel-sand mix) by special sanding trains, scrubbers and water jets, and long-term management of railside vegetation. On an adhesion railway, most locomotives will have a sand containment vessel, to apply sand onto the track; this is called "sanding".

Effect of Adhesion Limits

Adhesion is caused by friction, with maximum tangential force produced by a driving wheel before slipping given by:

$$F_{max} = \text{coefficient of friction} \times \text{Weight on wheel}$$

Usually the force needed to start sliding is greater than that needed to continue sliding. The former is concerned with static friction, referred colloquially to as "stiction", or "limiting friction", whilst the latter is dynamic friction, also called "sliding friction".

For steel on steel, the coefficient of friction can be as high as 0.78, under laboratory conditions, but typically on railways it is between 0.35 and 0.5, whilst under extreme conditions it can fall to as low as 0.05. Thus a 100-tonne locomotive could have a tractive effort of 350 kilonewtons, under the ideal conditions (assuming sufficient force can be produced by the engine), falling to a 50 kilonewtons under the worst conditions.

Steam locomotives suffer particularly badly from adhesion issues because power delivery is pulsed (especially in 2- or most 4-cylinder engines) and, on large locomotives, not all wheels are driven. The "factor of adhesion", being the weight on the driven wheels divided by the theoretical starting tractive effort, was generally designed to be a value of 4 or higher, reflecting a typical wheel-rail friction coefficient of 0.25. A locomotive with a factor of adhesion much lower than 4 would be highly prone to wheelslip. Other steam locomotive design factors significantly affecting traction include wheel size (smaller diameter wheels offer superior traction at the expense of top speed) and the sensitivity of the regulator.

All-weather Adhesion

The term *all-weather adhesion* is usually used in North America, and refers to the adhesion available during traction mode with 99% reliability in all weather conditions.

Toppling Conditions

The maximum speed a train can proceed around a turn is limited by the radius of turn, the position of the centre of mass of the units, the wheel gauge and whether the track is *superelevated* or *canted* .

Topples when:

$$\frac{mU^2}{R}$$

$$R = \frac{2hU^2}{gd}$$

h

d

mg

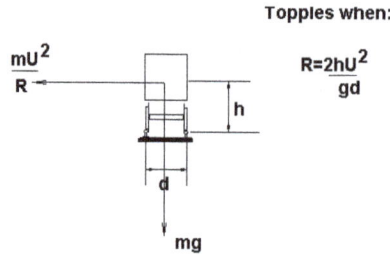

Toppling Limit on Turn Radius

Toppling limit on tight turn radius

Toppling will occur when the overturning moment due to the side force (centrifugal acceleration) is sufficient to cause the inner wheel to begin to lift off the rail. This may result in loss of adhesion - causing the train to slow, preventing toppling. Alternatively, the inertia may be sufficient to cause the train to continue to move at speed causing the vehicle to topple completely.

For a wheel gauge of 1.5 m, no canting, a centre of gravity height of 3 m and speed of 30 m/s (108 km/h), the radius of turn is 360 m. For a modern high speed train at 80 m/s, the toppling limit would be about 2.5 km. In practice, the minimum radius of turn is much greater than this, as contact between the wheel flanges and rail at high speed could cause significant damage to both. For very high speed, the minimum adhesion limit again appears appropriate, implying a radius of turn of about 13 km. In practice, curved lines used for high speed travel are *superelevated* or *canted* so that the turn limit is closer to 7 km.

During the 19th century, it was widely believed that coupling the drive wheels would compromise performance and was avoided on engines intended for express passenger service. With a single drive wheelset, the Herzian contact stress between the wheel and rail necessitated the largest diameter wheels that could be accommodated. The weight of locomotive was restricted by the stress on the rail and sandboxes were required, even under reasonable adhesion conditions.

Directional Stability and Hunting Instability

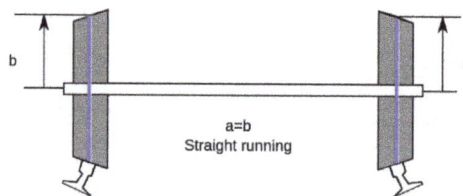

b

a

a=b
Straight running

Wheelset in the central position

It may be thought that the wheels are kept on the tracks by the flanges. However, close examination of a typical railway wheel reveals that the tread is burnished but the flange is not - the flanges rarely make contact with the rail and, when they do, most of the contact is sliding. The rubbing of a flange on the track dissipates large amounts of energy, mainly as heat but also including noise and, if sustained, would lead to excessive wheel wear.

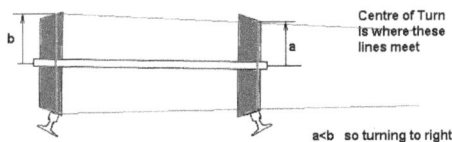

Effect of Lateral Displacement

The effect of lateral displacement

Centring is actually accomplished through shaping of the wheel. The tread of the wheel is slightly tapered. When the train is in the centre of the track, the region of the wheels in contact with the rail traces out a circle which has the same diameter for both wheels. The velocities of the two wheels are equal, so the train moves in a straight line.

If, however, the wheelset is displaced to one side, the diameters of the regions of contact, and hence the tangential velocities of the wheels at the running surfaces are different and the wheelset tends to steer back towards the centre. Also, when the train encounters an unbanked turn, the wheelset displaces laterally slightly, so that the outer wheel tread speeds up linearly, and the inner wheel tread slows down, causing the train to turn the corner. Some railway systems employ a flat wheel and track profile, relying on cant alone to reduce or eliminate flange contact.

Understanding how the train stays on the track, it becomes evident why Victorian locomotive engineers were averse to coupling wheelsets. This simple coning action is possible only with wheelsets where each can have some free motion about its vertical axis. If wheelsets are rigidly coupled together, this motion is restricted, so that coupling the wheels would be expected to introduce sliding, resulting in increased rolling losses. This problem was alleviated to a great extent by ensuring the diameter of all coupled wheels was very closely matched.

With perfect rolling contact between the wheel and rail, this coning behaviour manifests itself as a swaying of the train from side to side. In practice, the swaying is damped out below a critical speed, but is amplified by the forward motion of the train above the critical speed. This lateral swaying is known as hunting oscillation. The phenomenon of hunting was known by the end of the 19th century, although the cause was not fully understood until the 1920s and measures to eliminate it were not taken until the late 1960s. The limitation on maximum speed was imposed not by raw power but by encountering an instability in the motion.

The kinematic description of the motion of tapered treads on the two rails is insufficient to describe hunting well enough to predict the critical speed. It is necessary to deal with the forces involved. There are two phenomena which must be taken into account.

The first is the inertia of the wheelsets and vehicle bodies, giving rise to forces proportional to acceleration; the second is the distortion of the wheel and track at the point of contact, giving rise to elastic forces. The kinematic approximation corresponds to the case which is dominated by contact forces.

A fairly straightforward analysis of the kinematics of the coning action yields an estimate of the wavelength of the lateral oscillation:

$$\lambda = 2\pi\sqrt{\frac{rd}{2k}},$$

where d is the wheel gauge, r is the nominal wheel radius and k is the taper of the treads. For a given speed, the longer the wavelength and the lower the inertial forces will be, so the more likely it is that the oscillation will be damped out. Since the wavelength increases with reducing taper, increasing the critical speed requires the taper to be reduced, which implies a large minimum radius of turn.

A more complete analysis, taking account of the actual forces acting, yields the following result for the critical speed of a wheelset:

$$V^2 = \frac{Wrad^2}{k\left(4C + md^2\right)},$$

where W is the axle load for the wheelset, a is a shape factor related to the amount of wear on the wheel and rail, C is the moment of inertia of the wheelset perpendicular to the axle, m is the wheelset mass.

The result is consistent with the kinematic result in that the critical speed depends inversely on the taper. It also implies that the weight of the rotating mass should be minimised compared with the weight of the vehicle. The wheel gauge implicitly appears in both the numerator and denominator, implying that it has only a second-order effect on the critical speed.

The true situation is much more complicated, as the response of the vehicle suspension must be taken into account. Restraining springs, opposing the yaw motion of the wheelset, and similar restraints on bogies, may be used to raise the critical speed further. However, in order to achieve the highest speeds without encountering instability, a significant reduction in wheel taper is necessary, so there is little prospect of reducing the turn radius of high speed trains much below the current value of 7 km.

Forces on Wheels, Creep

The behaviour of adhesion railways is determined by the forces arising between two surfaces in contact. This may appear trivially simple from a superficial glance but it becomes extremely complex when studied to the depth necessary to predict useful results.

The first error to address is the assumption that wheels are round. A glance at the tyres of a parked car will immediately show that this is not true: the region in contact with the road is noticeably flattened, so that the wheel and road conform to each other over a region of contact. If this were not the case, the contact stress of a load being transferred through a point contact would be infinite. Rails and railway wheels are much stiffer than pneumatic tyres and tarmac but the same distortion takes place at the region of contact. Typically, the area of contact is elliptical, of the order of 15 mm across.

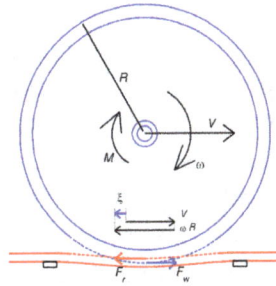

A torque M applied on the axle causes creepage: difference between forward velocity V and circumferential velocity ωR, , with resulting creep force F_w.

The distortion is small and localised but the forces which arise from it are large. In addition to the distortion due to the weight, both wheel and rail distort when braking and accelerating forces are applied and when the vehicle is subjected to side forces. These tangential forces cause distortion in the region where they first come into contact, followed by a region of slippage. The net result is that, during traction, the wheel does not advance as far as would be expected from rolling contact but, during braking, it advances further. This mix of elastic distortion and local slipping is known as "creep". The definition of creep in this context is:

$$\text{creep} = \frac{(\text{actual displacement} - \text{rolling displacement})}{(\text{rolling displacement})}$$

In analysing the dynamics of wheelsets and complete rail vehicles, the contact forces can be treated as linearly dependent on the creep (Kalker's linear theory, valid for small creepage) or more advanced theories can be used from frictional contact mechanics.

The forces which result in directional stability, propulsion and braking may all be traced to creep. It is present in a single wheelset and will accommodate the slight kinematic incompatibility introduced by coupling wheelsets together, without causing gross slippage, as was once feared.

Provided the radius of turn is sufficiently great (as should be expected for express passenger services), two or three linked wheelsets should not present a problem. However, 10 drive wheels (5 main wheelsets) are usually associated with heavy freight locomotives.

Sanding

On an adhesion railway, most locomotives will have a sand containment vessel. Properly dried sand can be dropped onto the rail to improve traction under slippery conditions. The sand is most often applied using compressed air via tower, crane, silo or train. When an engine slips, particularly when starting a heavy train, sand applied at the front of the driving wheels greatly aids in tractive effort causing the train to "lift", or to commence the motion intended by the engine driver.

Sanding however also has some negative effects. It can cause a "sandfilm", which consists of crushed sand, that is compressed to a film on the track where the wheels make contact. Together with some moisture on the track, which acts as a light adhesive and keeps the applied sand on the track, the wheels "bake" the crushed sand into a more solid layer of sand. Because the sand is applied to the first wheels on the locomotive, the following wheels may run, at least partially and for a limited time, on a layer of sand (sandfilm). While traveling this means that electric locomotives may lose contact to the track-ground, causing the locomotive to create EMI-emissions and currents through the couplers, or in standstill, when the locomotive is parked, "track-release-relays" may detect an empty track because the locomotive is electrically isolated from the track.

Booster

Some steam locomotives were fitted with booster engines on the rear trailing wheels. These were turned on as required at starting to give additional adhesive effort.

Road Slipperiness

A surface friction tester, used to measure road slipperiness

Road slipperiness (low skid resistance due to insufficient road friction) is the technical term for the cumulative effects of snow, ice, water, loose material and the texture of the road surface on the traction produced by the wheels of a vehicle.

Road slipperiness can be measured either in terms of the friction between a freely-spinning wheel and the ground, or the braking distance of a braking vehicle, and is related to the coefficient of friction between the tyre and the road surface.

Public works agencies spend a sizeable portion of their budget measuring and reducing road slipperiness. Even a small increase in slipperiness of a section of road can increase the accident rate of the section of road tenfold. Maintenance activities affecting slipperiness include drainage repair, snow removal and street sweeping. More intensive measures may include grinding or milling a surface that has worn smooth, a surface treatment such as a chipseal, or overlaying a new layer of asphalt.

A specific road safety problem is split friction or μ (mu) - split; when the friction significantly differs between the left and the right wheelpath. The road may then not be perceived as hazardous when accelerating, cruising or even braking softly, but in a case of hard braking, the difference in friction will cause the vehicle to start to rotate towards the side offering higher grip. Split friction may cause jack-knifing of articulated trucks, while trucks with towed trailers may experience trailer swing phenomena. Split friction may be caused by an improper road spot repair that results in high variance of texture (roads) and colour (thin ice on newly paved black spots thaws faster than ice on old greyish asphalt) across the road section.

Measurement

The two ways to measure road slipperiness are surface friction testing and stopping distance testing. Friction testing can use surface friction testers or portable friction testers, and involves allowing a freely moving object, usually a wheel, to move against the surface. By measuring the resistance experienced by the wheel, the friction between the ground and the wheel can be found.

Stopping distance testing involves performing an emergency stop in a test vehicle and measuring the distance required to come to a stop. This can be measured either from the length of the skid marks left by the vehicle, or by the "chalk-to-gun" method, where the brakes are connected to a small gun filled with chalk powder, which marks the point when the brakes were applied. This has the advantage of measuring the full stopping distance, while simply measuring the skid marks only measures the distance from the point where the wheels began to lock or slip.

Measurement of skidding resistance is not yet universally harmononised despite a number of attempts such as FEHRL's HERMES project. The European Standards Organisation (CEN)has been working on the topic for many years through its committee CEN/TC 227 - Road materials. Contributions to this were made through the FP7 Tyrosafe project which aims to raise awareness, to coordinate and prepare for European harmonisation and to optimise the assessment and management of essential tyre/road interaction parameters in order to increase safety and support greening of road transport. This project will provide a synopsis of the current state of scientific understanding and its current application in different standards. It will identify the needs for future research and propose a way forward in the context of the future objectives of road administrations in order to optimise three key properties of roads: skid resistance, rolling resistance and tyre/road noise emission.

Reduction

Dutch newsreel about the anti-skid training school of Rob Slotemaker

Road slipperiness can contribute to car accidents. In 1997, over 53,000 accidents were caused by slippery roads in the United Kingdom out of an estimated 4,000,000 accidents (or approximately 1.3 per cent) . A small change in road slipperiness can have a drastic effect on surface friction: decreasing the coefficient of friction from 0.45 to 0.35, equivalent to adding a dusting of wet snow, increased the accident rate by almost 1000%. As such, road agencies have a number of approaches to decreasing road slipperiness. Most roads are designed with a convex camber to provide sufficient drainage, thereby allowing surface water to drain out of the road. Trouble sections include entrances and exits of banked outercurves, where the cross slope is close to zero. Unless these sections have a longitudinal grade of at least 0.4–0.5%, the drainage gradient (resultant to crossfall and longitudinal grade) will be lower than 0.5% so water will not run off the road surface. Storm drains may be installed at regular intervals and modern paving materials are designed to provide high friction in most conditions. Permeable paving allows water to soak through the paving material, reducing slipperiness in very adverse conditions.

Road slipperiness can be prevented or delayed by proper pavement design. The aggregate used in the pavement should be selected with care, as certain aggregates such as dolomite may *polish*, or wear smooth under the action of tires. With asphalt pavements and surface treatments, using too much asphalt or asphalt emulsion can cause bleeding or *flushing*, a condition where excess asphalt rises to the top and fills in the road texture. Both problems increase slipperiness, especially when the pavement is wet.

Once lost, pavement texture can be restored with retexturing procedures such as diamond grinding of pavement, surface treatments such as chipsealing and resurfacing with asphalt concrete.

Snow and ice removal also decreases road slipperiness; snowploughs and snow blowers can remove the snow from the road surface while gritters drop road salt and sand, which both melts the snow and ice from the road surface, and provide a rougher surface to grip onto. However, in dry conditions, sand and salt on the road surface can themselves increase road slipperiness and pose a danger to road traffic, and therefore, roads are cleared by street sweepers after roadworks and gritting to make sure that all the loose material is cleared from the road surface.

Split Friction

Split friction (or μ (mu) - split) is a road condition that occurs when the friction significantly differs between the left and the right wheelpath.

The road may then not be perceived as hazardous when accelerating, cruising or even braking softly. But in a case of hard (emergency-)braking, the car will start to rotate over the wheelpath offering highest grip. Split friction may cause jack-knifing of articulated trucks, while trucks with towed trailers may experience trailer swing phenomena. Split friction may be caused by an improper road spot repair that results in high variance of texture and colour (thin ice on newly paved black spots thaws faster than ice on old greyish asphalt) across the road section.

To some extent, the risk for split friction can be measured with a road profilograph, scanning the pavement texture in both the left and right wheel paths.

A full analysis of the split friction issue requires a friction device that measures both the left and right wheel track at the same time. The ViaFriction skid device can be used for such a purpose.

Road Texture

Road surface textures are deviations from a planar and smooth surface, affecting the vehicle/tyre interaction. Pavement texture is divided into: microtexture with wavelengths from 0 mm to 0.5 millimetres (0.020 in), macrotexture with wavelengths from 0.5 millimetres (0.020 in) to 50 millimetres (2.0 in) and megatexture with wavelengths from 50 millimetres (2.0 in) to 500 millimetres (20 in).

Microtexture

Microtexture (MiTx) is the collaborative term for a material's crystallographic parameters and other aspects of micro-structure: such as morphology, including size and shape distributions; chemical composition; and crystal orientation and relationships

While vehicle suspension deflection and dynamic tire loads are affected by longer wavelength (roughness), road texture affects the interaction between the road surface and the tire footprint. Microtexture has wavelengths shorter than 0.5 mm. It relates to the surface of the binder, of the aggregate, and of contaminants such as rubber deposits from tires.

The mix of the road material contributes to dry road surface friction. Typically, road agencies do not monitor mix directly, but indirectly by brake friction tests. However, friction also depends on other surface properties, such as macro-texture.

Macrotexture

Macrotexture (MaTx) is partly a desired property and partly an undesired property. Short MaTx waves, about 5 mm, act as acoustical pores and reduce tyre/road noise. On the other hand, long wave MaTx increase noise. MaTx provide wet road friction, especially at high speeds. Excessive MaTx increases rolling resistance and thus fuel consumption and CO_2 emission contributing to global warming. Proper roads have MaTx of about 1 mm Mean Profile Depth.

Macrotexture is a family of wave-shaped road surface characteristics. While vehicle suspension deflection and dynamic tyre loads are affected by longer waves (roughness), road texture affects the interaction between the road surface and the tyre footprint. Macrotexture has wavelengths from 0.5 mm up to 50 mm.

Road agencies monitor macrotexture using measurements taken with highway speed laser or inertial profilometers.

Megatexture

Megatexture (MeTx) is the result of pavement wear and distress, causing noise and vibration. MeTx below 0.2 mm Root-Mean-Square is considered normal on proper roads.

Measurement

MaTx and MeTx are measured with laser/inertial profilographs. Since MiTx has so short waves, it is preferably measured by dry friction brake tests rather than by profiling. Profilographs that record texture in both left and right wheel paths can be used to identify road sections with hazardous split friction.

Profilograph

The profilograph is a device used to measure pavement surface roughness. In the early 20th century, profilographs were low speed rolling devices. Today many profilographs are advanced high speed systems with a laser based height sensor in combination with an inertial system that creates a large scale reference plane. It is used by construction crews or certified consultants to measure the roughness of in-service road networks, as well as before and after milling off ridges and paving overlays. Modern profilographs are fully computerized instruments.

The data collected by a profilograph is used to calculate the International Roughness Index (IRI), which is expressed in units of inches/mile or mm/m. IRI values range from 0 (equivalent to driving on a plate of glass) upwards to several hundred in/mile (a very rough road). The IRI value is used for road management to monitor road safety and quality issues.

Many road profilographs are also measuring the pavements cross slope, longitudinal gradient and rutting. Some profilographs take digital photos or videos while profiling the road. Most profilographs also record the position, using GPS technology. Yet another common measurement option is cracks. Some profilograph systems include a ground penetrating radar, used to record asphalt layer thickness.

Another type of profilograph system is for measuring the surface texture of a road and how it relates to the coefficient of friction and thus to skid resistance. Pavement texture is divided into three categories; megatexture, macrotexture, and microtexture. Microtexture cannot currently be measured directly, except in a laboratory. Megatexture is measured using a similar profiling method as when obtaining IRI values, while macrotexture is the measurement of the individual variations of the road within a small interval of a few centimeters. For example, a road which has gravel spread on top followed by an asphalt seal coat will have a high macrotexture, and a road built with concrete slabs will have low macrotexture. For this reason, concrete is often grooved or roughed up immediately after it is laid on the road bed to increase the friction between the tire and road.

Equipment to measure macrotexture currently consists of a distance measuring laser with an extremely small spot size (< 1 mm) and data acquisition systems capable of recording elevations spaced at 1 mm or less. The sample rate is generally over 32 kHz. Macrotexture data can be used to calculate the speed-dependent part of friction between typical car tires and the road surface in both dry and wet conditions. Microtexture affects friction as well.

Lateral friction and cross slope are the key reaction forces acting to keep a cornering vehicle in steady lateral position, while it is subject to exiting forces arising from speed and curvature. Cross slope and curvature can be measured with a road profilograph, and in combination with friction-related measurements can be used to identify improperly banked curves, which can increase the risk of motor vehicle accidents.

Road Profilometery

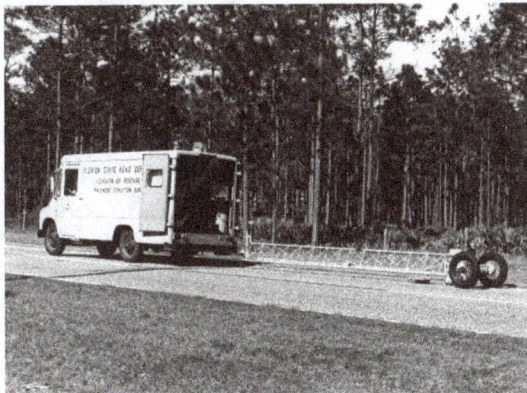

A van pulling a road profilometer (undated, before 1969).

Road pavement profilometers (aka profilographs, as used in the famous 1958-1960 AASHO Road Test) use a distance measuring laser (suspended approximately 30 cm from the pavement) in combination with an odometer and an inertial unit (normally an accelerometer to detect vehicle movement in the vertical plane) that establishes a moving reference plane to which the laser distances are integrated. The inertial compensation makes the profile data more or less independent of what speed the profilometer vehicle had during the measurements, with the assumption that the vehicle does not make large speed variations and the speed is kept above 25 km/h or 15 mph. The profilometer system collects data at normal highway speeds, sampling the surface elevations at intervals of 2–15 cm (1–6 in), and requires a high speed data acquisition system capable of obtaining measurements in the kilohertz range.

The data collected by a profilometer is used to calculate the International Roughness Index (IRI) which is expressed in units of inches/mile or mm/m. IRI values range from 0 (equivalent to driving on a plate of glass) upwards to several hundred in/mi (a very rough road). The IRI value is used for road management to monitor road safety and quality issues.

Many road profilers also measure the pavement's cross slope, curvature, longitudinal gradient and rutting. Some profilers take digital photos or videos while profiling the road. Most profilers also record the position, using GPS technology. Another quite common measurement option is cracks. Some profilometer systems include a ground penetrating radar, used to record asphalt layer thickness.

Another type of profilometer is for measuring the surface texture of a road and how it relates to the coefficient of friction and thus to skid resistance. Pavement texture is divided into three categories: megatexture, macrotexture, and microtexture. Microtexture cannot currently be measured directly, except in a laboratory. Megatexture is measured using a similar profiling method as when obtaining IRI values, while macrotexture is the measurement of the individual variations of the road within a small interval of a few centimeters. For example, a road which has gravel spread on top followed by an asphalt seal coat will have a high macrotexture, and a road built with concrete slabs will have low macrotexture. For this reason, concrete is often grooved or roughed up immediately after it is laid on the road bed to increase the friction between the tire and road.

Equipment to measure macrotexture currently consists of a distance measuring laser with an extremely small spot size (< 1 mm) and data acquisition systems capable of recording elevations spaced at a mm or less apart. The sample rate is generally over 32 kHz. Macrotexture data can be used to calculate the speed-depending part of the friction number between typical car tires and the road surface. The macrotexture also give information on the difference between dry and wet road friction. However, macrotexture cannot be used to calculate a relevant friction number, since also microtexture affects the friction.

Lateral friction and cross slope are the key reaction forces acting to keep a cornering vehicle in steady lateral position, while exposed to exciting forces from speed and curvature. Since friction is strongly dependent on macrotexture and texture, cross slope as well as curvature can be measured with a road profiler, so road profilers are very useful to identify improperly banked curves that may pose a risk to motor vehicles.

Rolling Resistance

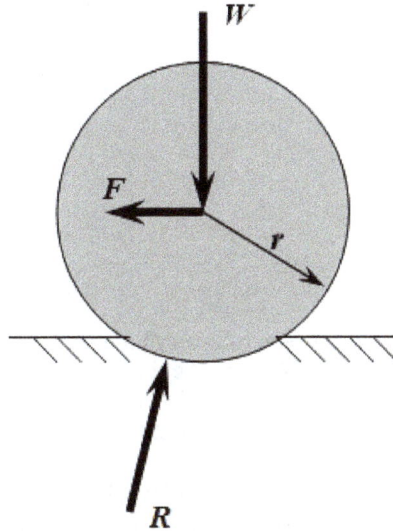

Hard wheel rolling on and deforming a soft surface, resulting in the reaction force R from the surface having a component that opposes the motion. (W is some vertical load on the axle, F is some towing force applied to the axle, r is the wheel radius, and both friction with the ground and friction at the axle are assumed to be negligible and so are not shown. The wheel is rolling to the left at constant speed.) Note that R is the resultant force from non-uniform pressure at the wheel-roadbed contact surface. This pressure is greater towards the front of the wheel due to hysteresis.

Rolling resistance, sometimes called rolling friction or rolling drag, is the force resisting the motion when a body (such as a ball, tire, or wheel) rolls on a surface. It is mainly caused by non-elastic effects; that is, not all the energy needed for deformation (or movement) of the wheel, roadbed, etc. is recovered when the pressure is removed. Two forms of this are hysteresis losses, and permanent (plastic) deformation of the object or the surface (e.g. soil). Another cause of rolling resistance lies in the slippage between the wheel and the surface, which dissipates energy. Note that only the last of these effects involves friction, therefore the name "rolling friction" is to an extent a misnomer.

In analogy with sliding friction, rolling resistance is often expressed as a coefficient times the normal force. This coefficient of rolling resistance is generally much smaller than the coefficient of sliding friction.

Any coasting wheeled vehicle will gradually slow down due to rolling resistance including that of the bearings, but a train car with steel wheels running on steel rails will roll farther than a bus of the same mass with rubber tires running on tarmac. Factors that contribute to rolling resistance are the (amount of) deformation of the wheels, the deformation of the roadbed surface, and movement below the surface. Additional contributing factors include wheel diameter, speed, load on wheel, surface adhesion, sliding, and relative micro-sliding between the surfaces of contact. The losses due to hysteresis also depend strongly on the material properties of the wheel or tire and the surface. For example, a rubber tire will have higher rolling resistance on a paved road than a steel railroad wheel on a steel rail. Also, sand on the ground will give more rolling resistance than concrete.

Primary Cause

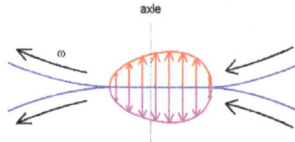

Asymmetrical pressure distribution between rolling cylinders due to viscoelastic material behavior (rolling to the right).

The primary cause of pneumatic tire rolling resistance is hysteresis:

A characteristic of a deformable material such that the energy of deformation is greater than the energy of recovery. The rubber compound in a tire exhibits hysteresis. As the tire rotates under the weight of the vehicle, it experiences repeated cycles of deformation and recovery, and it dissipates the hysteresis energy loss as heat. Hysteresis is the main cause of energy loss associated with rolling resistance and is attributed to the viscoelastic characteristics of the rubber.

        -- National Academy of Sciences

This main principle is illustrated in the figure of the rolling cylinders. If two equal cylinders are pressed together then the contact surface is flat. In the absence of surface friction, contact stresses are normal (i.e. perpendicular) to the contact surface. Consider a particle that enters the contact area at the right side, travels through the contact patch and leaves at the left side. Initially its vertical deformation is increasing, which is resisted by the hysteresis effect. Therefore an additional pressure is generated to avoid interpenetration of the two surfaces. Later its vertical deformation is decreasing. This is again resisted by the hysteresis effect. In this case this decreases the pressure that is needed to keep the two bodies separate.

The resulting pressure distribution is asymmetrical and is shifted to the right. The line of action of the (aggregate) vertical force no longer passes through the centers of the cylinders. This means that a moment occurs that tends to retard the rolling motion.

Materials that have a large hysteresis effect, such as rubber, which bounce back slowly, exhibit more rolling resistance than materials with a small hysteresis effect that bounce back more quickly and more completely, such as steel or silica. Low rolling resistance tires typically incorporate silica in place of carbon black in their tread compounds to reduce low-frequency hysteresis without compromising traction. Note that railroads also have hysteresis in the roadbed structure.

"Rolling Resistance" has Different Definitions

In the broad sense, specific "rolling resistance" (for vehicles) is the force per unit vehicle weight required to move the vehicle on level ground at a constant slow speed where aerodynamic drag (air resistance) is insignificant and also where there are no traction (motor) forces or brakes applied. In other words, the vehicle would be coasting if it were not for the force to maintain constant speed. An example of such usage for railroads is . This broad sense includes wheel bearing resistance, the energy dissipated by vibration and oscillation of both the roadbed and the vehicle, and sliding of the wheel on the roadbed surface (pavement or a rail).

But there is an even broader sense that would include energy wasted by wheel slippage due to the torque applied from the engine. This includes the increased power required due to the increased velocity of the wheels where the tangential velocity of the driving wheel(s) becomes greater than the vehicle speed due to slippage. Since power is equal to force times velocity and the wheel velocity has increased, the power required has increased accordingly.

The pure "rolling resistance" for a train is that which happens due to deformation and possible minor sliding at the wheel-road contact. For a rubber tire, an analogous energy loss happens over the entire tire, but it is still called "rolling resistance". In the broad sense, "rolling resistance" includes wheel bearing resistance, energy loss by shaking both the roadbed (and the earth underneath) and the vehicle itself, and by sliding of the wheel, road/rail contact. Railroad textbooks seem to cover all these resistance forces but do not call their sum "rolling resistance" (broad sense) as is done in this article. They just sum up all the resistance forces (including aerodynamic drag) and call the sum basic train resistance (or the like).

Since railroad rolling resistance in the broad sense may be a few times larger than just the pure rolling resistance reported values may be in serious conflict since they may be based on different definitions of "rolling resistance". The train's engines must, of course, provide the energy to overcome this broad-sense rolling resistance.

For tyres, rolling resistance is defined as the energy consumed by a tyre per unit distance covered. It is also called rolling friction or rolling drag. It is one of the forces that act to oppose the motion of a driver. The main reason for this is that when the tyres are in motion and touch the surface, the surface changes shape and causes deformation of the tyre.

For highway motor vehicles, there is obviously some energy dissipated in shaking the roadway (and the earth beneath it), the shaking of the vehicle itself, and the sliding of the tires. But, other than the additional power required due to torque and wheel bearing friction, non-pure rolling resistance doesn't seem to have been investigated, possibly because the "pure" rolling resistance of a rubber tire is several times higher than the neglected resistances.

Rolling Resistance Coefficient

The "rolling resistance coefficient" is defined by the following equation:

$$F = C_{rr}N$$

where

F is the rolling resistance force

C_{rr} is the dimensionless rolling resistance coefficient or coefficient of rolling friction (CRF), and

N is the normal force, the force perpendicular to the surface on which the wheel is rolling.

C_{rr} is the force needed to push (or tow) a wheeled vehicle forward (at constant speed on a level surface, or zero grade, with zero air resistance) per unit force of weight. It is assumed that all wheels are the same and bear identical weight. Thus: $C_{rr} = 0.01$ means that it would only take 0.01 pounds to tow a vehicle weighing one pound. For a 1000 pound vehicle, it would take 1000 times more tow force, i.e. 10 pounds. One could say that C_{rr} is in lb(tow-force)/lb(vehicle weight). Since this lb/lb is force divided by force, C_{rr} is dimensionless. Multiply it by 100 and you get the percent (%) of the weight of the vehicle required to maintain slow steady speed. C_{rr} is often multiplied by 1000 to get the parts per thousand, which is the same as kilograms (kg force) per metric ton (tonne = 1000 kg), which is the same as pounds of resistance per 1000 pounds of load or Newtons/kilo-Newton, etc. For the US railroads, lb/ton has been traditionally used; this is just $2000C_{rr}$. Thus, they are all just measures of resistance per unit vehicle weight. While they are all "specific resistances", sometimes they are just called "resistance" although they are really a coefficient (ratio)or a multiple thereof. If using pounds or kilograms as force units, mass is equal to weight (in earth's gravity a kilogram a mass weighs a kilogram and exerts a kilogram of force) so one could claim that C_{rr} is also the force per unit mass in such units. The SI system would use N/tonne (N/T, N/t), which is $1000gC_{rr}$ and is force per unit mass, where g is the acceleration of gravity in SI units (meters per second square).

The above shows resistance proportional to C_{rr} but does not explicitly show any variation with speed, loads, torque, surface roughness, diameter, tire inflation/wear, etc. because C_{rr} itself varies with those factors. It might seem from the above definition of C_{rr} that the rolling resistance is directly proportional to vehicle weight but it is not.

Measurement

There are at least two popular models for calculating rolling resistance.

1. "Rolling resistance coefficient (RRC). The value of the rolling resistance force divided by the wheel load. The Society of Automotive Engineers (SAE) has developed test practices to measure the RRC of tires. These tests (SAE J1269 and SAE J2452) are usually performed on new tires. When measured by using these standard test practices, most new passenger tires have reported RRCs ranging from 0.007 to 0.014." In the case of bicycle tires, values of 0.0025 to 0.005 are achieved. These coefficients are measured on rollers, with power meters on road surfaces, or with coast-down tests. In the latter two cases, the effect of air resistance must be subtracted or the tests performed at very low speeds.

2. The coefficient of rolling resistance b, which has the dimension of length, is approximately (due to the small-angle approximation of $cos(\theta) = 1$) equal to the value of the rolling resistance force times the radius of the wheel divided by the wheel load.

3. ISO 18164:2005 is used to test rolling resistance in Europe.

The results of these tests can be hard for the general public to obtain as manufacturers prefer to publicize "comfort" and "performance".

Physical Formulas

The coefficient of rolling resistance for a slow rigid wheel on a perfectly elastic surface, not adjusted for velocity, can be calculated by

$$C_{rr} = \sqrt{z/d}$$

where

z is the sinkage depth

d is the diameter of the rigid wheel

Empirical formula for C_{rr} for cast iron mine car wheels on steel rails.

$$C_{rr} = 0.0048(18/D)^{\frac{1}{2}}(100/W)^{\frac{1}{4}}$$

where

D is the wheel diameter in in.

W is the load on the wheel in lbs.

As an alternative to using C_{rr} one can use b, which is a different rolling resistance coefficient or coefficient of rolling friction with dimension of length. It is defined by the following formula:

$$F = \frac{Nb}{r}$$

where

F is the rolling resistance force

r is the wheel radius,

b is the rolling resistance coefficient or coefficient of rolling friction with dimension of length, and

N is the normal force

The above equation, where resistance is inversely proportional to radius r. seems to be based on the discredited "Coulomb's law" (Neither Coulomb's inverse square law nor Coulomb's law of friction). Equating this equation with the force per the #Rolling resistance coefficient, and solving for b, gives b = C_{rr}·r. Therefore, if a source gives rolling resistance coefficient (C_{rr}) as a dimensionless coefficient, it can be converted to b, having units of length, by multiplying C_{rr} by wheel radius r.

Rolling Resistance Coefficient Examples

Table of rolling resistance coefficient examples:

C_{rr}	b	Description
0.0003 to 0.0004		"Pure rolling resistance" Railroad steel wheel on steel rail
0.001 to 0.0015	0.1 mm	Hardened steel ball bearings on steel
0.0010 to 0.0024	0.5 mm	Railroad steel wheel on steel rail. Passenger rail car about 0.0020
0.0019 to 0.0065		Mine car cast iron wheels on steel rail
0.0022 to 0.005		Production bicycle tires at 120 psi (8.3 bar) and 50 km/h (31 mph), measured on rollers
0.0025		Special Michelin solar car/eco-marathon tires
0.005		Dirty tram rails (standard) with straights and curves
0.0045 to 0.008		Large truck (Semi) tires
0.0055		Typical BMX bicycle tires used for solar cars
0.0062 to 0.015		Car tire measurements
0.010 to 0.015		Ordinary car tires on concrete
0.0385 to 0.073		Stage coach (19th century) on dirt road. Soft snow on road for worst case.
0.3		Ordinary car tires on sand

For example, in earth gravity, a car of 1000 kg on asphalt will need a force of around 100 newtons for rolling (1000 kg × 9.81 m/s² × 0.01 = 98.1 N).

Depends on Diameter

Stagecoaches and Railroads (Diameter)

According to Dupuit (1837), rolling resistance (of wheeled carriages with wooden wheels with iron tires) is approximately inversely proportional to the square root of wheel diameter. This rule has been experimentally verified for cast iron wheels (8" - 24" diameter) on steel rail and for 19th century carriage wheels. But there are other tests on carriage wheels that do not agree. Theory of a cylinder rolling on an elastic roadway also gives this same rule These contradict earlier (1785) tests by Coulomb of rolling wooden cylinders where Coulomb reported that rolling resistance was inversely proportional to the diameter of the wheel (known as "Coulomb's law"). This disputed (or wrongly applied) -"Coulomb's law" is still found in handbooks, however.

Pneumatic Tires (Diameter)

For pneumatic tires on hard pavement, it is reported that the effect of diameter on rolling resistance is negligible (within a practical range of diameters).

Depends on Applied Torque

The driving torque T to overcome rolling resistance R_r and maintain steady speed on level ground (with no air resistance) can be calculated by:

$$T = \frac{V_s}{\Omega} R_r$$

where

V_s is the linear speed of the body (at the axle), and

Ω its rotational speed.

It is noteworthy that V_s / Ω is usually not equal to the radius of the rolling body.

All Wheels (Torque)

"Applied torque" may either be driving torque applied by a motor (often through a transmission) or a braking torque applied by brakes(including regenerative braking). Such torques results in energy dissipation (above that due to the basic rolling resistance of a freely rolling, non-driven, non-braked wheel). This additional loss is in part due to the fact that there is some slipping of the wheel, and for pneumatic tires, there is more flexing of the sidewalls due to the torque. Slip is defined such that a 2% slip means that the circumferential speed of the driving wheel exceeds the speed of the vehicle by 2%.

A small percentage slip can result in a much larger percentage increase in rolling resistance. For example, for pneumatic tires, a 5% slip can translate into a 200% increase in rolling resistance. This is partly because the tractive force applied during this slip is many times greater than the rolling resistance force and thus much more power per unit velocity is being applied (recall power = force x velocity so that power per unit of velocity is just force). So just a small percentage increase in circumferential velocity due to slip can translate into a loss of traction power which may even exceed the power loss due to basic (ordinary) rolling resistance. For railroads, this effect may be even more pronounced due to the low rolling resistance of steel wheels.

Railroad Steel Wheels (Torque)

In order to apply any traction to the wheels, some slippage of the wheel is required. For Russian trains climbing up a grade, this slip is normally 1.5% to 2.5%.

Slip (also known as creep) is normally roughly directly proportional to tractive effort. An exception is if the tractive effort is so high that the wheel is close to substantial slipping (more than just a few percent as discussed above), then slip rapidly increases with tractive effort and is no longer linear. With a little higher applied tractive effort the wheel spins out of control and the adhesion drops resulting in the wheel spinning even faster. This is the type of slipping that is observable by eye—the slip of say 2% for traction is only observed by instruments. Such rapid slip may result in excessive wear or damage.

Pneumatic Tires (Torque)

Rolling resistance greatly increases with applied torque. At high torques, which apply a tangential force to the road of about half the weight of the vehicle, the rolling resistance may triple (a 200% increase). This is in part due to a slip of about 5%. The rolling resistance increase with applied torque is not linear, but increases at a faster rate as the torque becomes higher.

Wheel Load Dependence
Railroad Steel Wheels (Load)

The rolling resistance coefficient, Crr, significantly decreases as the weight of the rail car per wheel increases. For example, an empty Russian freight car had about twice the Crr as loaded car (Crr=0.002 vs. Crr=0.001). This same "economy of scale" shows up in testing of mine rail cars. The theoretical Crr for a rigid wheel rolling on an elastic roadbed shows Crr inversely proportional to the square root of the load.

If Crr is itself dependent on wheel load per an inverse square-root rule, then for an increase in load of 2% only a 1% increase in rolling resistance occurs.

Pneumatic Tires (Load)

For pneumatic tires, the direction of change in Crr (rolling resistance coefficient) depends on whether or not tire inflation is increased with increasing load. It is reported that, if inflation pressure is increased with load according to an (undefined) "schedule", then a 20% increase in load decreases Crr by 3%. But, if the inflation pressure is not changed, then a 20% increase in load results in a 4% increase in Crr. Of course, this will increase the rolling resistance by 20% due to the increase in load plus 1.2 x 4% due to the increase in Crr resulting in a 24.8% increase in rolling resistance.

Depends on Curvature of Roadway

General

When a vehicle (motor vehicle or railroad train) goes around a curve, rolling resistance usually increases. If the curve is not banked so as to exactly counter the centrifugal force with an equal and opposing centripetal force due to the banking, then there will be a net unbalanced sideways force on the vehicle. This will result in increased rolling resistance. Banking is also known as "superelevation" or "cant". For railroads, this is called curve resistance but for roads it has (at least once) been called rolling resistance due to cornering.

Sound Effects

Rolling friction generates sound (vibrational) energy, as mechanical energy is converted to this form of energy due to the friction. One of the most common examples of rolling friction is the movement of motor vehicle tires on a roadway, a process which generates sound as a by-product. The sound generated by automobile and truck tires as they roll (especially noticeable at highway speeds) is mostly due to the percussion of the tire treads, and compression (and subsequent decompression) of air temporarily captured within the treads.

Factors that Contribute in Tires

Several factors affect the magnitude of rolling resistance a tire generates:

- As mentioned in the introduction: wheel radius, forward speed, surface adhesion, and relative micro-sliding.

- Material - different fillers and polymers in tire composition can improve traction while reducing hysteresis. The replacement of some carbon black with higher-priced silica–silane is one common way of reducing rolling resistance. The use of exotic materials including nano-clay has been shown to reduce rolling resistance in high performance rubber tires. Solvents may also be used to swell solid tires, decreasing the rolling resistance.

- Dimensions - rolling resistance in tires is related to the flex of sidewalls and the contact area of the tire For example, at the same pressure, wider bicycle tires flex less in the sidewalls as they roll and thus have lower rolling resistance (although higher air resistance).

- Extent of inflation - Lower pressure in tires results in more flexing of the sidewalls and higher rolling resistance. This energy conversion in the sidewalls increases resistance and can also lead to overheating and may have played a part in the infamous Ford Explorer rollover accidents.

- Over inflating tires (such a bicycle tires) may not lower the overall rolling resistance as the tire may skip and hop over the road surface. Traction is sacrificed, and overall rolling friction may not be reduced as the wheel rotational speed changes and slippage increases.

- Sidewall deflection is not a direct measurement of rolling friction. A high quality tire with a high quality (and supple) casing will allow for more flex per energy loss than a cheap tire with a stiff sidewall. Again, on a bicycle, a quality tire with a supple casing will still roll easier than a cheap tire with a stiff casing. Similarly, as noted by Goodyear truck tires, a tire with a "fuel saving" casing will benefit the fuel economy through many tread lives (i.e. retreading), while a tire with a "fuel saving" tread design will only benefit until the tread wears down.

- In tires, tread thickness and shape has much to do with rolling resistance. The thicker and more contoured the tread, the higher the rolling resistance Thus, the "fastest" bicycle tires have very little tread and heavy duty trucks get the best fuel economy as the tire tread wears out.

- Diameter effects seem to be negligible, provided the pavement is hard and the range of diameters is limited.

- Virtually all world speed records have been set on relatively narrow wheels, probably because of their aerodynamic advantage at high speed, which is much less important at normal speeds.

- Temperature: with both solid and pneumatic tires, rolling resistance has been found to decrease as temperature increases (within a range of temperatures: i.e. there is an upper limit to this effect) For a rise in temperature from 30 °C to 70 °C the rolling resistance decreased by 20-25%. It is claimed that racers heat their tire before racing.

Railroads: Components of Rolling Resistance

In a broad sense rolling resistance can be defined as the sum of components):

1. Wheel bearing torque losses.

2. Pure rolling resistance.

3. Sliding of the wheel on the rail.

4. Loss of energy to the roadbed (and earth).

5. Loss of energy to oscillation of railway rolling stock.

Wheel bearing torque losses can be measured as a rolling resistance at the wheel rim, Crr. Railroads normally use roller bearings which are either cylindrical (Russia) or tapered (United States). The specific rolling resistance in Russian bearings varies with both wheel loading and speed. Wheel bearing rolling resistance is lowest with high axle loads and intermediate speeds of 60–80 km/h with a Crr of 0.00013 (axle load of 21 tonnes). For empty freight cars with axle loads of 5.5 tonnes, Crr goes up to 0.00020 at 60 km/h but at a low speed of 20 km/h it increases to 0.00024 and at a high speed (for freight trains) of 120 km/h it is 0.00028. The Crr obtained above is added to the Crr of the other components to obtain the total Crr for the wheels.

Comparing Rolling Resistance of Highway Vehicles and Trains

The rolling resistance of steel wheels on steel rail of a train is far less than that of the rubber tires wheels of an automobile or truck. The weight of trains varies greatly; in some cases they may be much heavier per passenger or per net ton of freight than an automobile or truck, but in other cases they may be much lighter.

As an example of a very heavy passenger train, in 1975, Amtrak passenger trains weighed a little over 7 tonnes per passenger, which is much heavier than an average of a little over one ton per passenger for an automobile. This means that for an Amtrak passenger train in 1975, much of the energy savings of the lower rolling resistance was lost to its greater weight.

An example of a very light high-speed passenger train is the N700 Series Shinkansen, which weighs 715 tonnes and carries 1323 passengers, resulting in a per-passenger weight of about half a tonne. This lighter weight per passenger, combined with the lower rolling resistance of steel wheels on steel rail means that an N700 Shinkansen is much more energy efficient than a typical automobile.

In the case of freight, CSX ran an advertisement campaign in 2013 claiming that their freight trains move "a ton of freight 436 miles on a gallon of fuel", whereas some sources claim trucks move a ton of freight about 130 miles per gallon of fuel, indicating trains are more efficient overall.

Belt Friction

Belt friction is a term describing the friction forces between a belt and a surface, such as a belt wrapped around a bollard. When one end of the belt is being pulled only part of this force is transmitted to the other end wrapped about a surface. The friction force increases with the amount of wrap about a surface and makes it so the tension in the belt can be different at both ends of the belt. Belt friction can be modeled by the Belt friction equation.

In practice, the theoretical tension acting on the belt or rope calculated by the belt friction equation can be compared to the maximum tension the belt can support. This helps a designer of such a rig to know how many times the belt or rope must be wrapped around the pulley to prevent it from slipping. Mountain climbers and sailing crews demonstrate a standard knowledge of belt friction when accomplishing basic tasks.

Equation

The equation used to model belt friction is, assuming the belt has no mass and its material is a fixed composition:

$$T_2 = T_1 e^{\mu_s \beta}$$

where T_2 is the tension of the pulling side, T_1 is the tension of the resisting side, μ_s is the static friction coefficient, which has no units, and β is the angle, in radians, formed by the first and last spots the belt touches the pulley, with the vertex at the center of the pulley.

The tension on the pulling side of the belt and pulley has the ability to increase exponentially if the magnitude of the belt angle increases (e.g. it is wrapped around the pulley segment numerous times).

Generalization for a Rope Lying on an Arbitrary Orthotropic Surface

If a rope is laying in equilibrium under tangential forces on a rough orthotropic surface then three following conditions (all of them) are satisfied:

1. No separation – normal reaction N is positive for all points of the rope curve:

$N = -k_n T > 0$, where k_n is a normal curvature of the rope curve.

2. Dragging coefficient of friction μ_g and angle α are satisfying the following criteria for all points of the curve

$$-\mu_g < \tan \alpha < +\mu_g$$

3. Limit values of the tangential forces:

The forces at both ends of the rope T and T_0 are satisfying the following inequality

$$T_0 e^{-\int_s \omega ds} \leq T \leq T_0 e^{\int_s \omega ds}$$

with
$$\omega = \mu_\tau \sqrt{k_n^2 - \frac{k_g^2}{\mu_g^2}} = \mu_\tau k \sqrt{\cos^2 \alpha - \frac{\sin^2 \alpha}{\mu_g^2}},$$

where k_g is a geodesic curvature of the rope curve, k is a curvature of a rope curve, μ_τ is a coefficient of friction in the tangential direction.

If $\omega = const$ then $T_0 e^{-\mu_\tau ks \sqrt{\cos^2 \alpha - \frac{\sin^2 \alpha}{\mu_g^2}}} \leq T \leq T_0 e^{\mu_\tau ks \sqrt{\cos^2 \alpha - \frac{\sin^2 \alpha}{\mu_g^2}}}$ ..

This generalization has been obtained by Konyukhov A.,

Friction Coefficient

There are certain factors that help determine the value of the friction coefficient. These determining factors are:

- Belting material used – The age of the material also plays a part, where worn out and older material may become more rough or smoother, changing the sliding friction.

- Construction of the drive-pulley system – This involves strength and stability of the material used, like the pulley, and how greatly it will oppose the motion of the belt or rope.

- Conditions under which the belt and pulleys are operating – The friction between the belt and pulley may decrease substantially if the belt happens to be muddy or wet, as it may act as a lubricant between the surfaces. This also applies to extremely dry or warm conditions which will evaporate any water naturally found in the belt, nominally making friction greater.

- Overall design of the setup – The setup involves the initial conditions of the construction, such as the angle which the belt is wrapped around and geometry of the belt and pulley system.

Applications

An understanding of belt friction is essential for sailing crews and mountain climbers. Their professions require being able to understand the amount of weight a rope with a certain tension capacity can hold versus the amount of wraps around a pulley. Too many revolutions around a pulley make it inefficient to retract or release rope, and too few may cause the rope to slip. Misjudging the ability of a rope and capstan system to maintain the proper frictional forces may lead to failure and injury.

Superlubricity

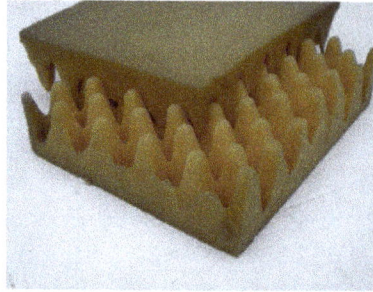

Foam in an egg carton which simulates the atomic surface structure of graphite, commensurable due to alignment in this photo

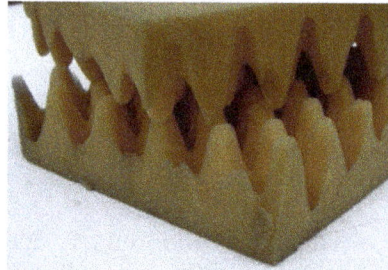

Incommensurable due to twisting, so the valleys and hills don't line up

Superlubricity is a regime of motion in which friction vanishes or very nearly vanishes. What is a "vanishing" friction level is not clear, which makes the term of the superlubricity to be quite vague. As an *ad hoc* definition, a kinetic coefficient of friction less than 0.001 can be adopted. This definition also requires further discussion and clarification.

Superlubricity may occur when two crystalline surfaces slide over each other in dry incommensurate contact. This effect, also called structural lubricity, was suggested in 1991 and verified with great accuracy between two graphite surfaces in 2004. The atoms in graphite are oriented in a hexagonal manner and form an atomic hill-and-valley landscape, which looks like an egg-crate. When the two graphite surfaces are in registry (every 60 degrees), the friction force is high. When the two surfaces are rotated out of registry, the friction is largely reduced. This is like two egg-crates which can slide over each other more easily when they are "twisted" with respect to each other.

Observation of superlubricity in microscale graphite structures was reported in 2012, by shearing a square graphite mesa a few micrometers across, and observing the self-retraction of the sheared layer. Such effects were also theoretically described for a model of graphene and nickel layers. This observation, which is reproducible even under ambient conditions, shifts interest in superlubricity from a primarily academic topic, accessible only under highly idealized conditions, to one with practical implications for micro and nanomechanical devices.

A state of ultralow friction can also be achieved when a sharp tip slides over a flat surface and the applied load is below a certain threshold. Such "superlubric" threshold depends on the tip-surface interaction and the stiffness of the materials in contact, as described by the Tomlinson model. The threshold can be significantly increased by exciting the sliding system at its resonance frequency, which suggests a practical way to limit wear in nanoelectromechanical systems.

Superlubricity was also observed between a gold AFM tip and Teflon substrate due to repulsive Van der Waals forces and hydrogen-bonded layer formed by glycerol on the steel surfaces. Formation of the hydrogen-bonded layer was also shown to lead to superlubricity between quartz glass surfaces lubricated by biological liquid obtained from mucilage of Brasenia schreberi.

The similarity of the term *superlubricity* with terms such as superconductivity and superfluidity is misleading; other energy dissipation mechanisms can lead to a finite (normally small) friction force.

Superlubricity at the Macroscale

In 2015, "Argonne scientists used Mira [supercomputer] to identify and improve a new mechanism for eliminating friction, which fed into the development of a hybrid material that exhibited superlubricity at the macroscale for the first time [..] simulating up to 1.2 million atoms for dry environments and up to 10 million atoms for humid environments [..] The researchers used the LAMMPS (Large-scale Atomic/Molecular Massively Parallel Simulator) code to carry out the computationally demanding reactive molecular dynamics simulations. [.. A] team of computational scientists [..] were able to overcome a performance bottleneck with the code's ReaxFF module, an add-on package that was needed to model the chemical reactions occurring in the system. [.. The team] optimized LAMMPS and its implementation of ReaxFF by adding OpenMP threading, replacing MPI point-to-point communication with MPI collectives in key algorithms, and leveraging MPI I/O. Altogether, these enhancements allowed the code to perform twice as fast as before."

"The research team is in the process of seeking a patent for the hybrid material, which could potentially be used for applications in dry environments, such as computer hard drives, wind turbine gears, and mechanical rotating seals for microelectromechanical and nanoelectromechanical systems."

Skin Friction Drag

Skin friction drag is a component of profile drag that occurs differently depending on the type of flow over the lifting body (laminar or turbulent). Just like any other form of drag, the coefficient of skin friction drag is calculated with various equations and measurements depending on the flow and then added to coefficients of other forms of drag to calculate total drag.

Flow and Effect on Skin Friction Drag

Laminar flow is when layers of the fluid move smoothly past each other in parallel lines. In nature, this kind of flow is rare. As the fluid flows over an object, it applies frictional forces on the surface of the object which works to impede forward movement of the object, in other words, create skin friction drag. Skin friction drag is often the major component of parasitic drag in objects experiencing laminar flow.

Turbulent flow has a fluctuating and irregular pattern of flow which is attributed to the formation of vortices. Turbulent flow suggests a faster rate of flow due to velocity increase and/or viscosity decrease relative to laminar flow. This results in a thinner boundary layer which, relative to laminar flow, depreciates the magnitude of friction force as the fluid flows over the object. This suggests that the total parasitic drag observed in turbulent flow is minimally impacted by skin friction drag.

Calculation

The calculating of skin friction drag is heavily based on the Reynolds number of the body. For reference, Reynolds number (Re) is calculated with:

$$\mathrm{Re} = \frac{vL}{v}$$

where:

- v is the velocity of the flow

- L is the length of the body that the flow travels across

- v is the kinematic viscosity of the fluid

Now that Reynolds number is known, the coefficient of skin friction drag can be calculated.

Laminar Flow

$$C_f = \frac{1.328}{\sqrt{\mathrm{Re}}} ,$$

Also known as the Blasius Friction law

Note: if measuring skin friction from a certain point on the body, replace the 'L' in the

Reynolds number equation with the distance from the leading edge that you want to measure (Re_x). Then use the following equation:

$$C_{fx} = \frac{0.664}{\sqrt{Re_x}}$$

Turbulent Flow

$$C_f = \frac{0.455}{log(Re)^{2.58}},$$

Also known as the Schlichting empirical formula

Drag

The total force on the body caused by skin friction drag in units of force can be calculated with:

$$F = C_f \frac{\rho_f v^2}{2} S_{wetted}$$

where ρ_f is the fluid density, v is the fluid velocity and S_{wetted} is the total surface area that is in contact with the fluid.

Parasitic Drag

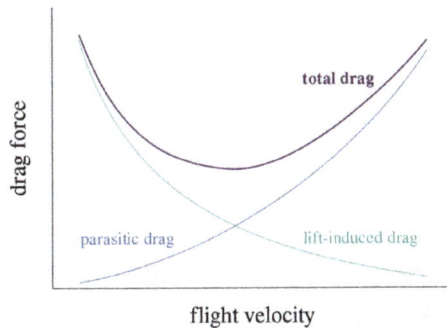

Drag curve for a body in steady flight

Parasitic drag is drag that results when an object is moved through a fluid medium. In the case of aerodynamic drag, the fluid medium is the atmosphere. Parasitic drag is a combination of form drag, skin friction drag and interference drag. The other components of total drag, induced drag, wave drag, and ram drag, are separate types of drag, and are not components of parasitic drag.

Description

In flight, induced drag results from the need to maintain a lift force, so that the craft can maintain level flight. It is greater at lower speeds where a high angle of attack is required. As speed increases, the induced drag decreases, but parasitic drag increases because the fluid is striking the object with greater force, and is moving across the object's surfaces at higher speed. As speed continues to increase into the transonic and supersonic regimes, wave drag grows in importance. Each of these drag components changes in proportion to the others based on speed. The combined overall drag curve therefore shows a minimum at some airspeed; an aircraft flying at this speed will be close to its optimal efficiency. Pilots will use this speed to maximize the gliding range in case of an engine failure. However, to maximize the gliding endurance, the aircraft's speed would have to be at the point of minimum power, which occurs at lower speeds than minimum drag.

At the point of minimum drag, $C_{D,o}$ (drag coefficient of aircraft when lift equals zero) is equal to $C_{D,i}$ (induced drag coefficient, or coefficient of drag created by lift). At the point of minimum power, $C_{D,o}$ is equal to one third times $C_{D,i}$. This can be proven by deriving the following equations:

$$F_{drag} = qA_sC_D$$

where:

$$q = \frac{1}{2}\rho u^2$$

is the dynamic pressure and

$$C_D = C_{D,o} + C_{D,i}$$

where

$$C_{D,i} = KC_L^2$$

Form Drag

Form drag or pressure drag arises because of the shape of the object. The general size and shape of the body are the most important factors in form drag; bodies with a larger presented cross-section will have a higher drag than thinner bodies; sleek ("streamlined") objects have lower form drag. Form drag follows the drag equation, meaning that it increases with velocity, and thus becomes more important for high-speed aircraft.

Form drag depends on the longitudinal section of the body. A prudent choice of body profile is essential for a low drag coefficient. Streamlines should be continuous, and separation of the boundary layer with its attendant vortices should be avoided.

Profile Drag

Profile drag is usually defined as the sum of form drag and skin friction. However, the term is often used synonymously with form drag.

Interference Drag

Interference drag results when airflow around one part of an object (such as a fuselage) must occupy the same space as the airflow around another part (such as a wing). The two competing airflows must speed up in order to pass through the restricted area; this speeding-up process requires extra energy and creates turbulence, resulting in a measurable increase in the form drag. This velocity increase is present at all airspeeds, but becomes even more important in the transonic range when the resulting velocity becomes sonic, producing shock waves.

Interference drag plays a role throughout the entire aircraft (e.g., nacelles, pylons, empennage) and its detrimental effect is always kept in mind by designers. Ideally, the pressure distributions on the intersecting bodies should complement each other's pressure distribution. If one body locally displays a negative pressure coefficient, the intersecting body should have a positive pressure coefficient. In reality, however, this is not always possible. Particular geometric characteristics on aircraft often show how designers have dealt with the issue of interference drag. A prime example is the wing-body fairing, which smooths the sharp angle between the wing and the fuselage. Another example is the junction between the horizontal and vertical tailplane in a T-tail. Often, an additional fairing (acorn) is positioned to reduce the added supervelocities. The position of the nacelle with respect to the wing is a third example of how interference-drag considerations dominate this geometric feature. For nacelles that are positioned beneath the wing, the lateral and longitudinal distance from the wing is dominated by interference-drag considerations. If there is little vertical space available between the wing and the nacelle (because of ground clearance) the nacelle is usually positioned much more in front of the wing. The NACA area rule is one approach to reducing transonic interference drag.

Skin Friction

Skin friction drag arises from the friction of the fluid against the "skin" of the object that is moving through it. Skin friction arises from the interaction between the fluid and the skin of the body, and is directly related to the wetted surface, the area of the surface of the body that is in contact with the fluid. As with other components of parasitic drag, skin friction follows the drag equation and rises with the square of the velocity.

The skin friction coefficient, C_f, is defined by

$$C_f \equiv \frac{\tau_w}{q},$$

where τ_w is the local wall shear stress, and q is the free-stream dynamic pressure. For boundary layers without a pressure gradient in the x direction, it is related to the momentum thickness as

$$C_f = 2\frac{d\theta}{dx}.$$

For comparison, the turbulent empirical relation known as the 1/7 Power Law (derived by Theodore von Kármán) is:

$$C_f = \frac{0.074}{Re^{0.2}},$$

where Re is the Reynolds number.

Skin friction is caused by viscous drag in the boundary layer around the object. The boundary layer at the front of the object is usually laminar and relatively thin, but becomes turbulent and thicker towards the rear. The position of the transition point depends on the shape of the object. There are two ways to decrease friction drag: the first is to shape the moving body so that laminar flow is possible, like an airfoil. The second method is to decrease the length and cross-section of the moving object as much as practicable. To do so, a designer can consider the fineness ratio, which is the length of the aircraft divided by its diameter at the widest point (L/D).

Triboelectric Effect

The triboelectric effect (also known as triboelectric charging) is a type of contact electrification in which certain materials become electrically charged after they come into frictional contact with a different material. Rubbing glass with fur, or a comb through the hair, can build up triboelectricity. Most everyday static electricity is triboelectric. The polarity and strength of the charges produced differ according to the materials, surface roughness, temperature, strain, and other properties.

The triboelectric effect is not very predictable, and only broad generalizations can be made. Amber, for example, can acquire an electric charge by contact and separation (or friction) with a material like wool. This property was first recorded by Thales of Miletus. The word "electricity" is derived from William Gilbert's initial coinage, "electra", which originates in the Greek word for amber. The prefix *tribo-* refers to 'friction', as in tribology. Other examples of materials that can acquire a significant charge when rubbed together include glass rubbed with silk, and hard rubber rubbed with fur.

The triboelectric effect is now considered to be very close to the phenomenon of adhesion, where two materials composed of different molecules tend to stick together on contact due to a form of chemical reaction. This is very close to a chemical bond; the adjacent dissimilar molecules exchange electrons. And when one material is physically moved away from the other, the bonding forces we experience are regarded by us as 'friction'. The result is that excess electrons are left behind in one material, while a deficit occurs in the other.

Triboelectric Series

John Carl Wilcke published the first triboelectric series in a 1757 paper on static charges. Materials are often listed in order of the polarity of charge separation when they are touched with another object. A material towards the bottom of the series, when touched to a material near the top of the series, will acquire a more negative charge. The farther away two materials are from each other on the series, the greater the charge transferred. Materials near to each other on the series may not exchange any charge, or may even exchange the opposite of what is implied by the list. This can be caused by rubbing, by contaminants or oxides, or other variables. Lists vary somewhat as to the exact order of some materials, since the relative charge varies for nearby materials. From actual tests, there is little or no measurable difference in charge affinity between metals, probably because the rapid motion of conduction electrons cancels such differences.

Cause

Although, the two materials only need to come into contact for electrons to be exchanged. After coming into contact, a chemical bond is formed between parts of the two surfaces, called adhesion, and charges move from one material to the other to equalize their electrochemical potential. This is what creates the net charge imbalance between the objects. When separated, some of the bonded atoms have a tendency to keep extra electrons, and some a tendency to give them away, though the imbalance will be partially destroyed by tunneling or electrical breakdown (usually corona discharge). In addition, some materials may exchange ions of differing mobility, or exchange charged fragments of larger molecules.

The triboelectric effect is related to friction only because they both involve adhesion. However, the effect is greatly enhanced by rubbing the materials together, as they touch and separate many times. For surfaces with differing geometry, rubbing may also lead to heating of protrusions, causing pyroelectric charge separation which may add to the existing contact electrification, or which may oppose the existing polarity. Surface nano-effects are not well understood, and the atomic force microscope has enabled rapid progress in this field of physics.

Sparks

Because the surface of the material is now electrically charged, either negatively or positively, any contact with an uncharged conductive object or with an object having substantially different charge may cause an electrical discharge of the built-up static electricity: a spark. A person simply walking across a carpet may build up a potential of many thousands of volts, enough to cause a spark one centimeter long or more. Low humidity in the ambient air increases the voltage at which electrical discharge occurs by increasing the ability of the insulating material to hold charge by decreasing the conductivity of the air, making it difficult for the charge build-up to dissipate gradually. Simply removing a nylon shirt or corset can also create sparks. Car travel can lead to a build-up of charge on the driver and passengers due to friction between the driver's clothes and the leather or plastic furnishings inside the vehicle. This charge can then be relaxed as a spark to the metal car body, fuel dispensers, or nearby door handles, etc. When the vehicle's body itself builds up a static charge (acting as a Faraday cage) it can relax through the carbon in the tires. If it remains charged when parked, sparks may jump from the door frame to occupants as they make contact with the ground.

This type of discharge is often harmless because the energy ($1/2V^2C$) of the spark is very small, being typically several tens of micro joules in cold dry weather, and much less than that in humid conditions. However, such sparks can ignite flammable vapours.

In Aircraft and Spacecraft

Aircraft flying in weather will develop a static charge from air friction on the airframe. The static can be discharged with static dischargers or static wicks.

NASA follows what they call the Triboelectrification Rule whereby they will cancel a launch if the launch vehicle is predicted to pass through certain types of clouds. Flying through high-level clouds can generate "P-static" (P for precipitation), which can create static around the launch vehicle that will interfere with radio signals sent by or to the vehicle. This may prevent transmitting of telemetry to the ground or, if the need arises, sending a signal to the vehicle, particularly critical signals for the flight termination system. When a hold is put in place due to the triboelectrification rule, it remains until Space Wing and observer personnel such as those in reconnaissance aircraft indicate that the skies are clear.

Risks and Counter-measures

Ignition

The effect is of considerable industrial importance in terms of both safety and potential damage to manufactured goods. Static discharge is a particular hazard in grain elevators owing to the danger of a dust explosion. The spark produced is fully able to ignite flammable vapours, for example, petrol, ether fumes as well as methane gas. For bulk fuel deliveries and aircraft fueling a grounding connection is made between the vehicle and the receiving tank prior to opening the tanks. When fueling vehicles at a retail station touching metal on the car before opening the gas tank or touching the nozzle may decrease one's risk of static ignition of fuel vapors.

In the Workplace

Means have to be provided to discharge carts which may carry such volatile liquids, flammable gasses, or oxygen in hospitals. Even where only a small charge is produced, it can result in dust particles being attracted to the rubbed surface. In the case of textile manufacture this can lead to a permanent grimy mark where the cloth comes in contact with dust accumulations held by a static charge. Dust attraction may be reduced by treating insulating surfaces with an antistatic cleaning agent.

Damage to Electronics

Some electronic devices, most notably CMOS integrated circuits and MOSFET transistors, can be accidentally destroyed by high-voltage static discharge. Such components are usually stored in a conductive foam for protection. Grounding oneself by touching the workbench, or using a special bracelet or anklet is standard practice while handling unconnected integrated circuits. Another way of dissipating charge is by using conducting materials such as carbon black loaded rubber mats in operating theatres, for example.

Devices containing sensitive components must be protected during normal use, installation, and disconnection, accomplished by designed-in protection at external connections where needed. Protection may be through the use of more robust devices or protective countermeasures at the device's external interfaces. These may be opto-isolators, less sensitive types of transistors, and static bypass devices such as metal oxide varistors.

References

- Beer, Ferdinand P.; Johnston, E. Russel, Jr. (1996). Vector Mechanics for Engineers (Sixth ed.). McGraw-Hill. p. 397. ISBN 0-07-297688-8.

- Meriam, J. L.; Kraige, L. G. (2002). Engineering Mechanics (fifth ed.). John Wiley & Sons. p. 328. ISBN 0-471-60293-0.

- Sambursky, Samuel (2014). The Physical World of Late Antiquity. Princeton University Press. pp. 65–66. ISBN 9781400858989.

- Dowson, Duncan (1997). History of Tribology (2nd ed.). Professional Engineering Publishing. ISBN 1-86058-070-X.

- Bhavikatti, S. S.; K. G. Rajashekarappa (1994). Engineering Mechanics. New Age International. p. 112. ISBN 978-81-224-0617-7. Retrieved 2007-10-21.

- Sheppard, Sheri; Tongue, Benson H.; Anagnos, Thalia (2005). Statics: Analysis and Design of Systems in Equilibrium. Wiley and Sons. p. 618. ISBN 0-471-37299-4. In general, for given contacting surfaces, $\mu k < \mu s$

- Meriam, James L.; Kraige, L. Glenn; Palm, William John (2002). Engineering Mechanics: Statics. Wiley and Sons. p. 330. ISBN 0-471-40646-5. Kinetic friction force is usually somewhat less than the maximum static friction force.

- Persson, B. N. J. (2000). Sliding friction: physical principles and applications. Springer. ISBN 978-3-540-67192-3. Retrieved 2016-01-23.

- Bigoni, D. Nonlinear Solid Mechanics: Bifurcation Theory and Material Instability. Cambridge University Press, 2012. ISBN 9781107025417.

- Nosonovsky, Michael (2013). Friction-Induced Vibrations and Self-Organization: Mechanics and Non-Equilibrium Thermodynamics of Sliding Contact. CRC Press. p. 333. ISBN 978-1466504011.

- Butt, Hans-Jürgen; Graf, Karlheinz and Kappl, Michael (2006) Physics and Chemistry of Interfaces, Wiley, ISBN 3-527-40413-9

- Bhandari, V.B. (2010). Design of machine elements. Tata McGraw-Hill. p. 472. ISBN 9780070681798. Retrieved 9 February 2016.

- Lane, Keith (2002). Automotive A-Z: Lane's Complete Dictionary of Automotive Terms. Veloce Publishing Plc. ISBN 978-1903706404. Retrieved 12 August 2015.

- Gwidon W. Stachowiak, Andrew William Batchelor, Engineering Tribology, Elsevier Publisher, 750 pages (2000) ISBN 0-7506-7304-4

- Hutchings, Ian M. (2016). "Leonardo da Vinci's studies of friction" (PDF). doi:10.1016/j.wear.2016.04.019.

- Hutchings, Ian M. (2016-08-15). "Leonardo da Vinci's studies of friction". Wear. 360–361: 51–66. doi:10.1016/j.wear.2016.04.019.

- Kirk, Tom (July 22, 2016). "Study reveals Leonardo da Vinci's 'irrelevant' scribbles mark the spot where he first recorded the laws of friction". phys.org. Retrieved 2016-07-26.

Surface Tension: Methods and Techniques

Surface tension is a characteristic that is found in a fluid surface; this quality allows the fluid to obtain the least surface area that is possible. The methods and techniques elucidated in this chapter are du Noüy ring method, Wilhelmy plate, spinning drop, maximum bubble pressure method and sessile drop technique.

Surface Tension

Surface tension is the elastic tendency of a fluid surface which makes it acquire the least surface area possible. Surface tension allows insects (e.g. water striders), usually denser than water, to float and stride on a water surface.

At liquid-air interfaces, surface tension results from the greater attraction of liquid molecules to each other (due to cohesion) than to the molecules in the air (due to adhesion). The net effect is an inward force at its surface that causes the liquid to behave as if its surface were covered with a stretched elastic membrane. Thus, the surface becomes under tension from the imbalanced forces, which is probably where the term "surface tension" came from. Because of the relatively high attraction of water molecules for each other through a web of hydrogen bonds, water has a higher surface tension (72.8 millinewtons per meter at 20 °C) compared to that of most other liquids. Surface tension is an important factor in the phenomenon of capillarity.

Surface tension has the dimension of force per unit length, or of energy per unit area. The two are equivalent, but when referring to energy per unit of area, it is common to use the term surface energy, which is a more general term in the sense that it applies also to solids.

In materials science, surface tension is used for either surface stress or surface free energy.

Causes

The cohesive forces among liquid molecules are responsible for the phenomenon of surface tension. In the bulk of the liquid, each molecule is pulled equally in every direction by neighboring liquid molecules, resulting in a net force of zero. The molecules at the surface do not have the *same* molecules on all sides of them and therefore are pulled inwards. This creates some internal pressure and forces liquid surfaces to contract to the minimal area.

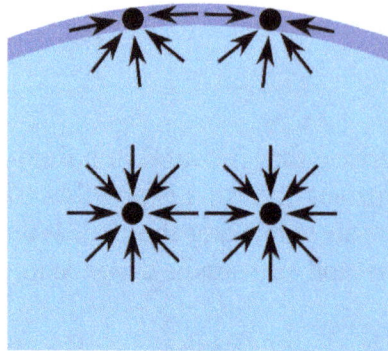

Diagram of the forces on molecules of a liquid

Surface tension is responsible for the shape of liquid droplets. Although easily deformed, droplets of water tend to be pulled into a spherical shape by the imbalance in cohesive forces of the surface layer. In the absence of other forces, including gravity, drops of virtually all liquids would be approximately spherical. The spherical shape minimizes the necessary "wall tension" of the surface layer according to Laplace's law.

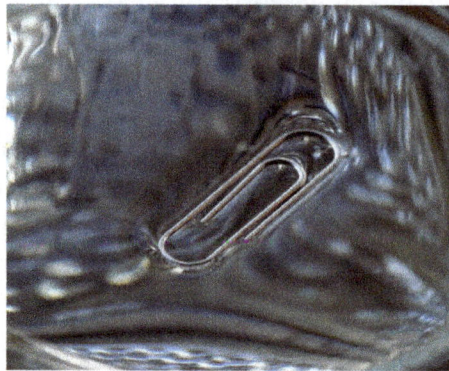

Surface tension preventing a paper clip from submerging.

Another way to view surface tension is in terms of energy. A molecule in contact with a neighbor is in a lower state of energy than if it were alone (not in contact with a neighbor). The interior molecules have as many neighbors as they can possibly have, but the boundary molecules are missing neighbors (compared to interior molecules) and therefore have a higher energy. For the liquid to minimize its energy state, the number of higher energy boundary molecules must be minimized. The minimized quantity of boundary molecules results in a minimal surface area. As a result of surface area minimization, a surface will assume the smoothest shape it can (mathematical proof that "smooth" shapes minimize surface area relies on use of the Euler–Lagrange equation). Since any curvature in the surface shape results in greater area, a higher energy will also result. Consequently, the surface will push back against any curvature in much the same way as a ball pushed uphill will push back to minimize its gravitational potential energy.

Effects of Surface Tension

Water

Several effects of surface tension can be seen with ordinary water:

A. Beading of rain water on a waxy surface, such as a leaf. Water adheres weakly to wax and strongly to itself, so water clusters into drops. Surface tension gives them their near-spherical shape, because a sphere has the smallest possible surface area to volume ratio.

A. Water beading on a leaf

B. Formation of drops occurs when a mass of liquid is stretched. The animation shows water adhering to the faucet gaining mass until it is stretched to a point where the surface tension can no longer keep the drop linked to the faucet. It then separates and surface tension forms the drop into a sphere. If a stream of water was running from the faucet, the stream would break up into drops during its fall. Gravity stretches the stream, then surface tension pinches it into spheres.

B. Water dripping from a tap

C. Flotation of objects denser than water occurs when the object is nonwettable and its weight is small enough to be borne by the forces arising from surface tension. For example, water striders use surface tension to walk on the surface of a pond by the following way. Nonwettability of leg of the water strider means no attraction between molecules of the leg and molecules of the water, so when the leg pushes down the water, the surface tension of the water only tries to recover its flatness from its deformation due to the leg. This behavior of the water push the water strider upward so it can stand on the surface of the water as long as its mass is small enough so that the water can support it. The surface of the water behaves like an elastic film: the insect's feet cause indentations in the water's surface, increasing its surface area and tendency of minimization of surface curvature (so area) of the water pushes the insect's feet upward.

C. Water striders stay atop the liquid because of surface tension

D. Separation of oil and water (in this case, water and liquid wax) is caused by a tension in the surface between dissimilar liquids. This type of surface tension is called "interface tension", but its chemistry is the same.

E. Photo showing the "tears of wine" phenomenon.

E. Tears of wine is the formation of drops and rivulets on the side of a glass containing an alcoholic beverage. Its cause is a complex interaction between the differing surface tensions of water and ethanol; it is induced by a combination of surface tension modification of water by ethanol together with ethanol evaporating faster than water.

D. Lava lamp with interaction between dissimilar liquids: water and liquid wax

Surfactants

Surface tension is visible in other common phenomena, especially when surfactants are used to decrease it:

- Soap bubbles have very large surface areas with very little mass. Bubbles in pure water are unstable. The addition of surfactants, however, can have a stabilizing effect on the bubbles. Note that surfactants actually reduce the surface tension of water by a factor of three or more.

- Emulsions are a type of colloid in which surface tension plays a role. Tiny fragments of oil suspended in pure water will spontaneously assemble themselves into much larger masses. But the presence of a surfactant provides a decrease in surface tension, which permits stability of minute droplets of oil in the bulk of water (or vice versa).

Physics

Physical Units

Surface tension, usually represented by the symbol γ, is measured in force per unit length. Its SI unit is newton per meter but the cgs unit of dyne per cm is also used.

$$\gamma = 1\frac{\text{dyn}}{\text{cm}} = 1\frac{\text{erg}}{\text{cm}^2} = 1\frac{\text{mN}}{\text{m}} = 0.001\frac{\text{N}}{\text{m}} = 0.001\frac{\text{J}}{\text{m}^2}$$

Surface Area Growth

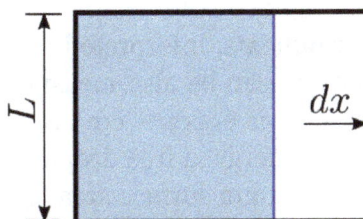

This diagram illustrates the force necessary to increase the surface area. This force is proportional to the surface tension.

Surface tension can be defined in terms of force or energy.

In terms of force: surface tension γ of a liquid is the force per unit length. In the illustration on the right, the rectangular frame, composed of three unmovable sides (black) that form a "U" shape, and a fourth movable side (blue) that can slide to the right. Surface tension will pull the blue bar to the left; the force F required to hold immobile the movable side is proportional to the length L of the movable side. Thus the ratio F/L depends only on the intrinsic properties of the liquid (composition, temperature, etc.), not on its geometry. For example, if the frame had a more complicated shape, the ratio F/L, with L the length of the movable side and F the force required to stop it from sliding, is found to be the same for all shapes. We therefore define the surface tension as

$$\gamma = \frac{1}{2}\frac{F}{L}.$$

The reason for the 1/2 is that the film has two sides, each of which contributes equally to the force; so the force contributed by a single side is $\gamma L = F/2$.

In terms of energy: surface tension γ of a liquid is the ratio of the change in the energy of the liquid, and the change in the surface area of the liquid (that led to the change in energy). This can be easily related to the previous definition in terms of force: if F is the force required to stop the side from *starting* to slide, then this is also the force that would keep the side in the state of *sliding at a constant speed* (by Newton's Second Law). But if the side is moving to the right (in the direction the force is applied), then the surface area of the stretched liquid is increasing while the applied force is doing work on the liquid. This means that increasing the surface area increases the energy of the film. The work done by the force F in moving the side by distance Δx is $W = F\Delta x$; at the same time the total area of the film increases by $\Delta A = 2L\Delta x$ (the factor of 2 is here because the liquid has two sides, two surfaces). Thus, multiplying both the numerator and the denominator of $\gamma = 1/2F/L$ by Δx, we get

$$\gamma = \frac{F}{2L} = \frac{F\Delta x}{2L\Delta x} = \frac{W}{\Delta A}.$$

This work W is, by the usual arguments, interpreted as being stored as potential energy. Consequently, surface tension can be also measured in SI system as joules per square meter and in the cgs system as ergs per cm². Since mechanical systems try to find a state of minimum potential energy, a free droplet of liquid naturally assumes a spherical shape, which has the minimum surface area for a given volume. The equivalence of measurement of energy per unit area to force per unit length can be proven by dimensional analysis.

Surface Curvature and Pressure

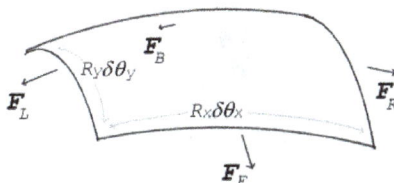

Surface tension forces acting on a tiny (differential) patch of surface. $\delta\theta_x$ and $\delta\theta_y$ indicate the amount of bend over the dimensions of the patch. Balancing the tension forces with pressure leads to the Young–Laplace equation

If no force acts normal to a tensioned surface, the surface must remain flat. But if the pressure on one side of the surface differs from pressure on the other side, the pressure difference times surface area results in a normal force. In order for the surface tension forces to cancel the force due to pressure, the surface must be curved. The diagram shows how surface curvature of a tiny patch of surface leads to a net component of surface tension forces acting normal to the center of the patch. When all the forces are balanced, the resulting equation is known as the Young–Laplace equation:

$$\Delta p = \gamma \left(\frac{1}{R_x} + \frac{1}{R_y} \right)$$

where:

- Δp is the pressure difference, known as the Laplace pressure.

- γ is surface tension.

- R_x and R_y are radii of curvature in each of the axes that are parallel to the surface.

The quantity in parentheses on the right hand side is in fact (twice) the mean curvature of the surface (depending on normalisation). Solutions to this equation determine the shape of water drops, puddles, menisci, soap bubbles, and all other shapes determined by surface tension (such as the shape of the impressions that a water strider's feet make on the surface of a pond). The table below shows how the internal pressure of a water droplet increases with decreasing radius. For not very small drops the effect is subtle, but the pressure difference becomes enormous when the drop sizes approach the molecular size. (In the limit of a single molecule the concept becomes meaningless.)

Δp for water drops of different radii at STP				
Droplet radius	1 mm	0.1 mm	1 μm	10 nm
Δp (atm)	0.0014	0.0144	1.436	143.6

Floating Objects

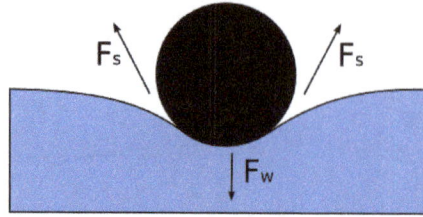

Cross-section of a needle floating on the surface of water. F_w is the weight and F_s are surface tension resultant forces.

When an object is placed on a liquid, its weight F_w depresses the surface, and if surface tension and downward force becomes equal than is balanced by the surface tension forces on either side F_s, which are each parallel to the water's surface at the points where it contacts the object. Notice that small moment in the body may cause the object sink as the angle of contact decreases surface tension decreases the horizontal components of the two F_s arrows point in opposite directions, so they cancel each other, but the vertical components point in the same direction and therefore add up to balance F_w. The object's surface must not be wettable for this to happen, and its weight must be low enough for the surface tension to support it.

$$F_w = 2F_s \cos\theta \iff \rho A_s L g = 2\gamma L z \cos\theta$$

Liquid Surface

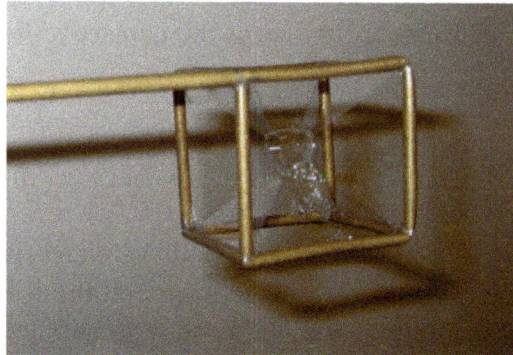

Minimal surface

To find the shape of the minimal surface bounded by some arbitrary shaped frame using strictly mathematical means can be a daunting task. Yet by fashioning the frame out of wire and dipping it in soap-solution, a locally minimal surface will appear in the resulting soap-film within seconds.

The reason for this is that the pressure difference across a fluid interface is proportional to the mean curvature, as seen in the Young-Laplace equation. For an open soap film, the pressure difference is zero, hence the mean curvature is zero, and minimal surfaces have the property of zero mean curvature.

Contact Angles

The surface of any liquid is an interface between that liquid and some other medium.The top surface of a pond, for example, is an interface between the pond water and the air. Surface tension, then, is not a property of the liquid alone, but a property of the liquid's interface with another medium. If a liquid is in a container, then besides the liquid/air interface at its top surface, there is also an interface between the liquid and the walls of the container. The surface tension between the liquid and air is usually different (greater than) its surface tension with the walls of a container. And where the two surfaces meet, their geometry must be such that all forces balance.

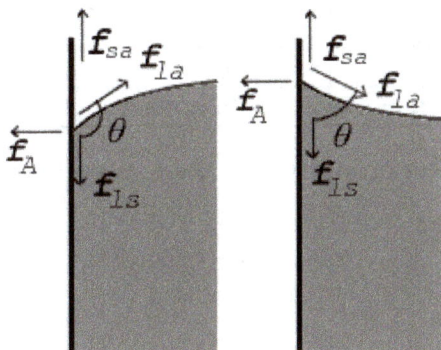

Forces at contact point shown for contact angle greater than 90° (left) and less than 90° (right)

Where the two surfaces meet, they form a contact angle, θ, which is the angle the tangent to the surface makes with the solid surface. The diagram to the right shows two examples. Tension forces are shown for the liquid–air interface, the liquid–solid interface, and the solid–air interface. The example on the left is where the difference between the liquid–solid and solid–air surface tension, $\gamma_{ls} - \gamma_{sa}$, is less than the liquid-air surface tension, γ_{la}, but is nevertheless positive, that is

$$\gamma_{la} > \gamma_{ls} - \gamma_{sa} > 0$$

In the diagram, both the vertical and horizontal forces must cancel exactly at the contact point, known as equilibrium. The horizontal component of f_{la} is canceled by the adhesive force, f_A.

$$f_A = f_{la} \sin \theta$$

The more telling balance of forces, though, is in the vertical direction. The vertical component of f_{la} must exactly cancel the force, f_{ls}.

$$f_{ls} - f_{sa} = -f_{la} \cos \theta$$

Liquid	Solid	Contact angle
water		
ethanol		
diethyl ether	soda-lime glass	
carbon tetrachloride	lead glass	0°
glycerol	fused quartz	
acetic acid		
water	paraffin wax	107°
	silver	90°
	soda-lime glass	29°
methyl iodide	lead glass	30°
	fused quartz	33°
mercury	soda-lime glass	140°
Some liquid-solid contact angles		

Since the forces are in direct proportion to their respective surface tensions, we also have:

$$\gamma_{ls} - \gamma_{sa} = -\gamma_{la} \cos \theta$$

where

- γ_{ls} is the liquid–solid surface tension,
- γ_{la} is the liquid–air surface tension,
- γ_{sa} is the solid–air surface tension,
- θ is the contact angle, where a concave meniscus has contact angle less than 90° and a convex meniscus has contact angle of greater than 90°.

This means that although the difference between the liquid-solid and solid–air surface tension, $\gamma_{ls} - \gamma_{sa}$, is difficult to measure directly, it can be inferred from the liquid–air surface tension, γ_{la}, and the equilibrium contact angle, θ, which is a function of the easily measurable advancing and receding contact angles .

This same relationship exists in the diagram on the right. But in this case we see that because the contact angle is less than 90°, the liquid–solid/solid–air surface tension difference must be negative:

$$\gamma_{la} > 0 > \gamma_{ls} - \gamma_{sa}$$

Special Contact Angles

Observe that in the special case of a water–silver interface where the contact angle is equal to 90°, the liquid–solid/solid–air surface tension difference is exactly zero.

Another special case is where the contact angle is exactly 180°. Water with specially prepared Teflon approaches this. Contact angle of 180° occurs when the liquid-solid surface tension is exactly equal to the liquid-air surface tension.

$$\gamma_{la} = \gamma_{ls} - \gamma_{sa} > 0 \qquad \theta = 180°$$

Methods of Measurement

Surface tension can be measured using the pendant drop method on a goniometer.

Because surface tension manifests itself in various effects, it offers a number of paths to its measurement. Which method is optimal depends upon the nature of the liquid being measured, the conditions under which its tension is to be measured, and the stability of its surface when it is deformed.

- Du Noüy ring method: The traditional method used to measure surface or interfacial tension. Wetting properties of the surface or interface have little influence on this measuring technique. Maximum pull exerted on the ring by the surface is measured.

- Du Noüy–Padday method: A minimized version of Du Noüy method uses a small diameter metal needle instead of a ring, in combination with a high sensitivity microbalance to record maximum pull. The advantage of this method is that very small sample volumes (down to few tens of microliters) can be measured with very high precision, without the need to correct for buoyancy (for a needle or rather, rod, with proper geometry). Further, the measurement can be performed very quickly, minimally in about 20 seconds. First commercial multichannel tensiometers [CMCeeker] were recently built based on this principle.

- Wilhelmy plate method: A universal method especially suited to check surface tension over long time intervals. A vertical plate of known perimeter is attached to a balance, and the force due to wetting is measured.

- Spinning drop method: This technique is ideal for measuring low interfacial tensions. The diameter of a drop within a heavy phase is measured while both are rotated.

- Pendant drop method: Surface and interfacial tension can be measured by this technique, even at elevated temperatures and pressures. Geometry of a drop is analyzed optically.

- Bubble pressure method (Jaeger's method): A measurement technique for determining surface tension at short surface ages. Maximum pressure of each bubble is measured.

- Drop volume method: A method for determining interfacial tension as a function of interface age. Liquid of one density is pumped into a second liquid of a different density and time between drops produced is measured.

- Capillary rise method: The end of a capillary is immersed into the solution. The height at which the solution reaches inside the capillary is related to the surface tension by the equation discussed below.

- Stalagmometric method: A method of weighting and reading a drop of liquid.

- Sessile drop method: A method for determining surface tension and density by placing a drop on a substrate and measuring the contact angle.

- Vibrational frequency of levitated drops: The natural frequency of vibrational oscillations of magnetically levitated drops has been used to measure the surface tension of superfluid ^{4}He. This value is estimated to be 0.375 dyn/cm at T = 0 K.

- Resonant oscillations of spherical and hemispherical liquid drop: The technique is based on measuring the resonant frequency of spherical and hemispherical pendant droplets driven in oscillations by a modulated electric field. The surface tension and viscosity can be evaluated from the obtained resonant curves.

Effects

Liquid in a Vertical Tube

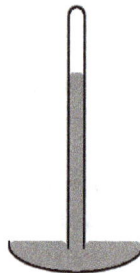

Diagram of a mercury barometer

An old style mercury barometer consists of a vertical glass tube about 1 cm in diameter partially filled with mercury, and with a vacuum (called Torricelli's vacuum) in the unfilled volume. Notice that the mercury level at the center of the tube is higher than at the edges, making the upper surface of the mercury dome-shaped. The center of mass of the entire column of mercury would be slightly lower if the top surface of the mercury were flat over the entire crossection of the tube. But the dome-shaped top gives slightly less surface area to the entire mass of mercury. Again the two effects combine to minimize the total potential energy. Such a surface shape is known as a convex meniscus.

We consider the surface area of the entire mass of mercury, including the part of the surface that is in contact with the glass, because mercury does not adhere to glass at all. So the surface tension of the mercury acts over its entire surface area, including where it is in contact with the glass. If instead of glass, the tube was made out of copper, the situation would be very different. Mercury aggressively adheres to copper. So in a copper tube, the level of mercury at the center of the tube will be lower than at the edges (that is, it would be a concave meniscus). In a situation where the liquid adheres to the walls of its container, we consider the part of the fluid's surface area that is in contact with the container to have *negative* surface tension. The fluid then works to maximize the contact surface area. So in this case increasing the area in contact with the container decreases rather than increases the potential energy. That decrease is enough to compensate for the increased potential energy associated with lifting the fluid near the walls of the container.

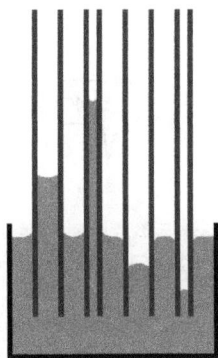

Illustration of capillary rise and fall. Red=contact angle less than 90°; blue=contact angle greater than 90°

If a tube is sufficiently narrow and the liquid adhesion to its walls is sufficiently strong, surface tension can draw liquid up the tube in a phenomenon known as capillary action. The height to which the column is lifted is given by:

$$h = \frac{2\gamma_{la}\cos\theta}{\rho g r}$$

where

- h is the height the liquid is lifted,

- γ_{la} is the liquid–air surface tension,

- ρ is the density of the liquid,

- r is the radius of the capillary,

- g is the acceleration due to gravity,

- θ is the angle of contact described above. If θ is greater than 90°, as with mercury in a glass container, the liquid will be depressed rather than lifted.

Puddles on a Surface

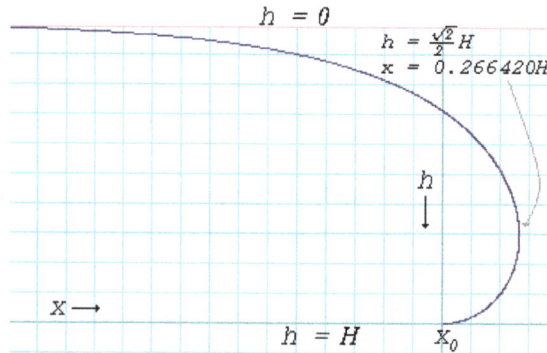

Profile curve of the edge of a puddle where the contact angle is 180°. The curve is given by the formula:

$$h = 2\sqrt{\frac{\gamma}{g\rho}}$$

where

Small puddles of water on a smooth clean surface have perceptible thickness.

Pouring mercury onto a horizontal flat sheet of glass results in a puddle that has a perceptible thickness. The puddle will spread out only to the point where it is a little under half a centimetre thick, and no thinner. Again this is due to the action of mercury's strong surface tension. The liquid mass flattens out because that brings as much of the mercury to as low a level as possible, but the surface tension, at the same time, is acting to reduce the total surface area. The result of the compromise is a puddle of a nearly fixed thickness.

The same surface tension demonstration can be done with water, lime water or even saline, but only on a surface made of a substance to which water does not adhere. Wax is such a substance. Water poured onto a smooth, flat, horizontal wax surface, say a waxed sheet of glass, will behave similarly to the mercury poured onto glass.

The thickness of a puddle of liquid on a surface whose contact angle is 180° is given by:

$$h = \sqrt{\frac{2\gamma_{la}\left(1-\cos\theta\right)}{g\rho}}.$$

where

- h is the depth of the puddle in centimeters or meters.

- γ is the surface tension of the liquid in dynes per centimeter or newtons per meter.

- g is the acceleration due to gravity and is equal to 980 cm/s² or 9.8 m/s²

- ρ is the density of the liquid in grams per cubic centimeter or kilograms per cubic meter

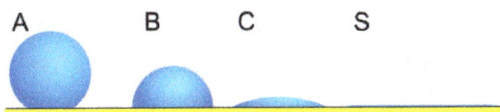

Illustration of how lower contact angle leads to reduction of puddle depth

In reality, the thicknesses of the puddles will be slightly less than what is predicted by the above formula because very few surfaces have a contact angle of 180° with any liquid. When the contact angle is less than 180°, the thickness is given by:

$$h = 2\sqrt{\frac{\gamma}{g\rho}}$$

For mercury on glass, $\gamma_{Hg} = 487$ dyn/cm, $\rho_{Hg} = 13.5$ g/cm³ and $\theta = 140°$, which gives h_{Hg} = 0.36 cm. For water on paraffin at 25 °C, $\gamma = 72$ dyn/cm, $\rho = 1.0$ g/cm³, and $\theta = 107°$ which gives $h_{H2O} = 0.44$ cm.

The formula also predicts that when the contact angle is 0°, the liquid will spread out into a micro-thin layer over the surface. Such a surface is said to be fully wettable by the liquid.

The Breakup of Streams into Drops

In day-to-day life all of us observe that a stream of water emerging from a faucet will break up into droplets, no matter how smoothly the stream is emitted from the faucet.

This is due to a phenomenon called the Plateau–Rayleigh instability, which is entirely a consequence of the effects of surface tension.

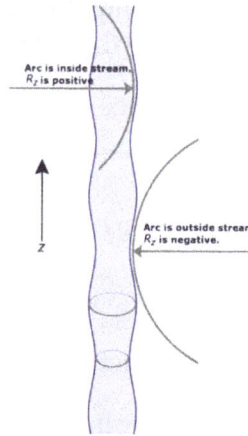

Intermediate stage of a jet breaking into drops. Radii of curvature in the axial direction are shown. Equation for the radius of the stream is $R(z) = R_o + A_k \cos(kz)$, where R_o is the radius of the unperturbed stream, A_k is the amplitude of the perturbation, z is distance along the axis of the stream, and k is the wave number

The explanation of this instability begins with the existence of tiny perturbations in the stream. These are always present, no matter how smooth the stream is. If the perturbations are resolved into sinusoidal components, we find that some components grow with time while others decay with time. Among those that grow with time, some grow at faster rates than others. Whether a component decays or grows, and how fast it grows is entirely a function of its wave number (a measure of how many peaks and troughs per centimeter) and the radii of the original cylindrical stream.

Thermodynamics

Thermodynamic Theories of Surface Tension

J.W. Gibbs developed the thermodynamic theory of capillarity based on the idea of surfaces of discontinuity. He introduced and studied thermodynamics of two-dimensional objects – surfaces. These surfaces have area, mass, entropy, energy and free energy. As stated above, the mechanical work needed to increase a surface area A is $dW = \gamma\, dA$. Hence at constant temperature and pressure, surface tension equals Gibbs free energy per surface area:

$$\gamma = \left(\frac{\partial G}{\partial A}\right)_{T,P,n}$$

where G is Gibbs free energy and A is the area.

Thermodynamics requires that all spontaneous changes of state are accompanied by a decrease in Gibbs free energy.

From this it is easy to understand why decreasing the surface area of a mass of liquid is always spontaneous ($G < 0$), provided it is not coupled to any other energy changes. It follows that in order to increase surface area, a certain amount of energy must be added.

Gibbs free energy is defined by the equation $G = H - TS$, where H is enthalpy and S is entropy. Based upon this and the fact that surface tension is Gibbs free energy per unit area, it is possible to obtain the following expression for entropy per unit area:

$$\left(\frac{\partial \gamma}{\partial T}\right)_{A,P} = -S^A$$

Kelvin's equation for surfaces arises by rearranging the previous equations. It states that surface enthalpy or surface energy (different from surface free energy) depends both on surface tension and its derivative with temperature at constant pressure by the relationship.

$$H^A = \gamma - T\left(\frac{\partial \gamma}{\partial T}\right)_P$$

Fifteen years after Gibbs, J.D. van der Waals developed the theory of capillarity effects based on the hypothesis of a continuous variation of density. He added to the energy density the term $c(\nabla \rho)^2$, where c is the capillarity coefficient and ρ is the density. For the multiphase *equilibria*, the results of the van der Waals approach practically coincide with the Gibbs formulae, but for modelling of the *dynamics* of phase transitions the van der Waals approach is much more convenient. The van der Waals capillarity energy is now widely used in the phase field models of multiphase flows. Such terms are also discovered in the dynamics of non-equilibrium gases.

Thermodynamics of Soap Bubbles

The pressure inside an ideal (one surface) soap bubble can be derived from thermodynamic free energy considerations. At constant temperature and particle number, $dT = dN = 0$, the differential Helmholtz energy is given by

$$dF = -P\,dV + \gamma\,dA$$

where P is the difference in pressure inside and outside of the bubble, and γ is the surface tension. In equilibrium, $dF = 0$, and so,

$$P\,dV = \gamma\,dA.$$

For a spherical bubble, the volume and surface area are given simply by

$$V = \tfrac{4}{3}\pi R^3 \quad \rightarrow \quad dV \approx 4\pi R^2\,dR,$$

and

$$A = 4\pi R^2 \quad \rightarrow \quad dA \approx 8\pi R\, dR.$$

Substituting these relations into the previous expression, we find

$$P = \frac{2}{R}\gamma,$$

which is equivalent to the Young–Laplace equation when $R_x = R_y$. For real soap bubbles, the pressure is doubled due to the presence of two interfaces, one inside and one outside.

Influence of Temperature

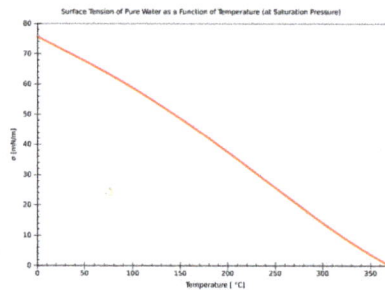

Temperature dependence of the surface tension between the liquid and vapor phases of pure water

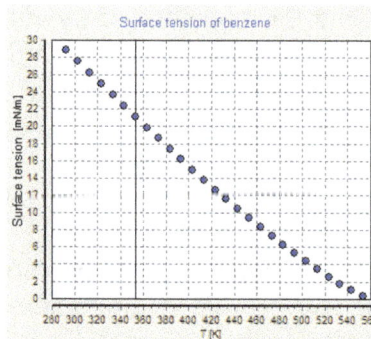

Temperature dependency of the surface tension of benzene

Surface tension is dependent on temperature. For that reason, when a value is given for the surface tension of an interface, temperature must be explicitly stated. The general trend is that surface tension decreases with the increase of temperature, reaching a value of 0 at the critical temperature. There are only empirical equations to relate surface tension and temperature:

- Eötvös:

$$\gamma V^{\frac{2}{3}} = k(T_c - T).$$

Here V is the molar volume of a substance, T_C is the critical temperature and k is a constant valid for almost all substances. A typical value is $k = 2.1{\times}10^{-7}$ J K^{-1} mol$^{-2/3}$. For water one can further use $V = 18$ ml/mol and $T_C = 647$ K (374 °C).

A variant on Eötvös is described by Ramay and Shields:

$$\gamma V^{\frac{2}{3}} = k\left(T_C - T - 6\right)$$

where the temperature offset of 6 kelvins provides the formula with a better fit to reality at lower temperatures.

- Guggenheim–Katayama:

$$\gamma = \gamma^{\circ}\left(1 - \frac{T}{T_C}\right)^{n}$$

γ° is a constant for each liquid and n is an empirical factor, whose value is 11/9 for organic liquids. This equation was also proposed by van der Waals, who further proposed that γ° could be given by the expression

$$K_2 T_C^{\frac{1}{3}} P_C^{\frac{2}{3}},$$

where K_2 is a universal constant for all liquids, and P_C is the critical pressure of the liquid (although later experiments found K_2 to vary to some degree from one liquid to another).

Both Guggenheim–Katayama and Eötvös take into account the fact that surface tension reaches 0 at the critical temperature, whereas Ramay and Shields fails to match reality at this endpoint.

Influence of Solute Concentration

Solutes can have different effects on surface tension depending on the nature of the surface and the solute:

- Little or no effect, for example sugar at water|air, most organic compounds at oil|air

- Increase surface tension, most inorganic salts at water|air

- Non-monotonic change, most inorganic acids at water|air

- Decrease surface tension progressively, as with most amphiphiles, e.g., alcohols at water|air

- Decrease surface tension until certain critical concentration, and no effect afterwards: surfactants that form micelles

What complicates the effect is that a solute can exist in a different concentration at the surface of a solvent than in its bulk. This difference varies from one solute–solvent combination to another.

Gibbs isotherm states that:

$$\Gamma = -\frac{1}{RT}\left(\frac{\partial \gamma}{\partial \ln C}\right)_{T,P}$$

- Γ is known as surface concentration, it represents excess of solute per unit area of the surface over what would be present if the bulk concentration prevailed all the way to the surface. It has units of mol/m²
- C is the concentration of the substance in the bulk solution.
- R is the gas constant and T the temperature

Certain assumptions are taken in its deduction, therefore Gibbs isotherm can only be applied to ideal (very dilute) solutions with two components.

Influence of Particle Size on Vapor Pressure

The Clausius–Clapeyron relation leads to another equation also attributed to Kelvin, as the Kelvin equation. It explains why, because of surface tension, the vapor pressure for small droplets of liquid in suspension is greater than standard vapor pressure of that same liquid when the interface is flat. That is to say that when a liquid is forming small droplets, the equilibrium concentration of its vapor in its surroundings is greater. This arises because the pressure inside the droplet is greater than outside.

$$P_v^{fog} = P_v^{\circ} e^{\frac{V 2\gamma}{RT r_k}}$$

Molecules on the surface of a tiny droplet (left) have, on average, fewer neighbors than those on a flat surface (right). Hence they are bound more weakly to the droplet than are flat-surface molecules.

- $P_v^{\,\circ}$ is the standard vapor pressure for that liquid at that temperature and pressure.

- V is the molar volume.

- R is the gas constant

- r_k is the Kelvin radius, the radius of the droplets.

The effect explains supersaturation of vapors. In the absence of nucleation sites, tiny droplets must form before they can evolve into larger droplets. This requires a vapor pressure many times the vapor pressure at the phase transition point.

This equation is also used in catalyst chemistry to assess mesoporosity for solids.

The effect can be viewed in terms of the average number of molecular neighbors of surface molecules.

The table shows some calculated values of this effect for water at different drop sizes:

P/P_0 for water drops of different radii at STP			
Droplet radius (nm)	1000	100	10	1
P/P_0	1.001	1.011	1.114	2.95

The effect becomes clear for very small drop sizes, as a drop of 1 nm radius has about 100 molecules inside, which is a quantity small enough to require a quantum mechanics analysis.

Surface Tension of Water and of Seawater

The two most abundant liquids on Earth are fresh water and seawater.

Surface Tension of Water

The surface tension of pure liquid water in contact with its vapor has been given by IAPWS as

$$\gamma_w = 235.8\left(1-\frac{T}{T_C}\right)^{1.256}\left(1-0.625\left(1-\frac{T}{T_C}\right)\right) \text{ mN m}^{-1}$$

where both T and the critical temperature, T_C = 647.098 K, are expressed in kelvin. The region of validity the entire vapor–liquid saturation curve, from the triple point (0.01 °C) to the critical point. It also provides reasonable results when extrapolated to metastable (supercooled) conditions, down to at least −25 °C. This formulation was originally adopted by IAPWS in 1976, and was adjusted in 1994 to conform to the International Temperature Scale of 1990.

The uncertainty of this formulation is given over the full range of temperature by IAPWS. For temperatures below 100 °C, the uncertainty is ±0.5%.

Surface Tension of Seawater

Nayar et al. published reference data for the surface tension of seawater over the salinity range of $20 \leq S \leq 131$ g/kg and a temperature range of $1 \leq t \leq 92$ °C at atmospheric pressure. The uncertainty of the measurements varied from 0.18 to 0.37 mN/m with the average uncertainty being 0.22 mN/m. This data is correlated by the following equation

$$\gamma_{sw} = \gamma_w \left(1 + 3.766 \times 10^{-4} S + 2.347 \times 10^{-6} St\right)$$

where γ_{sw} is the surface tension of seawater in mN/m, γ_w is the surface tension of water in mN/m, S is the reference salinity in g/kg, and t is temperature in degrees Celsius. The average absolute percentage deviation between measurements and the correlation was 0.19% while the maximum deviation is 0.60%.

The range of temperature and salinity encompasses both the oceanographic range and the range of conditions encountered in thermal desalination technologies.

Data Table

Surface tension of various liquids in dyn/cm against air
Mixture compositions denoted "%" are by mass
dyn/cm is equivalent to the SI units of mN/m (millinewton per meter)

Liquid	Temperature (°C)	Surface tension, γ
Acetic acid	20	27.60
Acetic acid (40.1%) + Water	30	40.68
Acetic acid (10.0%) + Water	30	54.56
Acetone	20	23.70
Diethyl ether	20	17.00
Ethanol	20	22.27
Ethanol (40%) + Water	25	29.63
Ethanol (11.1%) + Water	25	46.03
Glycerol	20	63.00
n-Hexane	20	18.40
Hydrochloric acid 17.7 M aqueous solution	20	65.95
Isopropanol	20	21.70
Liquid helium II	−273	0.37
Liquid nitrogen	−196	8.85
Mercury	15	487.00
Methanol	20	22.60

n-Octane	20	21.80
Sodium chloride 6.0 M aqueous solution	20	82.55
Sucrose (55%) + water	20	76.45
Water	0	75.64
Water	25	71.97
Water	50	67.91
Water	100	58.85
Toluene	25	27.73

Du Noüy Ring Method

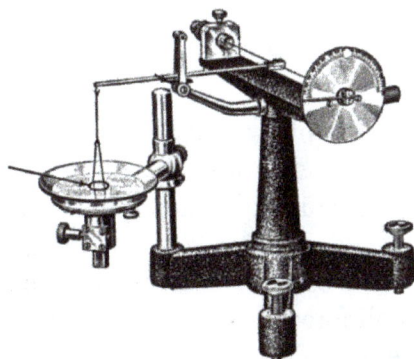

A du Noüy ring tensiometer. The arrow on the left points to the ring itself.

Close up of the ring drawn out of the liquid.

The du Noüy ring method is one technique by which the surface tension of a liquid can be measured. The method involves slowly lifting a ring, often made of platinum, from the surface of a liquid. The force, F, required to raise the ring from the liquid's surface is measured and related to the liquid's surface tension, γ:

$$F = 2\pi \cdot (r_i + r_a) \cdot \gamma$$

where r_i is the radius of the inner ring of the liquid film pulled and r_a is the radius of the outer ring of the liquid film.

This technique was proposed by the French physicist Pierre Lecomte du Noüy (1883–1947) in a paper published in 1925.

Du Noüy–Padday Method

The Du Noüy–Padday method is a minimized version of the Du Noüy ring method replacing the large platinum ring with a thin rod that is used to measure equilibrium surface tension or dynamic surface tension at an air–liquid interface. In this method, the rod is oriented perpendicular to the interface, and the force exerted on it is measured. Based on the work of Padday, this method finds wide use in the preparation and monitoring of Langmuir–Blodgett films, ink & coating development, pharmaceutical screening, and academic research.

Detailed Description

The Du Noüy Padday rod consists of a rod usually on the order of a few millimeters square making a small ring. The rod is often made from a composite metal material that may be roughened to ensure complete wetting at the interface. The rod is cleaned with water, alcohol and a flame or with strong acid to ensure complete removal of surfactants. The rod is attached to a scale or balance via a thin metal hook. The Padday method uses the maximum pull force method, i.e. the maximum force due to the surface tension is recorded as the probe is first immersed ca. one mm into the solution and then slowly withdrawn from the interface. The main forces acting on a probe are the buoyancy (due to the volume of liquid displaced by the probe) and the mass of the meniscus adhering to the probe. This is an old, reliable, and well-documented technique.

An important advantage of the maximum pull force technique is that the receding contact angle on the probe is effectively zero. The maximum pull force is obtained when the buoyancy force reaches its minimum,

The surface tension measurement used in the Padday devices based on the Du Noüy ring/maximum pull force method is explained further here:

The force acting on the probe can be divided into two components:

   i) Buoyancy stemming from the volume displaced by the probe, and

   ii) the mass of the meniscus of the liquid adhering to the probe.

The latter is in equilibrium with the surface tension force, i.e.

$$2\pi\tau_p\gamma\cos\theta = m_m g$$

where

- τ_p is the perimeter of the probe,

- m_m is the surface tension and the weight of the meniscus under the probe. In the situation considered here the volume displaced by the probe is included in the meniscus.

- θ is the contact angle between the probe and the solution that is measured, and is negligible for the majority of solutions with Kibron's probes.

Thus, the force measured by the balance is given by

$$F_p = m_m g + F_{buoayancy} = 2\tau_p + F_{buoyancy}$$

where

- F_p is the force acting on the probe and

- $F_{buoyancy}$ is the force due to buoyancy.

At the point of detachment the volume of the probe immersed in the solution vanishes, and thus, also the buoyancy term. This is observed as a maximum in the force curve, which relates to the surface tension through

$$\gamma = \frac{F_{max}}{2\tau_p}$$

The above derivation holds for ideal conditions. Non-idealities, e.g. from defect probe shape, are partly compensated in the calibration routine using a solution with known surface tension.

Advantages and Practice

Unlike a Du Noüy ring, no correction factors are required when calculating surface tensions. Due to its small size the rod can be used in high throughput instruments that use a 96-well plate to determine the surface tension. The small diameter of the rod allows its use in a small volume of liquid with 50 μl samples being used in some devices.

In addition, the rod also allows use for the Wilhelmy method because the rod is not completely removed during measurements. For this the dynamic surface tension can be used for accurate determination of surface kinetics on a wide range of timescales.

The Padday technique also offers low operator variance and does not need an anti-vibration table. This advantage over other devices allows the Padday devices to be used in the field easily. The rod when made of composite material is also less likely to bend and therefore cheaper than the more costly platinum rod offered in the Du Noüy method.

In a typical experiment, the rod is lowered using a manual or automatic device to the surface being analyzed until a meniscus is formed, and then raised so that the bottom edge of the rod lies on the plane of the undisturbed surface. One disadvantage of this technique is that it can not bury the rod into the surface to measure interfacial tension between two liquids.

Practical Uses

The practical uses of an instrument that uses a single probe are that it allows the for developing a high throughput device. A high throughput surface tension device can be used for formulation in real time for understanding the penetration of drugs in the blood–brain barrier (BBB), understanding the solubility of drugs, development of a screen to test a drugs toxicity, determining the physicochemical properties of oxidized phospholipids, and development of new surfactant/polymers.

Penetration of Drugs in the BBB

The physicochemical profiling of poorly soluble drug candidates performed using a HTS surface tension device. Allowed prediction of penetration through the blood–brain barrier.

Development of a Screen to Test a Drugs Toxicity

A correlation with drug-lipid-complexes were correlated with high-throughput surface tension device to predict phospholipidosis in particular cationic drugs.

Understanding the Solubility of Drugs

Drug solubility has previously been done by the shaker method. A 96-well high throughput device has allowed development of a new method to test drugs.

Oxidized Phospholipids

The physicochemical properties of oxidized lipids were characterized using a high throughput device. Since these oxidized lipids are expensive and only available in small quantities a surface tension device requiring only a small amount of volume is better.

Development of New Surfactant/Polymers

The surface tension profiles of the branched copolymers solutions were performed using a HTS surface tensiometer as a function polymer concentration to produce pH-triggered aggregation emulsion droplets.

Wilhelmy Plate

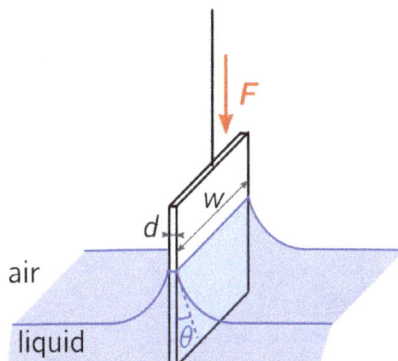

Illustration of Wilhelmy plate method. The magnitude of the capillary force F on the plate is proportional to the wetted perimeter, $l = 2w + 2d$, and to the surface tension γ of the liquid-air interface.

A Wilhelmy plate is a thin plate that is used to measure equilibrium surface or interfacial tension at an air–liquid or liquid–liquid interface. In this method, the plate is oriented perpendicular to the interface, and the force exerted on it is measured. Based on the work of Ludwig Wilhelmy, this method finds wide use in the preparation and monitoring of Langmuir–Blodgett films.

Detailed Description

The Wilhelmy plate consists of a thin plate usually on the order of a few square centimeters in area. The plate is often made from filter paper, glass or platinum which may be roughened to ensure complete wetting. In fact, the results of the experiment do not depend on the material used, as long as the material is wetted by the liquid. The plate is cleaned thoroughly and attached to a balance with a thin metal wire. The force on the plate due to wetting is measured using a tensiometer or microbalance and used to calculate the surface tension (γ) using the Wilhelmy equation:

$$\gamma = \frac{F}{l\cos(\theta)}$$

where l is the wetted perimeter ($2w + 2d$), w is the plate width, d is the plate thickness, and θ is the contact angle between the liquid phase and the plate. In practice the contact angle is rarely measured, instead either literature values are used, or complete wetting ($\theta = 0$) is assumed.

Advantages and Practice

If complete wetting is assumed (contact angle = 0), no correction factors are required to calculate surface tensions when using the Wilhelmy plate, unlike for a Du Noüy ring.

In addition, because the plate is not moved during measurements, the Wilhelmy plate allows accurate determination of surface kinetics on a wide range of timescales, and it displays low operator variance. In a typical plate experiment, the plate is lowered to the surface being analyzed until a meniscus is formed, and then raised so that the bottom edge of the plate lies on the plane of the undisturbed surface. If measuring a buried interface, the second (less dense) phase is then added on top of the undisturbed primary (denser) phase in such a way as to not disturb the meniscus. The force at equilibrium can then be used to determine the absolute surface or interfacial tension.

Spinning Drop Method

The spinning drop method (rotating drop method) is one of the methods used to measure interfacial tension. Measurements are carried out in a rotating horizontal tube which contains a dense fluid. A drop of a less dense liquid or a gas bubble is placed inside the fluid. Since the rotation of the horizontal tube creates a centrifugal force towards the tube walls, the liquid drop will start to deform into an elongated shape; this elongation stops when the interfacial tension and centrifugal forces are balanced. The surface tension between the two liquids (for bubbles: between the fluid and the gas) can then be derived from the shape of the drop at this equilibrium point. A device used for such measurements is called a "spinning drop tensiometer".

The spinning drop method is usually preferred for the accurate measurements of surface tensions below 10^{-2} mN/m. It refers to either using the fluids with low interfacial tension or working at very high angular velocities. This method is widely used in many different applications such as measuring the interfacial tension of polymer blends and copolymers.

Theory

An approximate theory was developed by Bernard Vonnegut in 1942 to measure the surface tension of the fluids, which is based on the principle that the interfacial tension and centrifugal forces are balanced at mechanical equilibrium. This theory assumes that the droplet's length L is much greater than its radius R, so that it may be approximated as a straight circular cylinder.

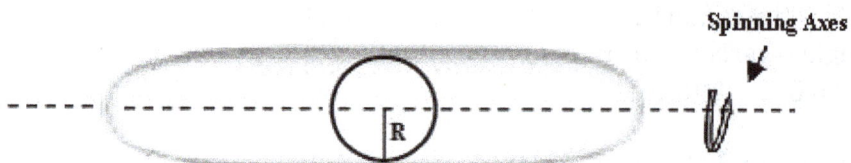

The relation between the surface tension and angular velocity of a droplet can be obtained in different ways. One of them involves considering the total mechanical energy of the droplet as the summation of its kinetic energy and its surface energy:

$$E = E_k + \gamma_s$$

The kinetic energy of a cylinder of length L and radius R rotating about its central axis is given by

$$E_k = \frac{1}{2} I \omega^2 = \frac{1}{4} m R^2 \omega^2$$

in which

$$I = \frac{1}{2} m R^2$$

is the moment of inertia of a cylinder rotating about its central axis and ω is its angular velocity. The surface energy of the droplet is given by

$$\gamma_s = 2\pi L R \sigma = \frac{2V}{R} \sigma$$

in which V is the constant volume of the droplet and σ is the interfacial tension. Then the total mechanical energy of the droplet is

$$E = E_k + \gamma_s = \frac{1}{4} \Delta\rho V R^2 \omega^2 + \frac{2V}{R} \sigma$$

in which $\Delta\rho$ is the difference between the densities of the droplet and of the surrounding fluid. At mechanical equilibrium, the mechanical energy is minimized, and thus

$$\frac{dE}{dR} = 0 = \frac{1}{2} \Delta\rho V R \omega^2 - \frac{2V}{R^2} \sigma$$

Substituting in

$$V = \pi L R^2$$

for a cylinder and then solving this relation for interfacial tension yields

$$\sigma = \frac{\Delta\rho \omega^2}{4} R^3$$

This equation is known as Vonnegut's expression. Interfacial tension of any liquid that gives a shape very close to a cylinder at steady state, can be estimated using this equation. The straight cylindrical shape will always develop for sufficiently high ω; this typically happens for $L/R > 4$. Once this shape has developed, further increasing ω will decrease R while increasing L keeping LR^2 fixed to meet conservation of volume.

New Developments After 1942

The full mathematical analysis on the shape of spinning drops was done by Princen and others. Progress in numerical algorithms and available computing resources turned solving the non linear implicit parameter equations to a pretty much 'common' task, which has been tackled by various authors and companies. The results are proving the Vonnegut restriction is no longer valid for the spinning drop method.

Comparison With Other Methods

Spinning drop method facilitates the measurement of surface tension compared to other widely used methods. Because in this method, measuring the contact angle is not required between a solid surface and the liquid. Another advantage of the method is that it is not necessary to estimate the curvature at the interface, which usually brings difficulties with taking the first and second derivatives describing the shape of the fluid drop.

On the other hand, this theory suggested by Vonnegut, is restricted with the rotational velocity. Spinning drop method is not expected to give accurate results for the high surface tension measurements, since the centrifugal force that is required to maintain a cylindrical shape of the drop is much higher in the case of liquids that has high interfacial tensions.

Comment on Comparison With Other Methods

We did state that the Vonnegut restriciton does not longer hold .. cause the mathematical exact solution is known .. right? Lemma: The possible precision of a spinning drop experiment is limited by

a, the precison of scaling b, the precison of rotation speed c, the precison of shape parmeter alpha (as in common papers) d, the precison of the density difference

and thus as accurate as any physical measurement is. Says: no limits in interfacial thension as long as the width to height ratio is not too close to 1. The same goes for similar methods like pendent / sessile drop too.

Maximum Bubble Pressure Method

In physics, the maximum bubble pressure method, or in short bubble measure method, is a technique to measure the surface tension of a liquid, with surfactants.

Background

When the liquid forms an interface with a gas phase, a molecule on the border has quite different physical properties due to the unbalance of attracting forces by the neighboring molecules. At the equilibrium state of the liquid, interior molecules are under the balanced forces with uniformly distributed adjacent molecules.

However, relatively fewer number of molecules in the gas phase above the interface than condensed liquid phase makes overall sum of forces applied to the surface molecule direct inside of the liquid and thus surface molecules tend to minimize their own surface area.

Such an inequality of molecular forces induces continuous movement of molecules from the inside to the surface, which means the surface molecules has extra energy, which is called surface free energy or potential energy, and such an energy acting on reduced unit area is defined as surface tension.

This is a frame work to interpret relevant phenomena which occurs surface or interface of materials and many methods to measure the surface tension has been developed.

Among the various ways to determine surface tension, Du Noüy ring method and Wilhelmy slide method are based on the separation of a solid object from the liquid surface, and Pendant drop method and Sessile drop or bubble method depend on the deformation of the spherical shape of a liquid drop.

Even though these methods are relatively simple and commonly used to determine the static surface tension, in case that the impurities are added to the liquid, measurement of surface tension based on the dynamic equilibrium should be applied since it takes more time to obtain a completely formed surface and this means that it is difficult to achieve the static equilibrium as a pure liquid does.

The most typical impurity to induce dynamic surface tension measurement is a surfactant molecule which has both of hydrophilic segment, generally called "head group" and hydrophobic segment, generally called "tail group" in a same molecule. Due to the characteristic molecular structure, surfactants migrate to the liquid surface bordering gas phase until an external force disperse the accumulated molecules from the interface or surface is fully occupied and thus cannot accommodate extra molecules. During this process, surface tension decrease as function of time and finally approach the equilibrium surface tension ($\sigma_{equilibrium}$). Such a process is illustrated in figure 1. (Image was reproduced from reference)

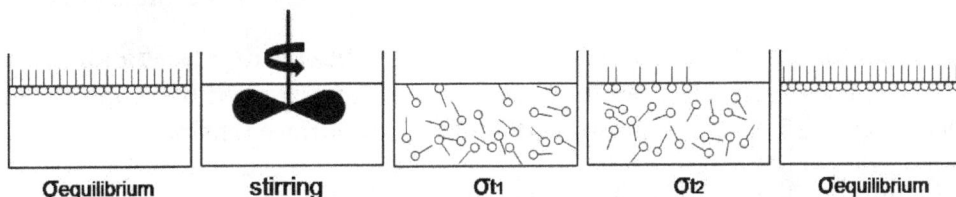

Figure 1 – Migration of surfactant molecules and change of surface tension ($\sigma_{t1} > \sigma_{t2} > \sigma_{equilibrium}$)

Maximum Bubble Pressure Method

One of the useful methods to determine the dynamic surface tension is measuring the "maximum bubble pressure method" or, simply, bubble pressure method.

Bubble pressure tensiometer produces gas bubbles (ex. air) at constant rate and blows them through a capillary which is submerged in the sample liquid and its radius is already known.

The pressure (P) inside of the gas bubble continues to increase and the maximum value is obtained when the bubble has the completely hemispherical shape whose radius is exactly corresponding to the radius of the capillary.

Figure 2 shows each step of bubble formation and corresponding change of bubble radius and each step is described below. (Image was reproduced from reference)

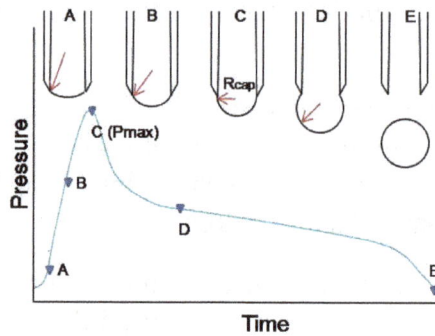

Figure 2 – Change of pressure during bubble formation plotted as a function of time.

A, B: A bubble appears on the end of the capillary. As the size increases, the radius of curvature of the bubble decreases.

C: At the point of the maximum bubble pressure, the bubble has a complete hemispherical shape whose radius is identical to the radius of the capillary denoted by Rcap. The surface tension can be determined using the Young–Laplace equation in the reduced form for spherical bubble shape within the liquid.

$$\sigma = \frac{\Delta P_{max} \times R_{cap}}{2}$$

(σ: surface tension, ΔP_{max}: maximum pressure drop, R_{cap}: radius of capillary)

D, E: After the maximum pressure, the pressure of the bubble decreases and the radius of the bubble increases until the bubble is detached from the end of a capillary and a new cycle begins. This is not relevant to determine the surface tension.

Currently developed and commercialized tensiometers monitors the pressure needed to form a bubble, the pressure difference between inside and outside the bubble, the radius of the bubble, and the surface tension of the sample are calculated in one time and a data acquisition is carried out via PC control.

Bubble pressure method is commonly used to measure the dynamic surface tension for the system containing surfactants or other impurities because it does not require contact angle measurement and has high accuracy even though the measurement is done rapidly. "Bubble pressure method" can be applied to measure the dynamic surface tension, particularly for the systems which contain surfactants. Moreover, this method is an appropriate technique to apply to biological fluids like serum because it does not require a large amount of liquid sample for the measurements.

Stalagmometric Method

The stalagmometric method is one of the most common methods for measuring surface tension which was developed by Mr 619. The principle is to measure the weight of drops of a fluid of interest falling from a capillary glass tube, and thereby calculate the surface tension of the fluid. We can determine the weight of the falling drops by counting them. From it we can determine the surface tension.

Stalagmometer

A stalagmometer, straight form.

A stalagmometer is a device for investigating surface tension using the stalagmometric method. It is also called a stactometer or stalogometer. The device is a capillary glass tube whose middle section is widened. The volume of a drop can be predetermined by the design of the stalagmometer. The lower end of the tube is narrowed to force the fluid to fall out of the tube as a drop. In an experiment, the drops of fluid flow slowly from the tube in a vertical direction. The drops hanging on the bottom of the tube start to fall when the volume of the drop reaches a maximum value that is dependent on the characteristics of the solution. At this moment, the weight of the drops is in equilibrium state with the surface tension. Based on Tate's law:

$$mg = 2\pi r\sigma$$

The drop falls when the weight (mg) is equal to the circumference (2πr) multiplied by the surface tension (σ). The surface tension can be calculated provided the radius of the tube (r) and mass of the fluid droplet (m) are known. Alternatively, since the surface tension is proportional to the weight of the drop, the fluid of interest may be compared to a reference fluid of known surface tension (typically water):

$$\frac{m_1}{\sigma_1} = \frac{m_2}{\sigma_2}$$

In the equation, m_1 and σ_1 represent the mass and surface tension of the reference fluid and m_2 and σ_2 the mass and surface tension of the fluid of interest. If we take water as a reference fluid,

$$\sigma = \sigma_{H_2O} \times \frac{m}{m_{H_2O}}$$

If the surface tension of water is known which is 72 dyne/cmsq, we can calculate the surface tension of the specific fluid from the equation. The more drops we weigh, the more precisely we can calculate the surface tension from the equation. The stalagmometer must be kept clean for meaningful readings. There are commercial tubes for stalagmometric method in three sizes: 2.5, 3.5, and 5.0 (ml). The 2.5-ml size is suitable for small volumes and low viscosity, that of 3.5 (ml) for relatively viscous fluids, and that of 5.0 (ml) for large volumes and high viscosity. The 2.5-ml size is suitable for most fluids.

Modified Method

The stalagmometric method was improved by S. V. Chichkanov and colleagues, who measured the weight of a fixed number of drops rather than counting the drops. This method for determining the surface tension may be more precise than the original method, especially for fluids whose surface is highly active.

Sessile Drop Technique

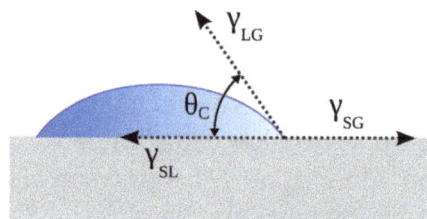

Fig 1: An illustration of the sessile drop technique with a liquid droplet partially wetting a solid substrate. θ_C is the contact angle, and γ_{SG}, γ_{LG}, γ_{SL} represent the solid–gas, gas–liquid, and liquid–solid interfaces, respectively.

The Sessile Drop Technique is a method used for the characterization of solid surface energies, and in some cases, aspects of liquid surface energies. The main premise of the method is that by placing a droplet of liquid with a known surface energy, the shape of the drop, specifically the contact angle, and the known surface energy of the liquid are the parameters which can be used to calculate the surface energy of the solid sample. The liquid used for such experiments is referred to as the probe liquid, and the use of several different probe liquids is required.

water droplet immersed in oil and resting on a brass surface

same fluids as above, but resting on a glass surface

Probe Liquid

The surface energy is measured in units of Joules per area, which is equivalent in the case of liquids to surface tension, measured in newtons per meter. The overall surface tension/energy of a liquid can be acquired through various methods using a tensiometer or using the pendant drop method technique and Maximum bubble pressure method.

The interface tension at the interface of the probe liquid and the solid surface can additionally be viewed as being the result of different types of intermolecular forces. As such, surface energies can be subdivided according to the various interactions that cause them, such as the surface energy due to dispersive (van der Waals) forces, hydrogen bonding, polar interactions, acid/base interactions, etc. It is often useful for the sessile drop technique to use liquids that are known to be incapable of some of those interactions . For example, the surface tension of all straight alkanes is said to be entirely dispersive, and all of the other components are zero. This is algebraically useful, as it eliminates a variable in certain cases, and makes these liquids essential testing materials.

The overall surface energy, both for a solid and a liquid, is assumed traditionally to simply be the sum of the components considered. For example, the equation describing the subdivision of surface energy into the contributions of dispersive interactions and polar interactions would be:

$$\sigma_S = \sigma_S^D + \sigma_S^P$$

$$\sigma_L = \sigma_L^D + \sigma_L^P$$

Where σ_S is the total surface energy of the solid, σ_S^D and σ_S^P are respectively the dispersive and polar components of the solid surface energy, σ_L is the total surface tension/surface energy of the liquid, and σ_L^D and σ_L^P are respectively the dispersive and polar components of the surface tension.

In addition to the tensiometer and pendant drop techniques, the sessile drop technique can be used in some cases to separate the known total surface energy of a liquid into its components. This is done by reversing the above idea with the introduction of a reference solid surface that is assumed to be incapable of polar interactions, such as Polytetrafluoroethylene (PTFE).

Contact Angle

The contact angle is defined as the angle made by the intersection of the liquid/solid interface and the liquid/air interface. It can be alternately described as the angle between solid sample's surface and the tangent of the droplet's ovate shape at the edge of the droplet. A high contact angle indicates a low solid surface energy or chemical affinity. This is also referred to as a low degree of wetting. A low contact angle indicates a high solid surface energy or chemical affinity, and a high or sometimes complete degree of wetting. For example, a contact angle of zero degrees will occur when the droplet has turned into a flat puddle; this is called complete wetting.

Measuring Contact Angle

Goniometer Method

The simplest way of measuring the contact angle is with a goniometer, which allows the user to measure the contact angle visually. The droplet is deposited by a syringe pointed vertically down onto the sample surface, and a high resolution camera captures the image, which can then be analyzed either by eye (with a protractor) or using image analysis software. The size of the droplet can be increased gradually so that it grows proportionally, and the contact angle remains congruent. By taking pictures incrementally as the droplet grows, the user can acquire a set of data to get a good average. If necessary, the receding contact angle can also be measured by depositing a droplet via syringe and recording images of the droplet being gradually sucked back up.

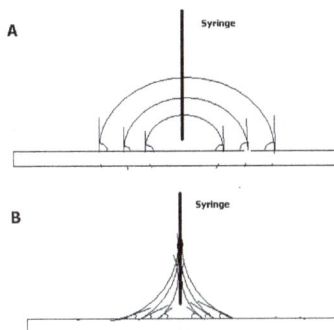

Fig 3: Sketch of the contact angle, as seen by a goniometer. In the top picture, the volume of the drop is being increased, and in the bottom it is being decreased. Each angle is measure of the same contact angle

Advantages and Disadvantages

The advantage of this method, aside from its relatively straightforward nature, is the fact that with a large enough solid surface, multiple droplets can be deposited in various locations on the sample to determine heterogeneity. The reproducibility of particular values of the contact angle will reflect the heterogeneity of the surface's energy properties. Conversely, the disadvantage is that if the sample is only large enough for one droplet, then it will be difficult to determine heterogeneity, or consequently to assume homogeneity. This is particularly true because conventional, commercially available goniometers do not swivel the camera/backlight set up relative to the stage, and thus can only show the contact angle at two points: the right and the left edge of the droplet. In addition to this, this measurement is hampered by its inherent subjectivity, as the placement of the lines is determined either by the user looking at the pictures or by the image analysis software's definition of the lines.

Wilhelmy Method

An alternative method for measuring the contact angle is the Wilhelmy method, which employs a sensitive force meter of some sort to measure a force that can be translated into a value of the contact angle. In this method, a small plate-shaped sample of the solid in question, attached to the arm of a force meter, is vertically dipped into a pool of the probe liquid (in actuality, the design of a stationary force meter would have the liquid being brought up, rather than the sample being brought down), and the force exerted on the sample by the liquid is measured by the force meter. This force is related to the contact angle by the following equation:

$$\cos\theta = (F - Fb)\,/\,I\sigma$$

Where F is the total force measured by the force meter, F_b is the force of buoyancy due to the solid sample displacing the liquid, I is the wetted length, and sigma is the known surface tension of the liquid.

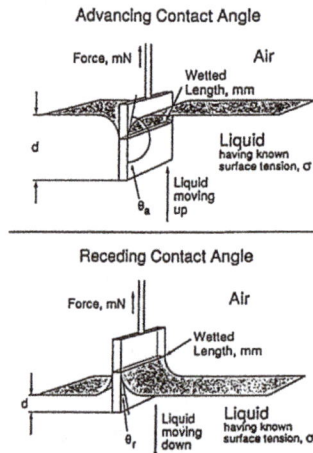

The Wilhelmy method for measuring contact angle. In the top picture a plate of the solid surface is lowered into a submerging liquid. The liquid pushes up on the solid sample with force due to the buoyancy and the surface tension, and these forces are measured by instruments attached to the arm above the sample, and depend on the length d, surface tension σ, and wetted length I (the perimeter of the sample along the line of contact of the air, liquid, and solid). In the bottom picture the sample is being raised and the liquid exerts a downward force

Advantages and Disadvantages

The advantage of this method is that it is fairly objective and the measurement yields data which is inherently averaged over the wetted length. Although this does not help determine heterogeneity, it does automatically give a more accurate average value. Its disadvantages, aside from being more complicated than the goniometer method, include the fact that sample of an appropriate size must be produced with a uniform cross section in the submersion direction, and the wetted length must be measured with some precision. In addition, this method is only appropriate if both sides of the sample are identical, otherwise the measured data will be a result of two completely different interactions.

Strictly speaking, this is not a sessile drop technique, as we are using a small submerging pool, rather than a droplet. However, the calculations described in the following sections, which were derived for the relation of the sessile drop contact angle to the surface energy, apply just as well.

Determining Surface Energy

While surface energy is conventionally defined as the work required to build a unit of area of a given surface, when it comes to its measurement by the sessile drop technique, the surface energy is not quite as well defined. The values obtained through the sessile drop technique depend not only on the solid sample in question, but equally on the properties of the probe liquid being used, as well as the particular theory relating the parameters mathematically to one another.

There are numerous such theories developed by various researchers. These methods differ in several regards, such as derivation and convention, but most importantly they differ in the number of components or parameters which they are equipped to analyze. The simpler methods containing fewer components simplify the system by lumping surface energy into one number, while more rigorous methods with more components are derived to distinguish between various components of the surface energy. Again, the total surface energy of solids and liquids depends on different types of molecular interactions, such as dispersive (van der Waals), polar, and acid/base interactions, and is considered to be the sum of these independent components. Some theories account for more of these phenomena than do other theories. These distinctions are to be considered when deciding which method is appropriate for the experiment at hand. The following are a few commonly used such theories.

One Component Theories

The Zisman Theory

The Zisman theory is the simplest commonly used theory, as it is a one component theory, and is best used for non-polar surfaces. This means that polymer surfaces that have been subjected to heat treatment, corona treatment, plasma cleaning, or polymers that contain heteroatoms do not lend themselves to this particular theory, as they tend to be at least somewhat polar. The Zisman theory also tends to be more useful in practice for surfaces with lower energies.

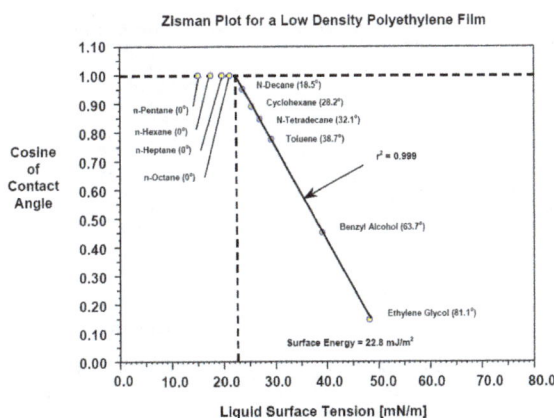

Fig 5: The Zisman plot for the surface energy of LDPE. Each plot point reflects the contact angle with a specified probe liquid. The line coefficient r²=.999 indicates a high degree of confidence.

The Zisman theory simply defines the surface energy as being equal to the surface energy of the highest surface energy liquid that wets the solid completely. That is to say, the droplet will disperse as much as possible, i.e. completely wetting the surface, for this liquid and any liquids with lower surface energies, but not for liquids with higher surface energies. Since this probe liquid could hypothetically be any liquid, including an imaginary liquid, the best way to determine the surface energy by the Zisman method is to acquire data points of contact angles for several probe liquids on the solid surface in question, and then plot the cosine of that angle against the known surface energy of the probe liquid. By constructing the Zisman plot, one can extrapolate the highest liquid surface energy, real or hypothetical, that would result in complete wetting of the sample with a contact angle of zero degrees.

Accuracy/Precision

The line coefficient (Fig 5) suggests that this is a fairly accurate result, however this is only the case for the pairing of that particular solid with those particular liquids. In other cases, the fit may not be so great (such is the case if we replace polyethylene with poly(methyl methacrylate), wherein the line coefficient of the plot results using the same list of liquids would be significantly lower). This shortcoming is a result of the fact that the Zisman theory treats the surface energy as one single parameter, rather than accounting for the fact that, for example, polar interactions are much stronger than dispersive ones, and thus the degree to which one is happening versus the other greatly affects the necessary calculations. As such, it is a simple but not particularly robust theory. Since the premise of this procedure is to determine the hypothetical properties of a liquid, the precision of the result depends on the precision to which the surface energy values of the probe liquids are known.

Two Component Theories

The Owens/Wendt Theory

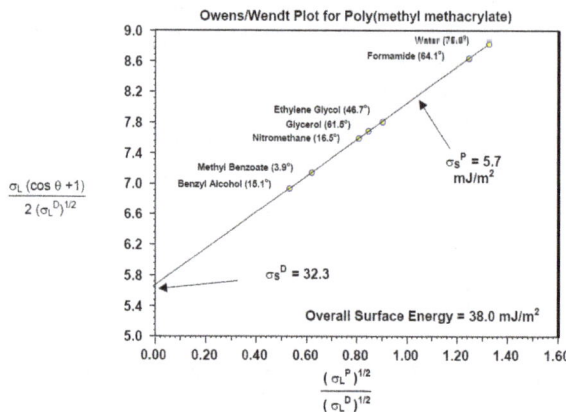

Fig: The Owens-Wendt plot for the surface energy of Poly(Methyl Methacrylate), with each data point reflecting the interaction of the solid with specified liquid. In this case, a high degree of confidence was indicated by a line coefficient of $r^2=.998$

The Owens/Wendt theory (after C. J. van Oss and John F. Wendt) divides the surface energy into two components: surface energy due to dispersive interactions and surface energy due to polar interactions. This theory is derived from the combination of Young's relation, which relates the contact angle to the surface energies of the solid and liquid and to the interface tension, and Good's equation (after R. J. Good), which relates the interface tension to the polar and dispersive components of the surface energy. The resulting principle equation is:

$$\frac{\sigma_L(\cos\theta+1)}{2\sqrt{\sigma_L^D}} = \frac{\sqrt{\sigma_S^P}\sqrt{\sigma_L^P}}{\sqrt{\sigma_L^D}} + \sqrt{\sigma_S^D}$$

Note that this equation has the form of y=mx+b, with:

$$y = \frac{\sigma_L(\cos\theta+1)}{2\sqrt{\sigma_L^D}} \; ; \; m = \sqrt{\sigma_S^P} \; ; \; x = \frac{\sqrt{\sigma_L^P}}{\sqrt{\sigma_L^D}} \; ; \; b = \sqrt{\sigma_S^D}$$

As such, the polar and dispersive components of the solid's surface energy are determined by the slope and intercept of the resulting graph. Of course, the problem at this point is that in order to make that graph, knowing the surface energy of the probe liquid is not enough, as it is necessary to know specifically how it breaks down into its polar and dispersive components as well.

To do this, one can simply reverse the procedure by testing the probe liquid against a standard reference solid that is not capable of polar interactions, such as PTFE. If the contact angle of a sessile drop of the probe liquid is measured on a PTFE surface with:

$$\sigma_S^P = 0$$

$$\sigma_S^D = 18.0\,\text{mN/m}$$

the principle equation reduces to:

$$\sigma_L^D = \frac{(\sigma_L(\cos\theta+1))^2}{72}$$

Since the total surface tension of the liquid is already known, this equation determines the dispersive component, and the difference between the total and dispersive components gives the polar component.

Accuracy/Precision

The accuracy and precision of this method is supported largely by the confidence level of the results for appropriate liquid/solid combinations (as seen, for example, in fig). The Owens/Wendt theory is typically applicable to surfaces with low charge and moderate polarity. Some good examples are polymers that contain heteroatoms, such as PVC, polyurethanes, polyamides, polyesters, polyacrylates, and polycarbonates

The Fowkes Theory

The Fowkes theory (after F. M. Fowkes) is derived in a slightly different way from the Owens/Wendt theory, although the Fowkes theory's principle equation is mathematically equivalent to that of Owens and Wendt:

$$\frac{\sigma_L (\cos \theta + 1)}{2} = \sqrt{\sigma_S^P} \sqrt{\sigma_L^P} + \sqrt{\sigma_S^D} \sqrt{\sigma_L^D}$$

Note that by dividing both sides of the equation by $\sqrt{\sigma_L^D}$, the Owens/Wendt principle equation is recovered. As such, one of the options for the proper determination of the surface energy components is the same.

In addition to that method, it is also possible to simply do tests using liquids with no polar component to their surface energies, and then liquids that do have both polar and dispersive components, and then linearize the equations. First, one performs the standard sessile drop contact angle measurement for the solid in question and a liquid with a polar components of zero ($\sigma_L^P = 0$; $\sigma_L = \sigma_L^D$) The second step is to use a second probe liquid that has both a dispersive and a polar component to its surface energy, and then solve for the unknowns algebraically. The Fowkes theory generally requires the use of only two probe liquids, as described above, and the recommended ones are diiodomethane, which should have no polar component due to its molecular symmetry, and water, which is commonly known to be a very polar liquid.

Accuracy/Precision

Though the principle equation is essentially identical to that of Owens and Wendt, the Fowkes theory in a larger sense has slightly different applications. Because it is derived from different principles than Owens/Wendt, the rest of the information that Fowkes theory is concerned with is related to adhesion. As such, it is more applicable to situations where adhesion occurs, and in general works better than does the Owens/Wendt theory when dealing with higher surface energies.

In addition, there is an extended Fowkes theory, rooted in the same principles, but dividing the total surface energy into a sum of three rather than two components: surface energy due to dispersive interactions, polar interactions, and hydrogen bonding.

The Wu Theory

The Wu theory (after Souheng Wu) is also essentially similar to the Owens/Wendt and Fowkes theories, in that it divides surface energy into a polar and a dispersive component. The primary difference is that Wu uses the harmonic means rather than the geometric means of the known surface tensions, and subsequently the use of more rigorous mathematics is employed.

Accuracy/Precision

The Wu theory provides more accurate results than do the other two component theories, particularly for high surface energies. It does, however, suffer from one complication: because of the mathematics involved, the Wu theory yields two results for each component, one being the true result, and one being simply a consequence of the mathematics. The challenge at this point lies in interpreting which is the true result. Sometimes this is as simple as eliminating the result that makes no physical sense (a negative surface energy) or the result that is clearly incorrect by virtue of being many orders of magnitude larger or smaller than it should be. Sometimes interpretation is more tricky.

The Schultz Theory

The Schultz theory (after D. L. Schultz) is applicable only for very high energy solids. Again, it is similar to the theories of Owens, Wendt, Fowkes, and Wu, but is designed for a situation where conventional measurement required for those theories is impossible. In the class of solids with sufficiently high surface energy, most liquids wet the surface completely with a contact angle of zero degrees, and thus no useful data can be gathered. The Schultz theory and procedure calls to deposit a sessile drop of probe liquid on the solid surface in question, but this is all done while the system is submerged in yet another liquid, rather than being done in the open air. As a result, the higher "atmospheric" pressure due to the surrounding liquid causes the probe liquid droplet to compress so that there is a measurable contact angle.

Accuracy/Precision

This method is designed to be robust where the other methods don't even provide any results in particular. As such, it is indispensable, since it is the only way to use the sessile drop technique on very high surface energy solids. Its major drawback is the fact that it is far more complex, both in its mathematics and experimentally. The Schultz theory requires one to account for many more factors, as there is now the unusual interaction of the probe liquid phase with the surrounding liquid phase. In addition, the set up of the camera and backlight become more complicated owing to the refractive properties of the surrounding liquid, not to mention the set up of the two-liquid system itself.

Three Component Theories

The Van Oss Theory

The van Oss theory separates the surface energy of solids and liquids into three components. It includes the dispersive surface energy, as before, and subdivides the polar component as being the sum of two more specific components: the surface energy due to acidic interactions (σ^+) and due to basic interactions (σ^-). The acid component theoretically describes a surface's propensity to have polar interactions with a second surface that has the ability to act basic by donating electrons. Conversely, the base component of the surface energy describes the propensity of a surface to have polar interactions with another surface that acts acidic by accepting electrons. The principle equation for this theory is:

$$\sigma_L(\cos\theta+1) = 2[\sqrt{\sigma_L^D \sigma_S^D} + \sqrt{\sigma_L^- \sigma_S^+} + \sqrt{\sigma_L^+ \sigma_S^-}]$$

Again, the best way to deal with this theory, much like the two component theories, is to use at least three liquids (more can be used to get more results for statistical purposes) – one with only a dispersive component to its surface energy ($\sigma_L = \sigma_L^D$), one with only a dispersive and an acidic or basic component ($\sigma_L = \sigma_L^D + \sigma_L^+$), and finally either a liquid with a dispersive and a basic or acidic component (whichever the second probe liquid did *not* have ($\sigma_L = \sigma_L^D + \sigma_L^\mp$)), or a liquid with all three components ($\sigma_L = \sigma_L^D + \sigma_L^+ + \sigma_L^-$) – and linearizing the results.

Accuracy/Precision

Being a three component theory, it is naturally more robust than other theories, particularly in cases where there is a great imbalance between the acid and base components of the polar surface energy. The van Oss theory is most suitable for testing the surface energies of inorganics, organometallics, and surface containing ions.

The most significant difficulty of applying the van Oss theory is the fact that there is not much of an agreement in regards to a set of reference solids that can be used to characterize the acid and base components of potential probe liquids. There are however some liquids that are generally agreed to have known dispersive/acid/base components to their surface energies. Two of them are listed in table 1.

List of common Probe Liquids

Table 1

Liquid	Total Surface Tension (mN/m)	Dispersive Component (mN/m)	Polar Component (mN/m)	Acid Component (mN/m)	Base Component (mN/m)
Formamide	58.0	39.0	19.0	2.28	39.6
Diiodomethane	50.8	50.8	0	0	0
Water	72.8	26.4	46.4	23.2	23.2

Potential Problems

The presence of surface active elements such as oxygen and sulfur will have a large impact on the measurements obtained with this technique. Surface active elements will exist in larger concentrations at the surface than in the bulk of the liquid, meaning that the total levels of these elements must be carefully controlled to a very low level. For example, the presence of only 50 ppm sulphur in liquid iron will reduce the surface tension by approximately 20%.

Practical Applications

The sessile drop technique has various applications for both materials engineering and straight characterization. In general, it is useful in determining the surface tension of liquids through the use of reference solids. There are various other specific applications which can be subdivided according to which of the above theories is most likely to be applicable to the circumstances:

The Zisman theory is mostly used for low energy surfaces and characterizes only the total surface energy. As such, it is probably most useful in cases that recall the conventional definition of surfaces, for example if a chemical engineer wants to know what the energy associated with fabricating a surface is. It may also be useful in cases where the surface energy has some effect on a spectroscopic technique being used on the solid in question.

The two component theories would most likely be applicable to materials engineering questions about the practical interactions of liquids and solids. The Fowkes theory, since it is more suited for higher energy solid surfaces, and since much of it is rooted in theories about adhesion, would likely be suited for the characterization of interactions where the solids and liquids have a high affinity for one another, such as, logically enough, adhesives and adhesive coatings. The Owens/Wendt theory, which deals in low energy solid surfaces, would be helpful in characterizing the interactions where the solids and liquids do *not* have a strong affinity for one another – for example, the effectiveness of waterproofing. Polyurethanes and PVC are good examples of waterproof plastics.

The Schultz theory is best used for the characterization of very high energy surfaces for which the other theories are ineffective, the most significant example being bare metals.

The van Oss theory is most suitable for cases in which acid/base interaction is an important consideration. Examples include pigments, pharmaceuticals, and paper. Specifically, notable examples include both paper used for the regular purpose of printing, and the more specialized case of litmus paper, which in itself is used to characterize acidity and basicity.

References

- Pierre-Gilles de Gennes; Françoise Brochard-Wyart; David Quéré (2002). Capillary and Wetting Phenomena—Drops, Bubbles, Pearls, Waves. Alex Reisinger. Springer. ISBN 0-387-00592-7.

- Gibbs, J.W. (2002) [1876-1878], "On the Equilibrium of Heterogeneous Substances", in Bumstead, H.A.; Van Nameeds, R.G., The Scientific Papers of J. Willard Gibbs, 1, Woodbridge, CT: Ox Bow Press, pp. 55–354, ISBN 0918024773

- G. Ertl, H. Knözinger and J. Weitkamp; Handbook of heterogeneous catalysis, Vol. 2, page 430; Wiley-VCH; Weinheim; 1997 ISBN 3-527-31241-2

- Gorban, A.N.; Karlin, I. V. (2016), "Beyond Navier–Stokes equations: capillarity of ideal gas", Contemporary Physics (Review article), doi:10.1080/00107514.2016.1256123

- US Geological Survey (July 2015). "Surface Tension (Water Properties) – USGS Water Science School". US Geological Survey. Retrieved November 6, 2015.

- International Association for the Properties of Water and Steam (June 2014). "Revised Release on Surface Tension of Ordinary Water Substance".

- K.G. Nayar; D. Panchanathan; G.H. McKinley & J.H. Lienhard V (November 2014). "Surface tension of seawater". J. Phys. Chem. Ref. Data. 43 (4): 43103. Bibcode:2014JPCRD..43d3103N. doi:10.1063/1.4899037.

- "Droplet oscillations driven by an electric field". Colloids and Surfaces A: Physicochemical and Engineering Aspects. 460: 351–354. doi:10.1016/j.colsurfa.2013.12.013.

- "Electrically Driven Resonant Oscillations of Pendant Hemispherical Liquid Droplet and Possibility to Evaluate the Surface Tension in Real Time". Zeitschrift für Physikalische Chemie. 227. doi:10.1524/zpch.2013.0420.

- "Oscillations of a Hanging Liquid Drop, Driven by Interfacial Dielectric Force". Zeitschrift für Physikalische Chemie. 225: 405–411. doi:10.1524/zpch.2011.0074.

Wear: An Integrated Study

Wear is the deformation that results from the exertion of force on a material. Abrasion, galling, adhesion, tribocorrosion and fretting are some of the topics that have been explained in the section. The chapter will provide an integrated understanding of the process of wear.

Wear

Wear is related to interactions between surfaces and specifically the removal and deformation of material on a surface as a result of mechanical action of the opposite surface.

In materials science, wear is erosion or sideways displacement of material from its "derivative" and original position on a solid surface performed by the action of another surface.

Wear of metals occurs by the plastic displacement of surface and near-surface material and by the detachment of particles that form wear debris. The size of the generated particles may vary from millimeter range down to an ion range. This process may occur by contact with other metals, nonmetallic solids, flowing liquids, or solid particles or liquid droplets entrained in flowing gasses.

Rear sprocket for a bicycle where the left is unused and the right has obvious wear from clockwise rotation.

Wear can also be defined as a process where interaction between two surfaces or bounding faces of solids within the working environment results in dimensional loss of one solid, with or without any actual decoupling and loss of material. Aspects of the working environment which affect wear include loads and features such as unidirectional sliding, reciprocating, rolling, and impact loads, speed, temperature, but also different types of counter-bodies such as solid, liquid or gas and type of contact ranging between single phase or multiphase, in which the last multiphase may combine liquid with solid particles and gas bubbles.

Stages of Wear

Under normal mechanical and practical procedures, the wear-rate normally changes through three different stages(ref.4):

- Primary stage or early run-in period, where surfaces adapt to each other and the wear-rate might vary between high and low.

- Secondary stage or mid-age process, where a steady rate of ageing is in motion. Most of the components operational life is comprised in this stage.

- Tertiary stage or old-age period, where the components are subjected to rapid failure due to a high rate of ageing.

The secondary stage is shortened with increasing severity of environmental conditions such as higher temperatures, strain rates, stress and sliding velocities etc. Note that, wear rate is strongly influenced by the operating conditions. Specifically, normal loads and sliding speeds play a pivotal role in determining wear rate. In addition, tribo-chemical reaction is also important in order to understand the wear behavior. Different oxide layers are developed during the sliding motion. The layers are originated from complex interaction among surface, lubricants, and environmental molecules. In general, a single plot, namely wear map. demonstrating wear rate under different loading condition is used for operation. This graph also represents dominating wear modes under different loading conditions (ref. 13).

In explicit wear tests simulating industrial conditions between metallic surfaces, there are no clear chronological distinction between different wear-stages due to big overlaps and symbiotic relations between various friction mechanisms. Surface engineering and treatments are used to minimize wear and extend the components working life.

Types

The study of the processes of wear is part of the discipline of tribology. The complex nature of wear has delayed its investigations and resulted in isolated studies towards specific wear mechanisms or processes. Some commonly referred to wear mechanisms (or processes) include:

1. Adhesive wear

2. Abrasive wear

3. Surface fatigue

4. Fretting wear

5. Erosive wear

6. Corrosion and oxidation wear

A number of different wear phenomena are also commonly encountered and presented in the literature. Impact-, cavitation-, diffusive- and corrosive- wear are all such examples. These wear mechanisms, however, do not necessarily act independently and wear mechanisms are not mutually exclusive. "Industrial Wear" are commonly described as incidence of multiple wear mechanisms occurring in unison. Another way to describe "Industrial Wear" is to define clear distinctions in how different friction mechanisms operate, for example distinguish between mechanisms with high or low energy density. Wear mechanisms and/or sub-mechanisms frequently overlap and occur in a synergistic manner, producing a greater rate of wear than the sum of the individual wear mechanisms.

Adhesive Wear

Adhesive wear can be found between surfaces during frictional contact and generally refers to unwanted displacement and attachment of wear debris and material compounds from one surface to another. Two separate mechanisms operate between the surfaces.

Friccohesity

Friccohesity defines actual changes in cohesive forces and their reproduction in form of kinetic or frictional forces in liquid when the clustering of the nano-particles scatter in medium for making smaller cluster or aggregates of different nanometer levels. The friccohesity basically is originated out of similar and dissimilar dipolar vectorization in liquid states. Since similar dipoles like water, water dipole develop their own patterns of stabilization so are called similar vectors leading to have specific force of attraction among them at surface. Similarly, the dissimilar vectors do have another patterns of optimization so their expression different with similar forces noted as cohesive forces along with exohesive (extrovert)forces leading to have distribution.

1. Adhesive wear is caused by relative motion, "direct contact" and plastic deformation which create wear debris and material transfer from one surface to another. In case of thermodynamic systems the friccohesity plays a master role and quantitatively explain a dissolution work to be undertaken or materialized by putting cohesive energy of solvent into distributing energy of mixtures allowing solute molecules to be distributed among the solvent phase with higher entropy. The solvent with higher cohesive forces gain higher potential energy and least entropy.

2. Cohesive adhesive forces, holds two surfaces together even though they are separated by a measurable distance, with or without any actual transfer of material.

The above description and distinction between "Adhesive wear" and its Counterpart "cohesive adhesive forces" are quite common. Usually cohesive surface forces and adhesive energy potentials between surfaces are examined as a special field in physics departments. The adhesive wear and material transfer due to direct contact and plastic deformation are examined in engineering science and in industrial research.

Two aligned surfaces may always cause material transfer and due to overlaps and symbiotic relations between relative motional "wear" and "chemical" cohesive attraction, the wear-categorization have been a source for discussion. Consequently, the definitions and nomenclature must evolve with the latest science and empiric observations.

Generally, adhesive wear occurs when two bodies slide over or are pressed into each other, which promote material transfer. This can be described as plastic deformation of very small fragments within the surface layers. The asperities or microscopic high points or surface roughness found on each surface, define the severity on how fragments of oxides are pulled off and adds to the other surface, partly due to strong adhesive forces between atoms but also due to accumulation of energy in the plastic zone between the asperities during relative motion.

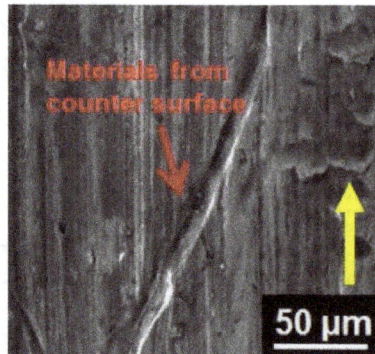

SEM micrograph of adhesive wear (transferred materials) on 52100 steel sample sliding against Al alloy.
(Yellow arrow indicate sliding direction)

The outcome can be a growing roughening and creation of protrusions (i.e., lumps) above the original surface, in industrial manufacturing referred to as galling, which eventually breaches the oxidized surface layer and connects to the underlying bulk material which enhance the possibility for a stronger adhesion and plastic flow around the lump. The geometry and the nominal sliding velocity of the lump defines how the flowing material will be transported and accelerated around the lump which is critical to define contact pressure and developed temperature during sliding. The mathematical function for acceleration of flowing material is thereby defined by the lumps surface contour.

It's clear, given these prerequisites, that contact pressure and developed temperature is highly dependent on the lumps geometry.Flow of material exhibits an increase in energy density, because initial phase transformation and displacement of material demand acceleration of material and high pressure.Low pressure is not compatible with plastic flow, only after deceleration may the flowing material be exposed to low pressure and quickly cooled. In other words, you can't deform a solid material using direct contact without applying a high pressure and somewhere along the process must acceleration and deceleration take place, i.e., high pressure must be applied on all sides of the deformed material. Flowing material will immediately exhibit energy loss and reduced ability to flow due to phase transformation, if ejected from high pressure into low pressure. This ability withholds the high pressure and energy density in the contact zone and decreases the amount of energy or friction force needed for further advancement when the sliding continues and partly explain the difference between the static and sliding coefficient of friction (μ) if the main fracture mechanisms are equal to the previous.

Adhesive wear is a common fault factor in industrial applications such as sheet metal forming (SMF) and commonly encountered in conjunction with lubricant failures and are often referred to as welding wear or galling due to the exhibited surface characteristics, phase transition and plastic flow followed by cooling. The type of mechanism and the amplitude of surface attraction, varies between different materials but are amplified by an increase in the density of "surface energy". Most solids will adhere on contact to some extent. However, oxidation films, lubricants and contaminants naturally occurring generally suppress adhesion. and spontaneous exothermic chemical reactions between surfaces generally produce a substance with low energy status in the absorbed species.

Abrasive Wear

Abrasive wear occurs when a hard rough surface slides across a softer surface. ASTM International (formerly American Society for Testing and Materials) defines it as the loss of material due to hard particles or hard protuberances that are forced against and move along a solid surface.

Abrasive wear is commonly classified according to the type of contact and the contact environment. The type of contact determines the mode of abrasive wear. The two modes of abrasive wear are known as two-body and three-body abrasive wear. Two-body wear occurs when the grits or hard particles remove material from the opposite surface. The common analogy is that of material being removed or displaced by a cutting or plowing operation. Three-body wear occurs when the particles are not constrained, and are free to roll and slide down a surface. The contact environment determines whether the wear is classified as open or closed. An open contact environment occurs when the surfaces are sufficiently displaced to be independent of one another

- Deep 'groove' like surface indicates abrasive wear over cast iron (yellow arrow indicate sliding direction)

There are a number of factors which influence abrasive wear and hence the manner of material removal. Several different mechanisms have been proposed to describe the manner in which the material is removed. Three commonly identified mechanisms of abrasive wear are:

1. Plowing

2. Cutting

3. Fragmentation

Plowing occurs when material is displaced to the side, away from the wear particles, resulting in the formation of grooves that do not involve direct material removal. The displaced material forms ridges adjacent to grooves, which may be removed by subsequent passage of abrasive particles. Cutting occurs when material is separated from the surface in the form of primary debris, or microchips, with little or no material displaced to the sides of the grooves. This mechanism closely resembles conventional machining. Fragmentation occurs when material is separated from a surface by a cutting process and the indenting abrasive causes localized fracture of the wear material. These cracks then freely propagate locally around the wear groove, resulting in additional material removal by spalling.

Abrasive wear can be measured as loss of mass by the Taber Abrasion Test according to ISO 9352 or ASTM D 4060.

Surface Fatigue

Surface fatigue is a process by which the surface of a material is weakened by cyclic loading, which is one type of general material fatigue. Fatigue wear is produced when the wear particles are detached by cyclic crack growth of microcracks on the surface. These microcracks are either superficial cracks or subsurface cracks.

Fretting Wear

Fretting wear is the repeated cyclical rubbing between two surfaces. Over a period of time fretting which will remove material from one or both surfaces in contact. It occurs typically in bearings, although most bearings have their surfaces hardened to resist the problem. Another problem occurs when cracks in either surface are created, known as fretting fatigue. It is the more serious of the two phenomena because it can lead to catastrophic failure of the bearing. An associated problem occurs when the small particles removed by wear are oxidised in air. The oxides are usually harder than the underlying metal, so wear accelerates as the harder particles abrade the metal surfaces further. Fretting corrosion acts in the same way, especially when water is present. Unprotected bearings on large structures like bridges can suffer serious degradation in behaviour, especially when salt is used the during winter to deice the highways carried by the bridges. The problem of fretting corrosion was involved in the Silver Bridge tragedy and the Mianus River Bridge accident.

Erosive Wear

Erosive wear can be defined as an extremely short sliding motion and is executed within a short time interval. Erosive wear is caused by the impact of particles of solid or liquid against the surface of an object. The impacting particles gradually remove material from the surface through repeated deformations and cutting actions. It is a widely encountered mechanism in industry. Due to the nature of the conveying process, piping systems are prone to wear when abrasive particles have to be transported.

The rate of erosive wear is dependent upon a number of factors. The material characteristics of the particles, such as their shape, hardness, impact velocity and impingement angle are primary factors along with the properties of the surface being eroded. The impingement angle is one of the most important factors and is widely recognized in literature. For ductile materials the maximum wear rate is found when the impingement angle is approximately 30°, whilst for non ductile materials the maximum wear rate occurs when the impingement angle is normal to the surface.

Corrosion and Oxidation Wear

This kind of wear occur in a variety of situations both in lubricated and unlubricated contacts. The fundamental cause of these forms of wear is chemical reaction between the worn material and the corroding medium. This kind of wear is a mixture of corrosion, wear and the synergistic term of corrosion-wear which is also called tribocorrosion.

Testing and Evaluation

Several standard test methods exist for different types of wear to determine the amount of material removal during a specified time period under well-defined conditions. The ASTM International Committee G-2 attempts to standardise wear testing for specific applications, which are periodically updated. The Society for Tribology and Lubrication Engineers (STLE) has documented a large number of frictional wear and lubrication tests. But all test methods have inbuilt limitations and do not give a true picture in every aspect. This can be attributed to the complex nature of wear, in particular "industrial wear", and the difficulties associated with accurately simulating wear processes. An attrition test is a test is carried out to measure the resistance of a granular material to wear.

A standard result review for wear tests, defined by the ASTM International and respective subcommittees such as Committee G-2, should be expressed as loss of material during wear in terms of volume. The volume loss gives a truer picture than weight loss, particularly when comparing the wear resistance properties of materials with large differences in density.

For example, a weight loss of 14 g in a sample of tungsten carbide + cobalt (density = 14000 kg/m³) and a weight loss of 2.7 g in a similar sample of aluminium alloy (density = 2700 kg/m³) both result in the same level of wear (1 cm³) when expressed as a volume loss. The inverse of volume loss can be used as a comparable index of wear resistance. Standard wear tests are only used for comparative material ranking of a specific test parameter as stipulated in the test method. For more realistic values of material deterioration in industrial applications it is necessary to conduct wear testing under conditions simulating the exact wear process.

The working life of an engineering component is expired when dimensional losses exceed the specified tolerance limits. Wear, along with other ageing processes such as fatigue and creep in association with stress concentration factors such as fracture toughness causes materials to progressively degrade, eventually leading to material failure at an advanced age. Wear in industrial applications is one of a limited number of fault factors in which an object loses its usefulness and the economic implication can be of enormous value to the industry.

Abrasion (Mechanical)

Abrasion is the process of scuffing, scratching, wearing down, marring, or rubbing away. It can be intentionally imposed in a controlled process using an abrasive. Abrasion can be an undesirable effect of exposure to normal use or exposure to the elements.

Abrasion in Stone Shaping

Ancient artists, working in stone, would use abrasion to create sculptures. The artist selected dense stones like carbonite and emory and rub them consistently against comparatively softer stones like limestone or granite. The artist would use different sizes and shapes of abrasives, or turn them in various ways as they rubbed, to create effects on the softer stone's surface. Water would be continuously poured over the surface to carry away particles. Abrasive technique in stone shaping was a long, tedious process that, with patience, resulted in eternal works of art in stone.

Models

The Archard equation is a simple model used to describe sliding wear and is based around the theory of asperity contact.

Abrasion Resistance

The resistance of materials and structures to abrasion can be measured by a variety of test methods. These often use a specified abrasive or other controlled means of abrasion. Under the conditions of the test, the results can be reported or can be compared items subjected to similar tests.

Such standardized measurements can produce two quantities: *abrasion rate* and *normalized abrasion rate* (also called *abrasion resistance index*). The former is the amount of mass lost per 1000 cycles of abrasion. The latter is the ratio of former with the known abrasion rate for some specific reference material.

One type of instrument used to get the quantities *abrasion rate* and *normalized abrasion rate*, is the abrasion scrub tester. This instrument is made up of a mechanical arm, liquid pump, and programmable electronics. The machine draws the mechanical arm with attached brush (or sandpaper, sponge, etc.) over the surface of the material that is being tested. The operator sets a pre-programmed number of passes for a repeatable and controlled result. The liquid pump can provide detergent or other liquids to the mechanical arm during testing to simulate washing and other normal uses.

The use of proper lubricants can help control abrasion in some instances. Some items can be covered with an abrasion resistant material. Controlling the cause of abrasion is sometimes an option.

Standards

ASTM

- ASTM B611 Test Method for Abrasive Wear Resistance of Cemented Carbides

- ASTM C131 Standard Test Method for Resistance to Degradation of Small-Size Coarse Aggregate by Abrasion and Impact in the Los Angeles Machine

- ASTM C448 Standard Test Methods for Abrasion Resistance of Porcelain Enamels

- ASTM C535 Standard Test Method for Resistance to Degradation of Large-Size Coarse Aggregate by Abrasion and Impact in the Los Angeles Machine

- ASTM C944 Standard Test Method for Abrasion Resistance of Concrete or Mortar Surfaces by the Rotating-Cutter Method

- ASTM C1027 Standard Test Method for Determining Visible Abrasion Resistance of Glazed Ceramic Tile

- ASTM C1353 Standard Test Method for Abrasion Resistance of Dimension Stone Subjected to Foot Traffic Using a Rotary Platform, Double-Head Abraser

- ASTM D968 Standard Test Methods for Abrasion Resistance of Organic Coatings by Falling Abrasive

- ASTM D1630 Standard Test Method for Rubber Property — Abrasion Resistance (Footwear Abrader)

- ASTM D 2228 Standard Test Method for Rubber Property - Relative Abrasion Resistance by the Pico Abrader Method

- ASTM D3389 Standard Test Method for Coated Fabrics Abrasion Resistance (Rotary Platform Abrader)

- ASTM D4060 Standard Test Method for Abrasion Resistance of Organic Coatings by the Taber Abraser

- ASTM D4158 Standard Guide for Abrasion Resistance of Textile Fabrics

- ASTM D4966 Standard Test Method for Abrasion Resistance of Textile Fabrics

- ASTM D5181 Standard Test Method for Abrasion Resistance of Printed Matter by the GA-CAT Comprehensive Abrasion Tester

- ASTM D5264 Standard Practice for Abrasion Resistance of Printed Materials by the Sutherland Rub Tester

- ASTM D5963 Standard Test Method for Rubber Property—Abrasion Resistance (Rotary Drum Abrader)

- ASTM D6279 Standard Test Method for Rub Abrasion Mar Resistance of High Gloss Coatings

- ASTM D7428 Standard Test Method for Resistance of Fine Aggregate to Degradation by Abrasion in the Micro-Deval Apparatus

- ASTM F1486 Standard Practice for Determination of Abrasion and Smudge Resistance of Images Produced from Office Products

- ASTM G56 Standard Test Method for Abrasiveness of Ink-Impregnated Fabric Printer Ribbons and Other Web Materials

- ASTM G65 Standard Test Method for Measuring Abrasion Using the Dry Sand/Rubber Wheel Apparatus

- ASTM G75 Standard Test Method for Determination of Slurry Abrasivity (Miller Number) and Slurry Abrasion Response of Materials (SAR Number)

- ASTM G81 Standard Test Method for Jaw Crusher Gouging Abrasion Test

- ASTM G105 Standard Test Method for Conducting Wet Sand/Rubber Wheel Abrasion Tests

- ASTM G132 Standard Test Method for Pin Abrasion Testing

- ASTM G171 Standard Test Method for Scratch Hardness of Materials Using a Diamond Stylus

- ASTM G174 Standard Test Method for Measuring Abrasion Resistance of Materials by Abrasive Loop Contact

DIN

- DIN 53516 Testing of Rubber and Elastomers; Determination of Abrasion Resistance

ISO

- ISO 4649 Rubber, vulcanized or thermoplastic -- Determination of abrasion resistance using a rotating cylindrical drum device

- ISO 9352 Plastics -- Determination of resistance to wear by abrasive wheels

- ISO 28080 Hardmetals -- Abrasion tests for hardmetals

- ISO 23794 Rubber, vulcanized or thermoplastic -- Abrasion testing -- Guidance

- ISO 21988:2006 Abrasion-resistant cast irons. Classification

- ISO 28080:2011 Hardmetals. Abrasion tests for hardmetals

- ISO 16282:2008 Methods of test for dense shaped refractory products. Determination of resistance to abrasion at ambient temperature

JSA

- JIS A 1121 Method of test for resistance to abrasion of coarse aggregate by use of the Los Angeles machine

- JIS A 1452 Method of abrasion test for building materials and part of building construction (falling sand method)

- JIS A 1453 Method of abrasion test for building materials and part of building construction (abrasive-paper method)

- JIS A 1509-5 Test methods for ceramic tiles -- Part 5: Determination of resistance to deep abrasion for unglazed floor tiles

- JIS A 1509-6 Test methods for ceramic tiles -- Part 6: Determination of resistance to surface abrasion for glazed floor tiles

- JIS C 60068-2 Environmental testing -- Part 2: Tests -- Test Xb: Abrasion of markings and letterings caused by rubbing of fingers and hands

- JIS H 8682-1 Test methods for abrasion resistance of anodic oxide coatings on aluminium and aluminium alloys -- Part 1: Wheel wear test

- JIS H 8682-2 Test methods for abrasion resistance of anodic oxide coatings on aluminium and aluminium alloys -- Part 2: Abrasive jet test

- JIS H 8682-3 Test methods for abrasion resistance of anodic oxide coatings on aluminium and aluminium alloys -- Part 3: Sand-falling abrasion resistance test

- JIS K 5600-5-8 Testing methods for paints -- Part 5: Mechanical property of film -- Section 8: Abrasion resistance (Rotating abrasive-paper-covered wheel method)

- JIS K 7204 Plastics -- Determination of resistance to wear by abrasive wheels

Galling

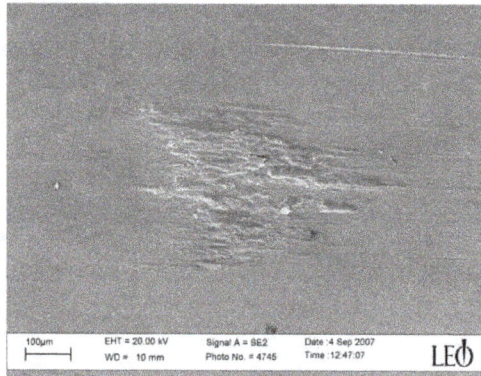

An electron microscope image shows transferred sheet material accumulated on a tool surface during sliding contact under controlled laboratory conditions. The outgrowth of material or localized, roughening and creation of protrusions on the tool surface is commonly referred to as a lump.

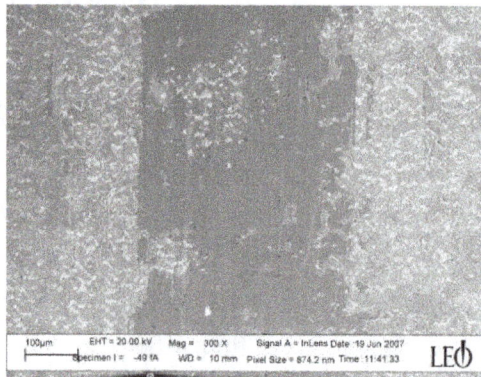

The damage on the metal sheet, wear mode, or characteristic pattern shows no breakthrough of the oxide surface layer, which indicates a small amount of adhesive material transfer and a flattening damage of the sheet's surface. This is the first stage of material transfer and galling build-up.

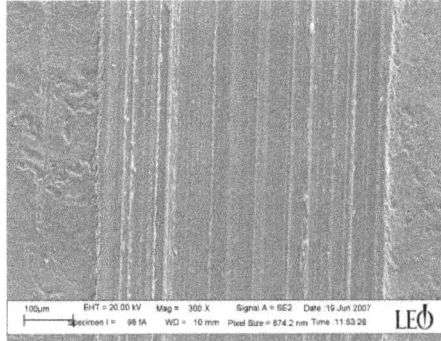

The damage on the metal sheet illustrates continuous lines or stripes, indicating a breakthrough of the oxide surface-layer.

Galling is a form of wear caused by adhesion between sliding surfaces. When a material galls, some of it is pulled with the contacting surface, especially if there is a large amount of force compressing the surfaces together. Galling is caused by a combination of friction and adhesion between the surfaces, followed by slipping and tearing of crystal structure beneath the surface. This will generally leave some material stuck or even friction welded to the adjacent surface, whereas the galled material may appear gouged with balled-up or torn lumps of material stuck to its surface.

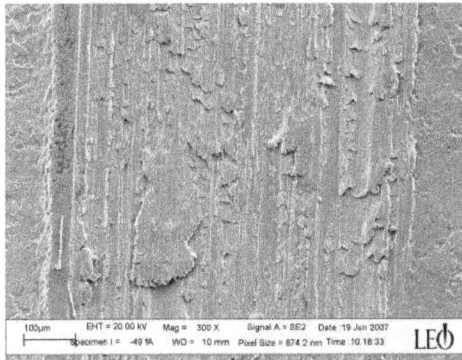

The damage on the metal sheet or characteristic pattern illustrates an "uneven surface", a change in the sheet material's plastic behaviour and involves a larger deformed volume compared to mere flattening of the surface oxides.

Galling is most commonly found in metal surfaces that are in sliding contact with each other. It is especially common where there is inadequate lubrication between the surfaces. However, certain metals will generally be more prone to galling, due to the atomic structure of their crystals. For example, aluminium is a metal that will gall very easily, whereas annealed (softened) steel is slightly more resistant to galling. Steel that is fully hardened is very resistant to galling.

Galling is a common problem in most applications where metals slide while in contact with other metals. This can happen regardless of whether the metals are the same or of different kinds. Metals such as brass are often chosen for bearings, bushings, and other sliding applications because of their resistance to galling, as well as other forms of mechanical abrasion.

Introduction

Galling is adhesive wear. Galling is caused by microscopic transfer of material between metallic surfaces, during transverse motion (sliding). Galling occurs frequently whenever metal surfaces are in contact, sliding against each other, especially with poor lubrication. Galling often occurs in high load, low speed applications, but also occurs in high-speed applications with very little load. Galling is a common problem in sheet metal forming, bearings and pistons in engines, hydraulic cylinders, air motors, and many other industrial operations. Galling is distinctive from gouging or scratching in that galling involves the visible transfer of material as it is adhesively pulled (mechanically spalled) from one surface, leaving it stuck to the other in the form of a raised lump (gall). Unlike other forms of wear, galling is usually not a gradual process, but occurs quickly and spreads rapidly as the raised lumps induce more galling.

Galling can often occur in screws and bolts, causing the threads to seize and tear free from either the fastener or the hole. In extreme cases, the bolt may lock up to the point where all turning force is used by the friction, which can lead to breakage of the fastener or the tool turning it. Threaded inserts of hardened steel are often used in metals like aluminium or stainless steel that can gall easily.

Galling requires two properties common to most metals, cohesion through metallic-bonding attractions and plasticity (the ability to deform without breaking). The tendency of a material to gall is affected by the ductility of the material. Typically, hardened materials are more resistant to galling whereas softer materials of the same type will gall more readily. The propensity of a material to gall is also affected by the specific arrangement of the atoms, because crystals arranged in a face-centered cubic (FCC) lattice will usually allow material-transfer to a greater degree than a body-centered cubic (BCC). This is because a face-centered cubic has a greater tendency to produce dislocations in the crystal lattice, which are defects that allow the lattice to shift, or "cross-slip," making the metal more prone to galling. However, if the metal has a high number of stacking faults (a difference in stacking sequence between atomic planes) it will be less apt to cross-slip at the dislocations. Therefore, a material's resistance to galling is usually determined by its stacking-fault energy. A material with high stacking-fault energy, such as aluminium or titanium, will be far more susceptible to galling than materials with low stacking-fault energy, like copper, bronze, or gold. Conversely, materials with a hexagonal close packed (HCP) structure, such as cobalt-based alloys, are extremely resistant to galling.

Galling occurs initially with material transfer from individual grains, on a microscopic scale, which become stuck or even diffusion welded to the adjacent surface. This transfer can be enhanced if one or both metals form a thin layer of hard oxides with high coefficients of friction, such as those found on aluminum or stainless-steel. As the lump grows it pushes against the adjacent material and begins forcing them apart, concentrating a majority of the friction heat-energy into a very small area. This in turn causes more adhesion and material build-up. The localized heat increases the plasticity of the galled surface, deforming the metal, until the lump breaks through the surface and begins plowing up large amounts of material from the galled surface. Methods of preventing galling include the use of lubricants like grease and oil, low-friction coatings and thin-film deposits like molybdenum disulfide or titanium nitride, and increasing the surface hardness of the metals using processes such as case hardening and induction hardening.

Mechanism

In engineering science and in other technical aspects, the term galling is widespread. The influence of acceleration in the contact zone between materials have been mathematically described and correlated to the exhibited friction mechanism found in the tracks during empiric observations of the galling phenomenon. Due to problems with previous incompatible definitions and test methods, better means of measurements in coordination with greater understanding of the involved frictional mechanisms, have led to the attempt to standardize or redefine the term galling to enable a more generalized use. ASTM International has formulated and established a common definition for the technical aspect of the galling phenomenon in the ASTM G40 standard: "Galling is a form of surface damage arising between sliding solids, distinguished by microscopic, usually localized, roughening and creation of protrusions (e.g.: lumps) above the original surface".

When two metallic surfaces are pressed against each other the initial interaction and the mating points are the asperities, or high points, found on each surface. An asperity may penetrate the opposing surface if there is a converging contact and relative movement. The contact between the surfaces initiates friction or plastic deformation and induces pressure and energy in a small area called the contact zone.

The elevation in pressure increases the energy density and heat level within the deformed area. This leads to greater adhesion between the surfaces which initiate material transfer, galling build-up, lump growth, and creation of protrusions above the original surface.

If the lump (or protrusion of transferred material to one surface) grows to a height of several micrometers, it may penetrate the opposing surface oxide-layer and cause damage to the underlying material. Damage in the bulk material is a prerequisite for plastic flow that is found in the deformed volume which surrounds the lump. The geometry and speed of the lump defines how the flowing material will be transported, accelerated, and decelerated around the lump. This material flow is critical when defining the contact pressure, energy density, and developed temperature during sliding. The mathematical function describing acceleration and deceleration of flowing material is thereby defined by the geometrical constraints, deduced or given by the lump's surface contour.

If the right conditions are met, such as geometric constraints of the lump, an accumulation of energy can cause a clear change in the materials contact and plastic behaviour; generally this increases adhesion and the friction force needed for further movement.

In sliding friction, increased compressive stress is proportionally equal to a rise in potential energy and temperature within the contact zone. The reasons for accumulation of energy during sliding can be a reduction of energy loss away from the contact zone, due to a small surface area on the surface boundary thus low heat conductivity. Another reason is the energy that is continuously forced into the metals, which is a product of acceleration and pressure. In cooperation, these mechanisms allow a constant accumulation of energy causing increased energy density and temperature in the contact zone during sliding.

The process and contact can be compared to cold welding or friction welding, because cold welding is not truly cold and the fusing points exhibit an increase in temperature and energy density derived from applied pressure and plastic deformation in the contact zone.

Incidence and Location

Galling is often found between metallic surfaces where direct contact and relative motion have occurred. Sheet metal forming, thread manufacturing and other industrial operations may include moving parts or contact surfaces made of stainless steel, aluminium, titanium, and other metals whose natural development of an external oxide layer through passivation increases their corrosion resistance but renders them particularly susceptible to galling.

In metalworking that involves cutting (primarily turning and milling), galling is often used to describe a wear phenomenon which occurs when cutting soft metal. The work material is transferred to the cutter and develops a "lump". The developed lump changes the contact behavior between the two surfaces, which usually increases adhesion, resistance to further cutting, and, due to created vibrations, can be heard as a distinct sound.

Galling often occurs with aluminium compounds and is a common cause of tool breakdown. Aluminium is a ductile metal, which means it possesses the ability for plastic flow with relative ease, which presupposes a relatively consistent and large plastic zone.

High ductility and flowing material can be considered a general prerequisite for excessive material transfer and galling because frictional heating is closely linked to the structure of plastic zones around penetrating objects.

Galling can occur even at relatively low loads and velocities, because it is the real energy-density in the system that induces a phase transition, which often leads to an increase in material transfer and higher friction.

Prevention

Generally there are two major frictional systems which affect adhesive wear or galling. In terms of prevention, they work in dissimilar ways and set different demands on the surface structure, alloys and crystal matrix used in the materials:

- Solid surface contact

- Lubricated contact

In solid surface contact or unlubricated conditions, the initial contact is characterised by interaction between asperities and the exhibition of two different sorts of attraction: cohesive surface-energy or the molecules connect and adhere the two surfaces together, notably even if they are separated by a measurable distance. Direct contact and plastic deformation generates another type of attraction through the constitution of a plastic zone with flowing material where induced energy, pressure and temperature allow bonding between the surfaces on a much larger scale than cohesive surface-energy.

In metallic compounds and sheet metal forming, the asperities are usually oxides and the plastic deformation mostly consists of brittle fracture, which presupposes a very small plastic zone. The accumulation of energy and temperature is low due to the discontinuity in the fracture mechanism. However, during the initial asperity/ asperity contact, wear debris or bits and pieces from the asperities adhere to the opposing surface, creating microscopic, usually localized, roughening and creation of protrusions (in effect lumps) above the original surface. The transferred wear debris and lumps penetrate the opposing oxide surface layer and cause damage to the underlying bulk material, plowing it forward. This allows continuous plastic deformation, plastic flow, and accumulation of energy and temperature. The prevention of adhesive material-transfer is accomplished by the following or similar approaches:

- Low temperature carburizing treatments such as Kolsterising can eliminate galling in austenitic stainless-steels by increasing surface hardness up to 1200 HV0.05 (depending on base material and surface conditions).

- Less cohesive or chemical attraction between surface atoms or molecules.

- Avoiding continuous plastic-deformation and plastic flow, for example through a thicker oxide layer on the subject material in sheet-metal forming (SMF).

- Coatings deposited on the SMF work tool, such as chemical vapor deposition (CVD) or physical vapor deposition (PVD) and titanium nitride (TiN) or diamond-like carbon coatings exhibit low chemical reactivity even in high energy frictional contact, where the subject material's protective oxide layer is breached, and the frictional contact is distinguished by continuous plastic deformation and plastic flow.

Lubricated contact sets other demands on the materials surface structure, and the main issue is to retain the protective lubrication thickness and avoid plastic deformation. This is important because plastic deformation raises the temperature of the oil or lubrication fluid and changes the viscosity. Any eventual material transfer or creation of protrusions above the original surface will also reduce the ability to retain a protective lubrication thickness. A proper protective lubrication thickness can be assisted or retained by:

- Surface cavities or small holes can create a favourable geometric situation for the oil to retain a protective lubrication thickness in the contact zone.

- Cohesive forces on the surface can increase the chemical attraction between the surface and lubricants, and enhance the lubrication thickness.

- Oil additives may reduce the tendency for galling or adhesive wear.

Clarification and Limitations

Galling should not be confused with other cases of attraction between surfaces which do not result in plastic deformation. These latter types of attraction involve adhesive surface forces or surface energy theories. Different energy potentials at the surfaces can develop adhesive bonds or cohesive forces that may hold the two surfaces together. Surface energy and the cohesive force phenomenon are not the same as galling, and are only partially correlated. Galling necessarily involves plastic deformation of at least one surface.

However, research generally fails to make a clear distinction between energy derived from plastic deformation and the cohesive surface-forces, and chemical attraction between atoms or surface molecules. The latter is likely to be involved in the initial material transfer, where only surface-oxide asperities are in contact. Difficulty comes in distinguishing these adhesive forces from more severe attractions, caused by accumulated energy and increased pressure from plastic deformation. Oxides are brittle and it is probable that most of the energy in the fracture mechanism is consumed in brittle fracture, but the created wear debris will instantaneously penetrate the opposing surface. This means that the transferred oxide material will instantly act as a penetrating body and the concentration of energy, pressure and frictional heating is immediate. Without this accumulation of energy, the tendency for material transfer will certainly decrease.

The formation and constitution (physique) of plastic zones around penetrating objects are arguably a prerequisite and the main factor for excessive material transfer, lump growth, and galling build-up even in the initial contact process.

Adhesion

Adhesion of a frog on a wet vertical glass surface.

Dew drops adhering to a spider web

IUPAC Definition

Process of attachment of a substance to the surface of another substance.

Adhesion is the tendency of dissimilar particles or surfaces to cling to one another (cohesion refers to the tendency of similar or identical particles/surfaces to cling to one another). The forces that cause adhesion and cohesion can be divided into several types. The intermolecular forces responsible for the function of various kinds of stickers and sticky tape fall into the categories of chemical adhesion, dispersive adhesion, and diffusive adhesion. In addition to the cumulative magnitudes of these intermolecular forces, there are certain emergent mechanical effects.

Surface Energy

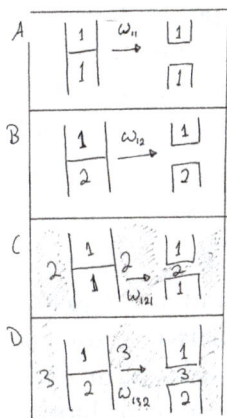

Diagram of various cases of cleavage, with each unique species labeled. A: $\gamma = (1/2)W_{11}$ B: $W_{12} = \gamma_1 + \gamma_2 - \gamma_{12}$ C: $\gamma_{12} = (1/2)W_{121} = (1/2)W_{212}$ D: $W_{12} + W_{33} - W_{13} - W_{23} = W_{132}$.

Surface energy is conventionally defined as the work that is required to build an area of a particular surface. Another way to view the surface energy is to relate it to the work required to cleave a bulk sample, creating two surfaces. If the new surfaces are identical, the surface energy γ of each surface is equal to half the work of cleavage, W: $\gamma = (1/2)W_{11}$.

If the surfaces are unequal, the Young-Dupré equation applies: $W_{12} = \gamma_1 + \gamma_2 - \gamma_{12}$, where γ_1 and γ_2 are the surface energies of the two new surfaces, and γ_{12} is the interfacial tension.

This methodology can also be used to discuss cleavage that happens in another medium: $\gamma_{12} = (1/2)W_{121} = (1/2)W_{212}$. These two energy quantities refer to the energy that is needed to cleave one species into two pieces while it is contained in a medium of the other species. Likewise for a three species system: $\gamma_{13} + \gamma_{23} - \gamma_{12} = W_{12} + W_{33} - W_{13} - W_{23} = W_{132}$, where W_{132} is the energy of cleaving species 1 from species 2 in a medium of species 3.

A basic understanding of the terminology of cleavage energy, surface energy, and surface tension is very helpful for understanding the physical state and the events that happen at a given surface, but as discussed below, the theory of these variables also yields some interesting effects that concern the practicality of adhesive surfaces in relation to their surroundings.

Mechanisms of Adhesion

There is no single theory covering adhesion, and particular mechanisms are specific to particular material scenarios. Five mechanisms of adhesion have been proposed to explain why one material sticks to another:

Mechanical Adhesion

Adhesive materials fill the voids or pores of the surfaces and hold surfaces together by interlocking. Other interlocking phenomena are observed on different length scales. Sewing is an example of two materials forming a large scale mechanical bond, velcro forms one on a medium scale, and some textile adhesives (glue) form one at a small scale.

Chemical Adhesion

Two materials may form a compound at the joint. The strongest joints are where atoms of the two materials swap or share electrons (known as ionic bonding or covalent bonding, respectively). A weaker bond is formed if a hydrogen atom in one molecule is attracted to an atom of nitrogen, oxygen, or fluorine in another molecule, a phenomenon called hydrogen bonding.

Chemical adhesion occurs when the surface atoms of two separate surfaces form ionic, covalent, or hydrogen bonds. The engineering principle behind chemical adhesion in this sense is fairly straightforward: if surface molecules can bond, then the surfaces will be bonded together by a network of these bonds. It bears mentioning that these attractive ionic and covalent forces are effective over only very small distances – less than a nanometer. This means in general not only that surfaces with the potential for chemical bonding need to be brought very close together, but also that these bonds are fairly brittle, since the surfaces then need to be kept close together.

Dispersive Adhesion

Cohesion causes water to form drops, surface tension causes them to be nearly spherical, and adhesion keeps the drops in place.

In dispersive adhesion, also known as physisorption, two materials are held together by van der Waals forces: the attraction between two molecules, each of which has a region of slight positive and negative charge. In the simple case, such molecules are therefore polar with respect to average charge density, although in larger or more complex molecules, there may be multiple "poles" or regions of greater positive or negative charge. These positive and negative poles may be a permanent property of a molecule (Keesom forces) or a transient effect which can occur in any molecule, as the random movement of electrons within the molecules may result in a temporary concentration of electrons in one region (London forces).

Water droplets are flatter on a Hibiscus flower which shows better adhesion.

In surface science, the term *adhesion* almost always refers to dispersive adhesion. In a typical solid-liquid-gas system (such as a drop of liquid on a solid surrounded by air) the contact angle is used to evaluate adhesiveness indirectly, while a Centrifugal Adhesion Balance allows for direct quantitative adhesion measurements. Generally, cases where the contact angle is low are considered of higher adhesion per unit area. This approach assumes that the lower contact angle corresponds to a higher surface energy. Theoretically, the more exact relation between contact angle and work of adhesion is more involved and is given by the Young-Dupre equation. The contact angle of the three-phase system is a function not only of dispersive adhesion (interaction between the molecules in the liquid and the molecules in the solid) but also cohesion (interaction between the liquid molecules themselves). Strong adhesion and weak cohesion results in a high degree of wetting, a lyophilic condition with low measured contact angles. Conversely, weak adhesion and strong cohesion results in lyophobic conditions with high measured contact angles and poor wetting.

London dispersion forces are particularly useful for the function of adhesive devices, because they don't require either surface to have any permanent polarity. They were described in the 1930s by Fritz London, and have been observed by many researchers. Dispersive forces are a consequence of statistical quantum mechanics. London theorized that attractive forces between molecules that cannot be explained by ionic or covalent interaction can be caused by polar moments within molecules. Multipoles could account for attraction between molecules having permanent multipole moments that participate in electrostatic interaction. However, experimental data showed that many of the compounds observed to experience van der Waals forces had no multipoles at all. London suggested that momentary dipoles are induced purely by virtue of molecules being in proximity to one another. By solving the quantum mechanical system of two electrons as harmonic oscillators at some finite distance from one another, being displaced about their respective rest positions and interacting with each other's fields, London showed that the energy of this system is given by:

$$E = 3h\nu - \frac{3}{4}\frac{h\nu\alpha^2}{R^6}$$

While the first term is simply the zero-point energy, the negative second term describes an attractive force between neighboring oscillators. The same argument can also be extended to a large number of coupled oscillators, and thus skirts issues that would negate the large scale attractive effects of permanent dipoles cancelling through symmetry, in particular.

The additive nature of the dispersion effect has another useful consequence. Consider a single such dispersive dipole, referred to as the origin dipole. Since any origin dipole is inherently oriented so as to be attracted to the adjacent dipoles it induces, while the other, more distant dipoles are not correlated with the original dipole by any phase relation (thus on average contributing nothing), there is a net attractive force in a bulk of such particles. When considering identical particles, this is called cohesive force.

When discussing adhesion, this theory needs to be converted into terms relating to surfaces. If there is a net attractive energy of cohesion in a bulk of similar molecules, then cleaving this bulk to produce two surfaces will yield surfaces with a dispersive surface energy, since the form of the energy remain the same. This theory provides a basis for the existence of van der Waals forces at the surface, which exist between any molecules having electrons. These forces are easily observed through the spontaneous jumping of smooth surfaces into contact. Smooth surfaces of mica, gold, various polymers and solid gelatin solutions do not stay apart when their separating becomes small enough – on the order of 1–10 nm. The equation describing these attractions was predicted in the 1930s by De Boer and Hamaker:

$$\frac{P}{area} = -\frac{A}{24\pi z^3}$$

where P is the force (negative for attraction), z is the separation distance, and A is a material-specific constant called the Hamaker constant.

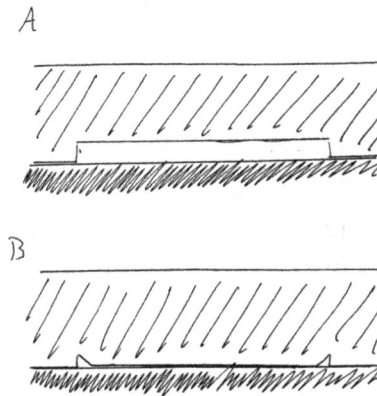

The two stages of PDMS microstructure collapse due to van der Waals attractions. The PDMS stamp is indicated by the hatched region, and the substrate is indicated by the shaded region. A) The PDMS stamp is placed on a substrate with the "roof" elevated. B) Van der Waals attractions make roof collapse energetically favorable for PDMS stamp.

The effect is also apparent in experiments where a polydimethylsiloxane (PDMS) stamp is made with small periodic post structures. The surface with the posts is placed face down on a smooth surface, such that the surface area in between each post is elevated above the smooth surface, like a roof supported by columns. Because of these attractive dispersive forces between the PDMS and the smooth substrate, the elevated surface – or "roof" – collapses down onto the substrate without any external force aside from the van der Waals attraction. Simple smooth polymer surfaces – without any micro-structures – are commonly used for these dispersive adhesive properties. Decals and stickers that adhere to glass without using any chemical adhesives are fairly common as toys and decorations and useful as removable labels because they do not rapidly lose their adhesive properties, as do sticky tapes that use adhesive chemical compounds.

It is important to note that these forces also act over very small distances – 99% of the work necessary to break van der Waals bonds is done once surfaces are pulled more than a nanometer apart. As a result of this limited motion in both the van der Waals and ionic/covalent bonding situations, practical effectiveness of adhesion due to either or both of these interactions leaves much to be desired. Once a crack is initiated, it propagates easily along the interface because of the brittle nature of the interfacial bonds.

As an additional consequence, increasing surface area often does little to enhance the strength of the adhesion in this situation. This follows from the aforementioned crack failure – the stress at the interface is not uniformly distributed, but rather concentrated at the area of failure.

Electrostatic Adhesion

Some conducting materials may pass electrons to form a difference in electrical charge at the join. This results in a structure similar to a capacitor and creates an attractive electrostatic force between the materials.

Diffusive Adhesion

Some materials may merge at the joint by diffusion. This may occur when the molecules of both materials are mobile and soluble in each other. This would be particularly effective with polymer chains where one end of the molecule diffuses into the other material. It is also the mechanism involved in sintering. When metal or ceramic powders are pressed together and heated, atoms diffuse from one particle to the next. This joins the particles into one.

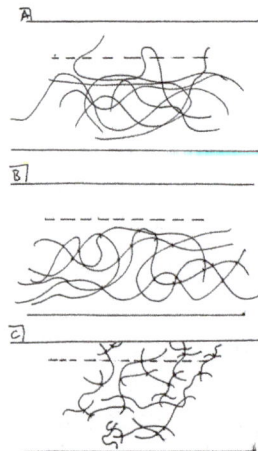

The interface is indicated by the dotted line. A) Non-crosslinked polymers are somewhat free to diffuse across the interface. One loop and two distal tails are seen diffusing. B) Crosslinked polymers not free enough to diffuse. C) "Scissed" polymers very free, with many tails extending across the interface.

Diffusive forces are somewhat like mechanical tethering at the molecular level. Diffusive bonding occurs when species from one surface penetrate into an adjacent surface while still being bound to the phase of their surface of origin. One instructive example is that of polymer-on-polymer surfaces. Diffusive bonding in polymer-on-polymer surfaces is the result of sections of polymer chains from one surface interdigitating with those of an adjacent surface. The freedom of movement of the polymers has a strong effect on their ability to interdigitate, and hence, on diffusive bonding. For example, cross-linked polymers are less capable of diffusion and interdigitation because they are bonded together at many points of contact, and are not free to twist into the adjacent surface. Uncrosslinked polymers (thermoplastics), on the other hand are freer to wander into the adjacent phase by extending tails and loops across the interface.

Another circumstance under which diffusive bonding occurs is "scission". Chain scission is the cutting up of polymer chains, resulting in a higher concentration of distal tails. The heightened concentration of these chain ends gives rise to a heightened concentration of polymer tails extending across the interface. Scission is easily achieved by ultraviolet irradiation in the presence of oxygen gas, which suggests that adhesive devices employing diffusive bonding actually benefit from prolonged exposure to heat/light and air. The longer such a device is exposed to these conditions, the more tails are scissed and branch out across the interface.

Once across the interface, the tails and loops form whatever bonds are favorable. In the case of polymer-on-polymer surfaces, this means more van der Waals forces. While these may be brittle, they are quite strong when a large network of these bonds is formed. The outermost layer of each surface plays a crucial role in the adhesive properties of such interfaces, as even a tiny amount of interdigitation – as little as one or two tails of 1.25 angstrom length – can increase the van der Waals bonds by an order of magnitude.

Strength

The strength of the adhesion between two materials depends on which of the above mechanisms occur between the two materials, and the surface area over which the two materials contact. Materials that wet against each other tend to have a larger contact area than those that do not. Wetting depends on the surface energy of the materials.

Low surface energy materials such as polyethylene, polypropylene, polytetrafluoroethylene and polyoxymethylene are difficult to bond without special surface preparation.

Other Effects

In concert with the primary surface forces described above, there are several circumstantial effects in play. While the forces themselves each contribute to the magnitude of the adhesion between the surfaces, the following play a crucial role in the overall strength and reliability of an adhesive device.

Stringing

Stringing is perhaps the most crucial of these effects, and is often seen on adhesive tapes. Stringing occurs when a separation of two surfaces is beginning and molecules at the interface bridge out across the gap, rather than cracking like the interface itself. The most significant consequence of this effect is the restraint of the crack. By providing the otherwise brittle interfacial bonds with some flexibility, the molecules that are stringing across the gap can stop the crack from propagating. Another way to understand this phenomenon is by comparing it to the stress concentration at the point of failure mentioned earlier. Since the stress is now spread out over some area, the stress at any given point has less of a chance of overwhelming the total adhesive force between the surfaces. If failure does occur at an interface containing a viscoelastic adhesive agent, and a crack does propagate, it happens by a gradual process called "fingering", rather than a rapid, brittle fracture. Stringing can apply to both the diffusive bonding regime and the chemical bonding regime. The strings of molecules bridging across the gap would either be the molecules that had earlier diffused across the interface or the viscoelastic adhesive, provided that there was a significant volume of it at the interface.

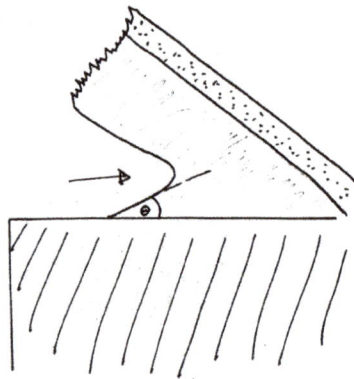

Fingering process. The hatched area is the receiving substrate, the dotted strip is the tape, and the shaded area in between is the adhesive chemical layer. The arrow indicates the direction of propagation for the fracture.

Microstructures

Technologically advanced adhesive devices sometimes make use of microstructures on surfaces, such as the periodic posts described above. These are biomimetic technologies inspired by the adhesive abilities of the feet of various arthropods and vertebrates (most notably, geckos). By intermixing periodic breaks into smooth, adhesive surfaces, the interface acquires valuable crack-arresting properties. Because crack initiation requires much greater stress than does crack propagation, surfaces like these are much harder to separate, as a new crack has to be restarted every time the next individual microstructure is reached.

Hysteresis

Hysteresis, in this case, refers to the restructuring of the adhesive interface over some period of time, with the result being that the work needed to separate two surfaces is greater than the work that was gained by bringing them together ($W > \gamma_1 + \gamma_2$). For the most part, this is a phenomenon associated with diffusive bonding. The more time is given for a pair of surfaces exhibiting diffusive bonding to restructure, the more diffusion will occur, the stronger the adhesion will become. The aforementioned reaction of certain polymer-on-polymer surfaces to ultraviolet radiation and oxygen gas is an instance of hysteresis, but it will also happen over time without those factors.

In addition to being able to observe hysteresis by determining if $W > \gamma_1 + \gamma_2$ is true, one can also find evidence of it by performing "stop-start" measurements. In these experiments, two surfaces slide against one another continuously and occasionally stopped for some measured amount of time. Results from experiments on polymer-on-polymer surfaces show that if the stopping time is short enough, resumption of smooth sliding is easy. If, however, the stopping time exceeds some limit, there is an initial increase of resistance to motion, indicating that the stopping time was sufficient for the surfaces to restructure.

Wettability and Adsorption

Some atmospheric effects on the functionality of adhesive devices can be characterized by following the theory of surface energy and interfacial tension. It is known that $\gamma_{12} = (1/2)W_{121} = (1/2)W_{212}$. If γ_{12} is high, then each species finds it favorable to cohere while in contact with a foreign species, rather than dissociate and mix with the other. If this is true, then it follows that when the interfacial tension is high, the force of adhesion is weak, since each species does not find it favorable to bond to the other. The interfacial tension of a liquid and a solid is directly related to the liquids wettability (relative to the solid), and thus one can extrapolate that cohesion increases in non-wetting liquids and decreases in wetting liquids. One example that verifies this is polydimethyl siloxane rubber, which has a work of self-adhesion of 43.6 mJ/m² in air, 74 mJ/m² in water (a nonwetting liquid) and 6 mJ/m² in methanol (a wetting liquid).

This argument can be extended to the idea that when a surface is in a medium with which binding is favorable, it will be less likely to adhere to another surface, since the medium is taking up the potential sites on the surface that would otherwise be available to adhere to another surface. Naturally this applies very strongly to wetting liquids, but also to gas molecules that could adsorb onto the surface in question, thereby occupying potential adhesion sites. This last point is actually fairly intuitive: Leaving an adhesive exposed to air too long gets it dirty, and its adhesive strength will decrease. This is observed in the experiment: when mica is cleaved in air, its cleavage energy, W_{121} or $W_{mica/air/mica}$, is smaller than the cleavage energy in vacuum, $W_{mica/vac/mica}$, by a factor of 13.

Lateral Adhesion

Lateral adhesion is the adhesion associated with sliding one object on a substrate such as sliding a drop on a surface. When the two objects are solids, either with or without a liquid between them, the lateral adhesion is described as friction. However, the behavior of lateral adhesion between a drop and a surface is tribologically very different from friction between solids, and the naturally adhesive contact between a flat surface and a liquid drop makes the lateral adhesion in this case, an individual field. Lateral adhesion can be measured using the Centrifugal Adhesion Balance (CAB), which uses a combination of centrifugal and gravitational forces to decouple the normal and lateral forces in the problem.

Fatigue (Material)

In materials science, fatigue is the weakening of a material caused by repeatedly applied loads. It is the progressive and localised structural damage that occurs when a material is subjected to cyclic loading. The nominal maximum stress values that cause such damage may be much less than the strength of the material typically quoted as the ultimate tensile stress limit, or the yield stress limit.

Fatigue occurs when a material is subjected to repeated loading and unloading. If the loads are above a certain threshold, microscopic cracks will begin to form at the stress concentrators such as the surface, persistent slip bands (PSBs), and grain interfaces. Eventually a crack will reach a critical size, the crack will propagate suddenly, and the structure will fracture. The shape of the structure will significantly affect the fatigue life; square holes or sharp corners will lead to elevated local stresses where fatigue cracks can initiate. Round holes and smooth transitions or fillets will therefore increase the fatigue strength of the structure.

Fatigue Life

ASTM defines *fatigue life*, N_f, as the number of stress cycles of a specified character that a specimen sustains before failure of a specified nature occurs. For some materials, notably steel and titanium, there is a theoretical value for stress amplitude below which the material will not fail for any number of cycles, called a fatigue limit, endurance limit, or fatigue strength.

Engineers have used any of three methods to determine the fatigue life of a material: the stress-life method, the strain-life method, and the linear-elastic fracture mechanics method. One method to predict fatigue life of materials is the Uniform Material Law (UML). UML was developed for fatigue life prediction of aluminium and titanium alloys by the end of 20th century and extended to high-strength steels, and cast iron.

Characteristics of Fatigue

- In metal alloys, and for the simplifying case when there are no macroscopic or microscopic discontinuities, the process starts with dislocation movements at the microscopic level, which eventually form persistent slip bands that become the nucleus of short cracks.

- Macroscopic and microscopic discontinuities (at the crystalline grain scale) as well as component design features which cause stress concentrations (holes, keyways, sharp changes of load direction etc.) are common locations at which the fatigue process begins.

- Fatigue is a process that has a degree of randomness (stochastic), often showing considerable scatter even in seemingly identical sample in well controlled environments.

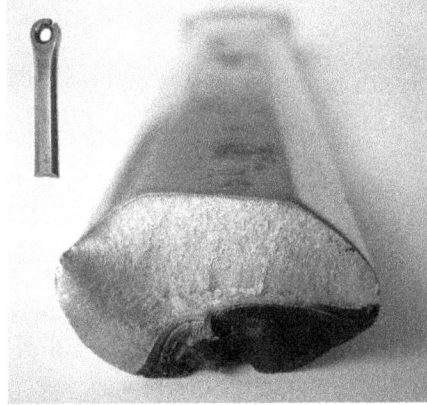

Fracture of an aluminium crank arm. Dark area of striations: slow crack growth. Bright granular area: sudden fracture.

- Fatigue is usually associated with tensile stresses but fatigue cracks have been reported due to compressive loads.

- The greater the applied stress range, the shorter the life.

- Fatigue life scatter tends to increase for longer fatigue lives.

- Damage is cumulative. Materials do not recover when rested.

- Fatigue life is influenced by a variety of factors, such as temperature, surface finish, metallurgical microstructure, presence of oxidizing or inert chemicals, residual stresses, scuffing contact (fretting), etc.

- Some materials (e.g., some steel and titanium alloys) exhibit a theoretical fatigue limit below which continued loading does not lead to fatigue failure.

- High cycle fatigue strength (about 10^4 to 10^8 cycles) can be described by stress-based parameters. A load-controlled servo-hydraulic test rig is commonly used in these tests, with frequencies of around 20–50 Hz. Other sorts of machines—like resonant magnetic machines—can also be used, to achieve frequencies up to 250 Hz.

- Low cycle fatigue (loading that typically causes failure in less than 10^4 cycles) is associated with localized plastic behavior in metals; thus, a strain-based parameter should be used for fatigue life prediction in metals. Testing is conducted with constant strain amplitudes typically at 0.01–5 Hz.

Timeline of Early Fatigue Research History

- 1837: Wilhelm Albert publishes the first article on fatigue. He devised a test machine for conveyor chains used in the Clausthal mines.

- 1839: Jean-Victor Poncelet describes metals as being 'tired' in his lectures at the military school at Metz.

- 1842: William John Macquorn Rankine recognises the importance of stress concentrations in his investigation of railroad axle failures. The Versailles train crash was caused by axle fatigue.

- 1843: Joseph Glynn reports on the fatigue of an axle on a locomotive tender. He identifies the keyway as the crack origin.

- 1848: The Railway Inspectorate reports one of the first tyre failures, probably from a rivet hole in tread of railway carriage wheel. It was likely a fatigue failure.

- 1849: Eaton Hodgkinson is granted a "small sum of money" to report to the UK Parliament on his work in "ascertaining by direct experiment, the effects of continued changes of load upon iron structures and to what extent they could be loaded without danger to their ultimate security".

- 1854: Braithwaite reports on common service fatigue failures and coins the term *fatigue*.

- 1860: Systematic fatigue testing undertaken by Sir William Fairbairn and August Wöhler.

- 1870: Wöhler summarises his work on railroad axles. He concludes that cyclic stress range is more important than peak stress and introduces the concept of *endurance limit*.

Micrographs showing how surface fatigue cracks grow as material is further cycled. From Ewing & Humfrey, 1903

- 1903: Sir James Alfred Ewing demonstrates the origin of fatigue failure in microscopic cracks.

- 1910: O. H. Basquin proposes a log-log relationship for S-N curves, using Wöhler's test data.

- 1945: A. M. Miner popularises Palmgren's (1924) linear damage hypothesis as a practical design tool.

- 1954: The world's first commercial jetliner, the de Havilland Comet, suffers disaster as three planes break up in mid-air, causing de Havilland and all other manufacturers to redesign high altitude aircraft and in particular replace square apertures like windows with oval ones.

- 1954: L. F. Coffin and S. S. Manson explain fatigue crack-growth in terms of plastic strain in the tip of cracks.

- 1961: P. C. Paris proposes methods for predicting the rate of growth of individual fatigue cracks in the face of initial scepticism and popular defence of Miner's phenomenological approach.

- 1968: Tatsuo Endo and M. Matsuishi devise the rainflow-counting algorithm and enable the reliable application of Miner's rule to random loadings.

- 1970: W. Elber elucidates the mechanisms and importance of crack closure in slowing the growth of a fatigue crack due to the wedging effect of plastic deformation left behind the tip of the crack.

High-cycle Fatigue

Historically, most attention has focused on situations that require more than 10^4 cycles to failure where stress is low and deformation is primarily elastic.

Stress-cycle (S-N) Curve

In high-cycle fatigue situations, materials performance is commonly characterized by an *S-N curve*, also known as a *Wöhler curve* . This is a graph of the magnitude of a cyclic stress (*S*) against the logarithmic scale of cycles to failure (*N*).

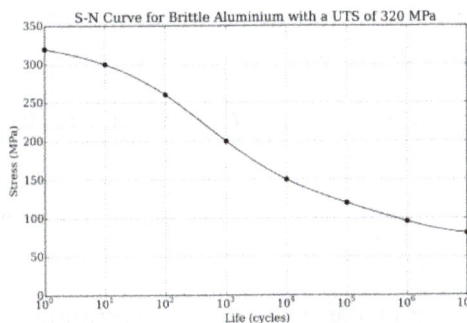

S-N curve for a brittle aluminium with an ultimate tensile strength of 320 MPa

S-N curves are derived from tests on samples of the material to be characterized (often called *coupons*) where a regular sinusoidal stress is applied by a testing machine which also counts the number of cycles to failure. This process is sometimes known as *coupon testing*. Each coupon test generates a point on the plot though in some cases there is a *runout* where the time to failure exceeds that available for the test. Analysis of fatigue data requires techniques from statistics, especially survival analysis and linear regression.

The progression of the *S-N curve* can be influenced by many factors such as corrosion, temperature, residual stresses, and the presence of notches. The Goodman-Line is a method used to estimate the influence of the mean stress on the fatigue strength.

Probabilistic Nature of Fatigue

As coupons sampled from a homogeneous frame will display a variation in their number of cycles to failure, the S-N curve should more properly be a Stress-Cycle-Probability (S-N-P) curve to capture the probability of failure after a given number of cycles of a certain stress. Probability distributions that are common in data analysis and in design against fatigue include the log-normal distribution, extreme value distribution, Birnbaum–Saunders distribution, and Weibull distribution.

Complex Loadings

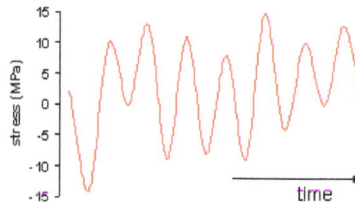

Spectrum loading

In practice, a mechanical part is exposed to a complex, often random, sequence of loads, large and small. In order to assess the safe life of such a part:

1. Complex loading is reduced to a series of simple cyclic loadings using a technique such as rainflow analysis;

2. A histogram of cyclic stress is created from the rainflow analysis to form a fatigue damage spectrum;

3. For each stress level, the degree of cumulative damage is calculated from the S-N curve; and

4. The effect of the individual contributions are combined using an algorithm such as *Miner's rule*.

For Multiaxial Loading

Since S-N curves are typically generated for uniaxial loading, some equivalence rule is needed whenever the loading is multiaxial. For simple, proportional loading histories (lateral load in a constant ratio with the axial), Sines rule may be applied. For more complex situations, such as nonproportional loading, Critical plane analysis must be applied.

Miner's Rule

In 1945, M A Miner popularised a rule that had first been proposed by A. Palmgren in 1924. The rule, variously called *Miner's rule* or the *Palmgren-Miner linear damage hypothesis*, states that where there are k different stress magnitudes in a spectrum, S_i ($1 \le i \le k$), each contributing $n_i(S_i)$ cycles, then if $N_i(S_i)$ is the number of cycles to failure of a constant stress reversal S_i (determined by uni-axial fatigue tests), failure occurs when:

$$\sum_{i=1}^{k} \frac{n_i}{N_i} = C$$

C is experimentally found to be between 0.7 and 2.2. Usually for design purposes, C is assumed to be 1. This can be thought of as assessing what proportion of life is consumed by a linear combination of stress reversals at varying magnitudes.

Though Miner's rule is a useful approximation in many circumstances, it has several major limitations:

1. It fails to recognise the probabilistic nature of fatigue and there is no simple way to relate life predicted by the rule with the characteristics of a probability distribution. Industry analysts often use design curves, adjusted to account for scatter, to calculate $N_i(S_i)$.

2. The sequence in which high vs. low stress cycles are applied to a sample in fact affect the fatigue life, for which Miner's Rule does not account. In some circumstances, cycles of low stress followed by high stress cause more damage than would be predicted by the rule. It does not consider the effect of an overload or high stress which may result in a compressive residual stress that may retard crack growth. High stress followed by low stress may have less damage due to the presence of compressive residual stress.

Paris' Law

In Fracture mechanics, Anderson, Gomez, and Paris derived relationships for the stage II crack growth with cycles N, in terms of the cyclical component ΔK of the Stress Intensity Factor K

$$\frac{da}{dN} = C(\Delta K)^m$$

where a is the crack length and m is typically in the range 3 to 5 (for metals), which states that the rate of crack growth with respect to the cycles of load applied is a function of the stress intensity factor; this is named Paris' law.

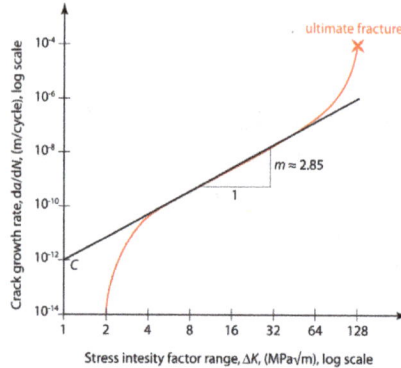

Typical fatigue crack growth rate graph

This relationship was later modified (by Forman, 1967) to make better allowance for the mean stress, by introducing a factor depending on (1-R) where R= min stress/max stress, in the denominator.

Goodman Relation

In the presence of a steady stress superimposed on the cyclic loading, the Goodman relation can be used to estimate a failure condition. It plots stress amplitude against mean stress with the fatigue limit and the ultimate tensile strength of the material as the two extremes. Alternative failure criteria include Soderberg and Gerber.

Low-cycle Fatigue

Where the stress is high enough for plastic deformation to occur, the accounting of the loading in terms of stress is less useful and the strain in the material offers a simpler and more accurate description. This type of fatigue is normally experienced by components which undergo a relatively small number of straining cycles. Low-cycle fatigue is usually characterised by the *Coffin-Manson relation* (published independently by L. F. Coffin in 1954 and S. S. Manson in 1953):

$$\frac{\Delta \varepsilon_p}{2} = \varepsilon_{f'}(2N)^c$$

where,

- $\Delta \varepsilon_p /2$ is the plastic strain amplitude;

- ε_f' is an empirical constant known as the *fatigue ductility coefficient*, the failure strain for a single reversal;

- $2N$ is the number of reversals to failure (N cycles);

- c is an empirical constant known as the *fatigue ductility exponent*, commonly ranging from -0.5 to -0.7 for metals in time independent fatigue. Slopes can be considerably steeper in the presence of creep or environmental interactions.

A similar relationship for materials such as Zirconium is used in the nuclear industry.

Fatigue and Fracture Mechanics

The account above is purely empirical and, though it allows life prediction and design assurance, life improvement or design optimisation can be enhanced using Fracture mechanics. Fatigue of materials can be described as having four stages.

1. Crack nucleation,

2. Stage I crack-growth,

3. Stage II crack-growth, and

4. Ultimate ductile failure.

Design Against Fatigue

Dependable design against fatigue-failure requires thorough education and supervised experience in structural engineering, mechanical engineering, or materials science. There are four principal approaches to life assurance for mechanical parts that display increasing degrees of sophistication:

1. Design to keep stress below threshold of fatigue limit (infinite lifetime concept);

2. Fail-safe, graceful degradation, and fault-tolerant design: Instruct the user to replace parts when they fail. Design in such a way that there is no single point of failure, and so that when any one part completely fails, it does not lead to catastrophic failure of the entire system.

3. Safe-life design: Design (conservatively) for a fixed life after which the user is instructed to replace the part with a new one (a so-called *lifed* part, finite lifetime concept, or "safe-life" design practice); planned obsolescence and disposable product are variants that design for a fixed life after which the user is instructed to replace the entire device;

4. Damage tolerant design: Instruct the user to inspect the part periodically for cracks and to replace the part once a crack exceeds a critical length. This approach usually uses the technologies of nondestructive testing and requires an

accurate prediction of the rate of crack-growth between inspections. The designer sets some aircraft maintenance checks schedule frequent enough that parts are replaced while the crack is still in the "slow growth" phase. This is often referred to as damage tolerant design or "retirement-for-cause".

Stopping Fatigue

Fatigue cracks that have begun to propagate can sometimes be stopped by drilling holes, called *drill stops*, in the path of the fatigue crack. This is not recommended as a general practice because the hole represents a stress concentration factor which depends on the size of the hole and geometry, though the hole is typically less of a stress concentration than the removed tip of the crack. The possibility remains of a new crack starting in the side of the hole. It is always far better to replace the cracked part entirely.

Material Change

Changes in the materials used in parts can also improve fatigue life. For example, parts can be made from better fatigue rated metals. Complete replacement and redesign of parts can also reduce if not eliminate fatigue problems. Thus helicopter rotor blades and propellers in metal are being replaced by composite equivalents. They are not only lighter, but also much more resistant to fatigue. They are more expensive, but the extra cost is amply repaid by their greater integrity, since loss of a rotor blade usually leads to total loss of the aircraft. A similar argument has been made for replacement of metal fuselages, wings and tails of aircraft.

Peening Treatment of Welds and Metal Components

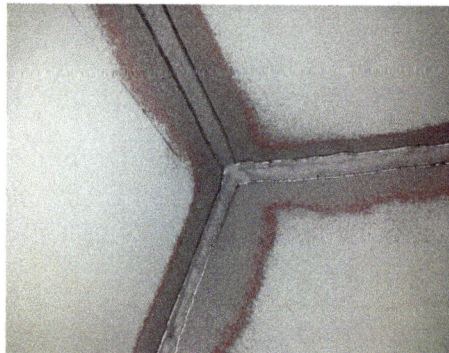

Example of a HFMI treated steel highway bridge to avoid fatigue along the weld transition.

Increases in fatigue life and strength are proportionally related to the depth of the compressive residual stresses imparted by surface enhancement processes such as shot peening but particularly by laser peening. Shot peening imparts compressive residual stresses approximately 0.005 inches deep, laser peening imparts compressive residual stresses from 0.040 to 0.100 inches deep, or deeper. Laser peening provide significant fatigue life extension through shock wave mechanics which plastically deform the surface of the metal component changing the material properties. Laser peening can be applied to existing parts without redesign requirements or incorporated into new designs to allow for lighter materials or thinner designs to achieve comparable engineering results.

High Frequency Mechanical Impact (HFMI) Treatment of Welds

The durability and life of dynamically loaded, welded steel structures are determined often by the welds, particular by the weld transitions. By selective treatment of weld transitions with the High Frequency Mechanical Impact (HFMI) treatment method, the durability of many designs can be increased significantly. This method is universally applicable, requires only specific equipment and offers high reproducibility and a high degree of quality control.

Deep Cryogenic Treatment

The use of Deep Cryogenic treatment has been shown to increase resistance to fatigue failure. Springs used in industry, auto racing and firearms have been shown to last up to six times longer when treated. Heat checking, which is a form of thermal cyclic fatigue has been greatly delayed.

Notable Fatigue Failures

Versailles Train Crash

Versailles train disaster

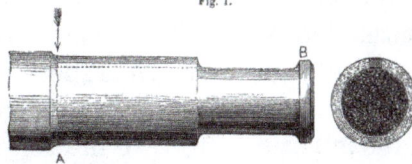

Drawing of a fatigue failure in an axle by Joseph Glynn, 1843

Following the King's fête celebrations at the Palace of Versailles, a train returning to Paris crashed in May 1842 at Meudon after the leading locomotive broke an axle. The carriages behind piled into the wrecked engines and caught fire. At least 55 passengers were killed trapped in the carriages, including the explorer Jules Dumont d'Urville.

This accident is known in France as the "Catastrophe ferroviaire de Meudon". The accident was witnessed by the British locomotive engineer Joseph Locke and widely reported in Britain. It was discussed extensively by engineers, who sought an explanation.

The derailment had been the result of a broken locomotive axle. Rankine's investigation of broken axles in Britain highlighted the importance of stress concentration, and the mechanism of crack growth with repeated loading. His and other papers suggesting a crack growth mechanism through repeated stressing, however, were ignored, and fatigue failures occurred at an ever increasing rate on the expanding railway system. Other spurious theories seemed to be more acceptable, such as the idea that the metal had somehow "crystallized". The notion was based on the crystalline appearance of the fast fracture region of the crack surface, but ignored the fact that the metal was already highly crystalline.

de Havilland Comet

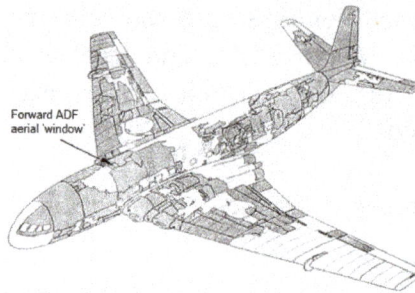

The recovered (shaded) parts of the wreckage of *G-ALYP* and the site (arrowed) of the failure

Two de Havilland Comet passenger jets broke up in mid-air and crashed within a few months of each other in 1954. As a result, systematic tests were conducted on a fuselage immersed and pressurised in a water tank. After the equivalent of 3,000 flights, investigators at the Royal Aircraft Establishment (RAE) were able to conclude that the crash had been due to failure of the pressure cabin at the forward Automatic Direction Finder window in the roof. This 'window' was in fact one of two apertures for the aerials of an electronic navigation system in which opaque fibreglass panels took the place of the window 'glass'. The failure was a result of metal fatigue caused by the repeated pressurisation and de-pressurisation of the aircraft cabin. Also, the supports around the windows were riveted, not bonded, as the original specifications for the aircraft had called for. The problem was exacerbated by the punch rivet construction technique employed. Unlike drill riveting, the imperfect nature of the hole created by punch riveting caused manufacturing defect cracks which may have caused the start of fatigue cracks around the rivet.

The Comet's pressure cabin had been designed to a safety factor comfortably in excess of that required by British Civil Airworthiness Requirements (2.5 times the cabin proof test pressure as opposed to the requirement of 1.33 times and an ultimate load of 2.0

times the cabin pressure) and the accident caused a revision in the estimates of the safe loading strength requirements of airliner pressure cabins.

The fuselage roof fragment of *G-ALYP* on display in the Science Museum in London, showing the two ADF windows at-which the initial failure occurred.

In addition, it was discovered that the stresses around pressure cabin apertures were considerably higher than had been anticipated, especially around sharp-cornered cut-outs, such as windows. As a result, all future jet airliners would feature windows with rounded corners, greatly reducing the stress concentration. This was a noticeable distinguishing feature of all later models of the Comet. Investigators from the RAE told a public inquiry that the sharp corners near the Comets' window openings acted as initiation sites for cracks. The skin of the aircraft was also too thin, and cracks from manufacturing stresses were present at the corners.

Alexander L. Kielland Oil Platform Capsizing

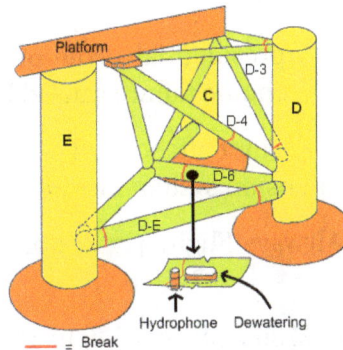

Fractures on the right side of the Alexander L. Kielland rig

The *Alexander L. Kielland* was a Norwegian semi-submersible drilling rig that capsized whilst working in the Ekofisk oil field in March 1980 killing 123 people. The capsizing was the worst disaster in Norwegian waters since World War II. The rig, located approximately 320 km east of Dundee, Scotland, was owned by the Stavanger Drilling Company of Norway and was on hire to the United States company Phillips Petroleum at the time of the disaster. In driving rain and mist, early in the evening of 27 March 1980 more than 200 men were off duty in the accommodation on the *Alexander L. Kielland*. The wind was gusting to 40 knots with waves up to 12 m high. The rig had just been winched away from the *Edda* production platform. Minutes before 18:30 those on board felt a 'sharp crack' followed by 'some kind of trembling'. Suddenly the rig heeled over 30° and then stabilised. Five of the six anchor cables had broken, with one remaining cable preventing the rig from capsizing. The list continued to increase and at 18.53 the remaining anchor cable snapped and the rig turned upside down.

A year later in March 1981, the investigative report concluded that the rig collapsed owing to a fatigue crack in one of its six bracings (bracing D-6), which connected the collapsed D-leg to the rest of the rig. This was traced to a small 6 mm fillet weld which joined a non-load-bearing flange plate to this D-6 bracing. This flange plate held a sonar device used during drilling operations. The poor profile of the fillet weld contributed to a reduction in its fatigue strength. Further, the investigation found considerable amounts of lamellar tearing in the flange plate and cold cracks in the butt weld. Cold cracks in the welds, increased stress concentrations due to the weakened flange plate, the poor weld profile, and cyclical stresses (which would be common in the North Sea), seemed to collectively play a role in the rig's collapse.

Others

- The 1862 Hartley Colliery Disaster was caused by the fracture of a steam engine beam and killed 220 people.

- The 1919 Great Molasses Flood has been attributed to a fatigue failure.

- The 1948 Northwest Airlines Flight 421 crash due to fatigue failure in a wing spar root

- The 1957 "Mt. Pinatubo", presidential plane of Philippine President Ramon Magsaysay, crashed due to engine failure caused by metal fatigue.

- The 1965 capsize of the UK's first offshore oil platform, the Sea Gem, was due to fatigue in part of the suspension system linking the hull to the legs.

- The 1968 Los Angeles Airways Flight 417 lost one of its main rotor blades due to fatigue failure.

- The 1968 MacRobertson Miller Airlines Flight 1750 that lost a wing due to improper maintenance leading to fatigue failure

- The 1977 Dan-Air Boeing 707 crash caused by fatigue failure resulting in the loss of the right horizontal stabilizer

- The 1980 LOT Flight 7 that crashed due to fatigue in an engine turbine shaft resulting in engine disintegration leading to loss of control

- The 1985 Japan Airlines Flight 123 crashed after the aircraft lost its vertical stabilizer due to faulty repairs on the rear bulkhead.

- The 1988 Aloha Airlines Flight 243 suffered an explosive decompression due to fatigue failure.

- The 1989 United Airlines Flight 232 lost its tail engine due to fatigue failure in a fan disk hub.

- The 1992 El Al Flight 1862 lost both engines on its right-wing due to fatigue failure in the pylon mounting of the #3 Engine.

- The 1998 Eschede train disaster was caused by fatigue failure of a single composite wheel.

- The 2000 Hatfield rail crash was likely caused by rolling contact fatigue.

- The 2000 recall of 6.5 million Firestone tires on Ford Explorers originated from fatigue crack growth leading to separation of the tread from the tire.

- The 2002 China Airlines Flight 611 had disintegrated in-flight due to fatigue failure.

- The 2005 Chalk's Ocean Airways Flight 101 lost its right wing due to fatigue failure brought about by inadequate maintenance practices.

- The 2009 Viareggio train derailment due to fatigue failure.

Tribocorrosion

Tribocorrosion is a material degradation process due to the combined effect of corrosion and wear. The name tribocorrosion expresses the underlying disciplines of tribology and corrosion. Tribology is concerned with the study of friction, lubrication and wear and corrosion is concerned with the chemical and electrochemical interactions between a material, normally a metal, and its environment. As a field of research tribocorrosion is relatively new, but tribocorrosion phenomena have been around ever since machines and installations are being used.

Wear is a mechanical material degradation process occurring on rubbing or impacting surfaces, while corrosion involves chemical or electrochemical reactions of the material. Corrosion may accelerate wear and wear may accelerate corrosion. One then speaks of corrosion accelerated wear or wear accelerated corrosion. Both these phenomena, as well as fretting corrosion (which results from small amplitude oscillations between contacting surfaces) fall into the broader category of tribocorrosion. Erosion-corrosion is another tribocorrosion phenomenon involving mechanical and chemical effects: impacting particles or fluids erode a solid surface by abrasion, chipping or fatigue while simultaneously the surface corrodes.

Phenomena in Different Engineering Fields

Tribocorrosion occurs in many engineering fields. It reduces the life-time of pipes, valves and pumps, of waste incinerators, of mining equipment or of medical implants, and it can affect the safety of nuclear reactors or of transport systems. On the other hand, tribocorrosion phenomena can also be applied to good use, for example in the chemical-mechanical planarization of wafers in the electronics industry or in metal grinding and cutting in presence of aqueous emulsions. Keeping this in mind, we may define tribocorrosion in a more general way independently of the notion of usefulness or damage or of the particular type of mechanical interaction: Tribocorrosion concerns the irreversible transformation of materials or of their function as a result of simultaneous mechanical and chemical/electrochemical interactions between surfaces in relative motion.

Biotribocorrosion

Biotribocorrosion covers the science of surface transformations resulting from the interactions of mechanical loading and chemical/electrochemical reactions that occur between elements of a tribological system exposed to biological environments. It has been studied for aritificial joint prostheses. It is important to understand material degradation processes for joint implants to achieve longer service life and better safety issues for such devices.

Passive Metals

While tribocorrosion phenomena may affect many materials, they are most critical for metals, especially the normally corrosion resistant so-called passive metals. The vast majority of corrosion resistant metals and alloys used in engineering (stainless steels, titanium, aluminium etc.) fall into this category. These metals are thermodynamically unstable in the presence of oxygen or water and they derive their corrosion resistance from the presence at the surface of a thin oxide film, called the passive film, which acts as a protective barrier between the metal and its environment. Passive films are usually just a few atomic layers thick. Nevertheless, they can provide excellent corrosion protection because if damaged accidentally they spontaneously self-heal by metal oxidation. However, when a metal surface is subjected to severe rubbing or to a stream of impacting particles the passive film damage becomes continuous and extensive. The self-healing process may no longer be effective and in addition it requires a high rate of metal oxidation. In other words, the underlying metal will strongly corrode before the protective passive film is reformed, if at all. In such a case, the total material loss due to tribocorrosion will be much higher than the sum of wear and corrosion one would measure in experiments with the same metal where only wear or only corrosion takes place. The example illustrates the fact that the rate of tribocorrosion is not simply the addition of the rate of wear and the rate of corrosion but it is strongly affected by synergistic and antagonistic effects between mechanical and chemical mechanisms. To study such effects in the laboratory, one most often uses mechanical wear testing rigs which are equipped with an electrochemical cell. This permits one to control independently the mechanical and chemical parameters. For example, by imposing a given potential to the rubbing metal one can simulate the oxidation potential of the environment and in addition, under certain conditions, the current flow is a measure of the instantaneous corrosion rate. Volume loss due to electrochemical dissolution can be measured by Faraday's laws of electrolysis and subtracted from total volume loss in tribocorrosion so the sum of mechanical wear loss and the synergies can be calculated. For a deeper understanding tribocorrosion experiments are supplemented by detailed microscopic and analytical studies of the contacting surfaces.

At high temperatures, the more rapid generation of oxide due to a combination of temperature and tribological action during sliding wear can generate potentially wear resistant oxide layers known as 'glazes'. Under such circumstances, tribocorrosion can be used potentially in a beneficial way.

Fretting

Fretting refers to wear and sometimes corrosion damage at the asperities of contact surfaces. This damage is induced under load and in the presence of repeated relative surface motion, as induced for example by vibration. The ASM Handbook on Fatigue and Fracture defines fretting as: "*A special wear process that occurs at the contact area between two materials under load and subject to minute relative motion by vibration or some other force.*" Fretting tangibly downgrades the surface layer quality producing increased surface roughness and micropits; which reduces the fatigue strength of the components.

The amplitude of the relative sliding motion is often in the order from micrometers to millimeters, but can be as low as 3 to 4 nanometers.

The contact movement causes mechanical wear and material transfer at the surface, often followed by oxidation of both the metallic debris and the freshly exposed metallic surfaces. Because the oxidized debris is usually much harder than the surfaces from which it came, it often acts as an abrasive agent that increases the rate of both fretting and a mechanical wear called false brinelling.

Steel

Fretting damage in steel can be identified by the presence of a pitted surface and fine 'red' iron oxide dust resembling cocoa powder. Strictly this debris is not 'rust' as its production requires no water. The particles are much harder than the steel surfaces in contact, so abrasive wear is inevitable; however, particulates are not required to initiate fret.

Products Affected

Fretting examples include wear of drive splines on driveshafts, wheels at the lug bolt interface, and cylinder head gaskets subject to differentials in thermal expansion coefficients.

There is currently a focus on fretting research in the aerospace industry. The dovetail blade-root connection and the spline coupling of gas turbine aero engines experience fretting.

Fretting Fatigue

Fretting decreases fatigue strength of materials operating under cycling stress. This can result in *fretting fatigue*, whereby fatigue cracks can initiate in the fretting zone. Afterwards, the crack propagates into the material. Lap joints, common on airframe surfaces, are a prime location for fretting corrosion. This is also known as frettage or fretting corrosion.

Mitigation

The fundamental way to prevent fretting is to design for no relative motion of the surfaces at the contact. Surface roughness plays an important role as fretting normally occurs by the contact of the asperities of the mating surfaces. Lubricants are often employed to mitigate fretting because they reduce friction and inhibit oxidation.

Soft materials often exhibit higher susceptibility to fretting than hard materials of a similar type. The hardness ratio of the two sliding materials also has an effect on fretting wear. However, softer materials such as polymers can show the opposite effect when they capture hard debris which becomes embedded in their bearing surfaces. They then act as a very effective abrasive agent, wearing down the harder metal with which they are in contact.

References

- Chattopadhyay, R. (2001). Surface Wear - Analysis, Treatment, and Prevention. OH, USA: ASM International. ISBN 0-87170-702-0.

- Chattopadhyay, R. (2004). Advanced Thermally Assisted Surface Engineering Processes. MA, USA: Kluwer Academic Publishers. ISBN 1-4020-7696-7.

- Stephens, Ralph I.; Fuchs, Henry O. (2001). Metal Fatigue in Engineering (Second ed.). John Wiley & Sons, Inc. p. 69. ISBN 0-471-51059-9.

- Joseph E. Shigley; Charles R. Mischke; Richard G. Budynas. Mechanical Engineering Design (7th ed.). McGraw Hill Higher Education. ISBN 9780072520361.

- The Alexander L. Kielland accident, Report of a Norwegian public commission appointed by royal decree of March 28, 1980, presented to the Ministry of Justice and Police March, 1981 ISBN B0000ED27N

- CAR, Duarte; FJ, de Souza; VF, dos Santos (January 2016). "Mitigating elbow erosion with a vortex chamber". Powder Technology. 288: 6–25.

- ANSBERRY, CLARE (Feb 5, 2001). "In Firestone Tire Study, Expert Finds Vehicle Weight Was Key in Failure". Wall Street Journal. Retrieved 6 September 2016.

- Surface Hardening of Stainless Steels by Kolsterising by Gümpel P. -- University of Applied Science, Konstanz Germany AIJSTPME (2012) 5(1): 11-18 (PDF)

- "Terminology for biorelated polymers and applications (IUPAC Recommendations 2012)" (PDF). Pure and Applied Chemistry. 84 (2): 377–410. 2012. doi:10.1351/PAC-REC-10-12-04.

- Can Yıldırım, Halid; Marquis, Gary. "Fatigue strength improvement factors for high strength steel welded joints treated by high frequency mechanical impact". International Journal of Fatigue. 44: 168–176. doi:10.1016/j.ijfatigue.2012.05.002.

- Korkmaz, S. (2011). "A Methodology to Predict Fatigue Life of Cast Iron: Uniform Material Law for Cast Iron". Journal of Iron and Steel Research, International. 18: 8. doi:10.1016/S1006-706X(11)60102-7.

6

Friction and Surface Tension Reducing Agents

The friction and surface tension reducing agents that have been explained in the chapter are bearing, ball bearing, composite bearing, wetting, surface finishing, false brinelling and gear. Bearing is a mechanical device that is used to restrict relative motion to particular desired motion. It also helps in reducing friction between parts that are in motion. The chapter strategically incorporates the main components and key concepts of friction and surface tension reducing agents, providing a complete understanding.

Bearing (Mechanical)

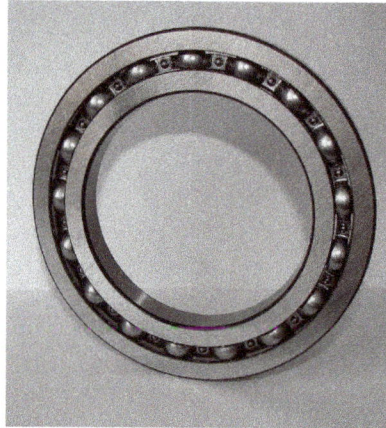

Ball bearing

A bearing is a machine element that constrains relative motion to only the desired motion, and reduces friction between moving parts. The design of the bearing may, for example, provide for free linear movement of the moving part or for free rotation around a fixed axis; or, it may *prevent* a motion by controlling the vectors of normal forces that bear on the moving parts. Most bearings facilitate the desired motion by minimizing friction. Bearings are classified broadly according to the type of operation, the motions allowed, or to the directions of the loads (forces) applied to the parts.

Rotary bearings hold rotating components such as shafts or axles within mechanical systems, and transfer axial and radial loads from the source of the load to the structure supporting it. The simplest form of bearing, the *plain bearing*, consists of a shaft rotating in a hole. Lubrication is often used to reduce friction. In the *ball bearing* and *roller bearing*, to prevent sliding friction, rolling elements such as rollers or balls with a circular cross-section are located between the races or journals of the bearing assembly. A wide variety of bearing designs exists to allow the demands of the application to be correctly met for maximum efficiency, reliability, durability and performance.

The term "bearing" is derived from the verb "to bear"; a bearing being a machine element that allows one part to bear (i.e., to support) another. The simplest bearings are bearing surfaces, cut or formed into a part, with varying degrees of control over the form, size, roughness and location of the surface. Other bearings are separate devices installed into a machine or machine part. The most sophisticated bearings for the most demanding applications are very precise devices; their manufacture requires some of the highest standards of current technology.

History

Tapered roller bearing

Drawing of Leonardo da Vinci (*1452-1519*) Study of a ball bearing

The invention of the rolling bearing, in the form of wooden rollers supporting, or bearing, an object being moved is of great antiquity, and may predate the invention of the wheel.

Though it is often claimed that the Egyptians used roller bearings in the form of tree trunks under sleds, this is modern speculation. They are depicted in their own drawings in the tomb of Djehutihotep as moving massive stone blocks on sledges with liquid-lubricated runners which would constitute a plain bearing. There are also Egyptian drawings of bearings used with hand drills.

The earliest recovered example of a rolling element bearing is a wooden ball bearing supporting a rotating table from the remains of the Roman Nemi ships in Lake Nemi, Italy. The wrecks were dated to 40 BC.

Leonardo da Vinci incorporated drawings of ball bearings in his design for a helicopter around the year 1500. This is the first recorded use of bearings in an aerospace design. However, Agostino Ramelli is the first to have published sketches of roller and thrust bearings. An issue with ball and roller bearings is that the balls or rollers rub against each other causing additional friction which can be reduced by enclosing the balls or rollers within a cage. The captured, or caged, ball bearing was originally described by Galileo in the 17th century.

The first practical caged-roller bearing was invented in the mid-1740s by horologist John Harrison for his H3 marine timekeeper. This uses the bearing for a very limited oscillating motion but Harrison also used a similar bearing in a truly rotary application in a contemporaneous regulator clock.

Industrial Era

The first modern recorded patent on ball bearings was awarded to Philip Vaughan, a British inventor and ironmaster who created the first design for a ball bearing in Carmarthen in 1794. His was the first modern ball-bearing design, with the ball running along a groove in the axle assembly.

Bearings played a pivotal role in the nascent Industrial Revolution, allowing the new industrial machinery to operate efficiently. For example, they saw use for holding wheel and axle to greatly reduce friction over that of dragging an object by making the friction act over a shorter distance as the wheel turned.

The first plain and rolling-element bearings were wood closely followed by bronze. Over their history bearings have been made of many materials including ceramic, sapphire, glass, steel, bronze, other metals and plastic (e.g., nylon, polyoxymethylene, polytetrafluoroethylene, and UHMWPE) which are all used today.

Watch makers produce "jeweled" watches using sapphire plain bearings to reduce friction thus allowing more precise time keeping.

Even basic materials can have good durability. As examples, wooden bearings can still be seen today in old clocks or in water mills where the water provides cooling and lubrication.

Early Timken tapered roller bearing with notched rollers

The first patent for a radial style ball bearing was awarded to Jules Suriray, a Parisian bicycle mechanic, on 3 August 1869. The bearings were then fitted to the winning bicycle ridden by James Moore in the world's first bicycle road race, Paris-Rouen, in November 1869.

In 1883, Friedrich Fischer, founder of FAG, developed an approach for milling and grinding balls of equal size and exact roundness by means of a suitable production machine and formed the foundation for creation of an independent bearing industry.

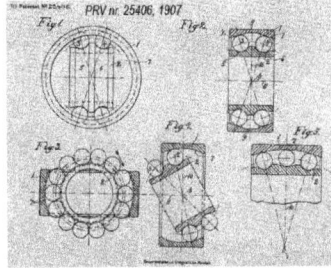

Wingquist original patent of self-aligning ball bearing

The modern, self-aligning design of ball bearing is attributed to Sven Wingquist of the SKF ball-bearing manufacturer in 1907, when he was awarded Swedish patent No. 25406 on its design.

Henry Timken, a 19th-century visionary and innovator in carriage manufacturing, patented the tapered roller bearing in 1898. The following year he formed a company to produce his innovation. Over a century the company grew to make bearings of all types, including specialty steel and an array of related products and services.

Erich Franke invented and patented the wire race bearing in 1934. His focus was on a bearing design with a cross section as small as possible and which could be integrated into the enclosing design. After World War II he founded together with Gerhard Heydrich the company Franke & Heydrich KG (today Franke GmbH) to push the development and production of wire race bearings.

Richard Stribeck's extensive research on ball bearing steels identified the metallurgy of the commonly used 100Cr6 (AISI 52100) showing coefficient of friction as a function of pressure.

Designed in 1968 and later patented in 1972, Bishop-Wisecarver's co-founder Bud Wisecarver created vee groove bearing guide wheels, a type of linear motion bearing consisting of both an external and internal 90-degree vee angle.

In the early 1980s, Pacific Bearing's founder, Robert Schroeder, invented the first bi-material plain bearing which was size interchangeable with linear ball bearings. This bearing had a metal shell (aluminum, steel or stainless steel) and a layer of Teflon-based material connected by a thin adhesive layer.

Today ball and roller bearings are used in many applications which include a rotating component. Examples include ultra high speed bearings in dental drills, aerospace bearings in the Mars Rover, gearbox and wheel bearings on automobiles, flexure bearings in optical alignment systems, bicycle wheel hubs, and air bearings used in Coordinate-measuring machines.

Common

By far, the most common bearing is the plain bearing, a bearing which uses surfaces in rubbing contact, often with a lubricant such as oil or graphite. A plain bearing may or may not be a discrete device. It may be nothing more than the bearing surface of a hole with a shaft passing through it, or of a planar surface that bears another (in these cases, not a discrete device); or it may be a layer of bearing metal either fused to the substrate (semi-discrete) or in the form of a separable sleeve (discrete). With suitable lubrication, plain bearings often give entirely acceptable accuracy, life, and friction at minimal cost. Therefore, they are very widely used.

However, there are many applications where a more suitable bearing can improve efficiency, accuracy, service intervals, reliability, speed of operation, size, weight, and costs of purchasing and operating machinery.

Thus, there are many types of bearings, with varying shape, material, lubrication, principle of operation, and so on.

Types

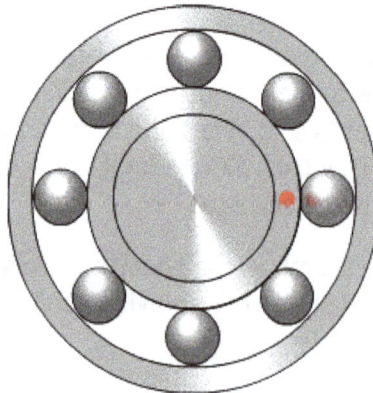

Image of ball bearing (without a cage). The inner ring rotates and the outer ring is stationary.

There are at least 6 common types of bearing, each of which operates on different principles:

- Plain bearing, consisting of a shaft rotating in a hole. There are several specific styles: bushing, journal bearing, sleeve bearing, rifle bearing, composite bearing.

- Rolling-element bearing, in which rolling elements placed between the turning and stationary races prevent sliding friction. There are two main types

- o   Ball bearing, in which the rolling elements are spherical balls

- o   Roller bearing, in which the rolling elements are cylindrical rollers

- Jewel bearing, a plain bearing in which one of the bearing surfaces is made of an ultrahard glassy jewel material such as sapphire to reduce friction and wear

- Fluid bearing, a noncontact bearing in which the load is supported by a gas or liquid,

- Magnetic bearing, in which the load is supported by a magnetic field

- Flexure bearing, in which the motion is supported by a load element which bends.

Motions

Common motions permitted by bearings are:

- axial rotation e.g. shaft rotation

- linear motion e.g. drawer

- spherical rotation e.g. ball and socket joint

- hinge motion e.g. door, elbow, knee

Friction

Reducing friction in bearings is often important for efficiency, to reduce wear and to facilitate extended use at high speeds and to avoid overheating and premature failure of the bearing. Essentially, a bearing can reduce friction by virtue of its shape, by its material, or by introducing and containing a fluid between surfaces or by separating the surfaces with an electromagnetic field.

- By shape, gains advantage usually by using spheres or rollers, or by forming flexure bearings.

- By material, exploits the nature of the bearing material used. (An example would be using plastics that have low surface friction.)

- By fluid, exploits the low viscosity of a layer of fluid, such as a lubricant or as a pressurized medium to keep the two solid parts from touching, or by reducing the normal force between them.

- By fields, exploits electromagnetic fields, such as magnetic fields, to keep solid parts from touching.

- Air pressure exploits air pressure to keep solid parts from touching.

Combinations of these can even be employed within the same bearing. An example of this is where the cage is made of plastic, and it separates the rollers/balls, which reduce friction by their shape and finish.

Loads

Bearing design varies depending on the size and directions of the forces that they are required to support. Forces can be predominately radial, axial (thrust bearings), or bending moments perpendicular to the main axis.

Speeds

Different bearing types have different operating speed limits. Speed is typically specified as maximum relative surface speeds, often specified ft/s or m/s. Rotational bearings typically describe performance in terms of the product DN where D is the mean diameter (often in mm) of the bearing and N is the rotation rate in revolutions per minute.

Generally there is considerable speed range overlap between bearing types. Plain bearings typically handle only lower speeds, rolling element bearings are faster, followed by fluid bearings and finally magnetic bearings which are limited ultimately by centripetal force overcoming material strength.

Play

Some applications apply bearing loads from varying directions and accept only limited play or "slop" as the applied load changes. One source of motion is gaps or "play" in the bearing. For example, a 10 mm shaft in a 12 mm hole has 2 mm play.

Allowable play varies greatly depending on the use. As example, a wheelbarrow wheel supports radial and axial loads. Axial loads may be hundreds of newtons force left or right, and it is typically acceptable for the wheel to wobble by as much as 10 mm under the varying load. In contrast, a lathe may position a cutting tool to ±0.02 mm using a ball lead screw held by rotating bearings. The bearings support axial loads of thousands of newtons in either direction, and must hold the ball lead screw to ±0.002 mm across that range of loads

Stiffness

A second source of motion is elasticity in the bearing itself. For example, the balls in a ball bearing are like stiff rubber, and under load deform from round to a slightly flattened shape. The race is also elastic and develops a slight dent where the ball presses on it.

The stiffness of a bearing is how the distance between the parts which are separated by the bearing varies with applied load. With rolling element bearings this is due to the strain of the ball and race. With fluid bearings it is due to how the pressure of the fluid varies with the gap (when correctly loaded, fluid bearings are typically stiffer than rolling element bearings).

Service Life

Fluid and magnetic bearings

Fluid and magnetic bearings can have practically indefinite service lives. In practice, there are fluid bearings supporting high loads in hydroelectric plants that have been in nearly continuous service since about 1900 and which show no signs of wear.

Rolling element bearings

Rolling element bearing life is determined by load, temperature, maintenance, lubrication, material defects, contamination, handling, installation and other factors. These factors can all have a significant effect on bearing life. For example, the service life of bearings in one application was extended dramatically by changing how the bearings were stored before installation and use, as vibrations during storage caused lubricant failure even when the only load on the bearing was its own weight; the resulting damage is often false brinelling. Bearing life is statistical: several samples of a given bearing will often exhibit a bell curve of service life, with a few samples showing significantly better or worse life. Bearing life varies because microscopic structure and contamination vary greatly even where macroscopically they seem identical.

L10 Life

Bearings are often specified to give an "L10" life (outside the USA, it may be referred to as "B10" life.) This is the life at which ten percent of the bearings in that application can be expected to have failed due to classical fatigue failure (and not any other mode of failure like lubrication starvation, wrong mounting etc.), or, alternatively, the life at which ninety percent will still be operating. The L10 life of the bearing is theoretical life and may not represent service life of the bearing. Bearings are also rated using C_0 (static loading) value. This is the basic load rating as a reference, and not an actual load value.

Plain bearings

For plain bearings some materials give much longer life than others. Some of the John Harrison clocks still operate after hundreds of years because of the *lignum vitae* wood employed in their construction, whereas his metal clocks are seldom run due to potential wear.

Flexure bearings

Flexure bearings rely on elastic properties of material. Flexure bearings bend a piece of material repeatedly. Some materials fail after repeated bending, even at low loads, but careful material selection and bearing design can make flexure bearing life indefinite.

Short-life bearings

Although long bearing life is often desirable, it is sometimes not necessary. Tedric A. Harris describes a bearing for a rocket motor oxygen pump that gave several hours life, far in excess of the several tens of minutes life needed.

Composite Bearings

Depending on the customized specifications (backing material and PTFE compounds), composite bearings can operate up to 30 years without maintenance.

External Factors

The service life of the bearing is affected by many parameters that are not controlled by the bearing manufacturers. For example, bearing mounting, temperature, exposure to external environment, lubricant cleanliness and electrical currents through bearings etc. The disruption from PWM inverter which are generating high frequency motor-bearing currents can be suppressed by inductive absorbers like CoolBLUE cores, which need to be put over the three phases giving a high frequency impedance against the common mode or motorbearing currents.

The temperature and terrain of the micro-surface will determine the amount of friction by the touching of solid parts.

Certain elements and fields reduce friction, while increasing speeds.

Strength and mobility help determine the amount of load the bearing type can carry.

Alignment factors can play a damaging role in wear and tear, yet overcome by computer aid signaling and non-rubbing bearing types, such as magnetic levitation or air field pressure.

Maintenance and Lubrication

Many bearings require periodic maintenance to prevent premature failure, but many others require little maintenance. The latter include various kinds of fluid and magnetic bearings, as well as rolling-element bearings that are described with terms including *sealed bearing* and *sealed for life*. These contain seals to keep the dirt out and the grease in. They work successfully in many applications, providing maintenance-free operation. Some applications cannot use them effectively.

Nonsealed bearings often have a grease fitting, for periodic lubrication with a grease gun, or an oil cup for periodic filling with oil. Before the 1970s, sealed bearings were not encountered on most machinery, and oiling and greasing were a more common activity than they are today. For example, automotive chassis used to require "lube jobs" nearly as often as engine oil changes, but today's car chassis are mostly sealed for life.

From the late 1700s through mid 1900s, industry relied on many workers called oilers to lubricate machinery frequently with oil cans.

Factory machines today usually have *lube systems*, in which a central pump serves periodic charges of oil or grease from a reservoir through *lube lines* to the various *lube points* in the machine's bearing surfaces, bearing journals, pillow blocks, and so on. The timing and number of such *lube cycles* is controlled by the machine's computerized control, such as PLC or CNC, as well as by manual override functions when occasionally needed. This automated process is how all modern CNC machine tools and many other modern factory machines are lubricated. Similar lube systems are also used on non-automated machines, in which case there is a hand pump that a machine operator is supposed to pump once daily (for machines in constant use) or once weekly. These are called *one-shot systems* from their chief selling point: one pull on one handle to lube the whole machine, instead of a dozen pumps of an alemite gun or oil can in a dozen different positions around the machine.

The oiling system inside a modern automotive or truck engine is similar in concept to the lube systems mentioned above, except that oil is pumped continuously. Much of this oil flows through passages drilled or cast into the engine block and cylinder heads, escaping through ports directly onto bearings, and squirting elsewhere to provide an oil bath. The oil pump simply pumps constantly, and any excess pumped oil continuously escapes through a relief valve back into the sump.

Many bearings in high-cycle industrial operations need periodic lubrication and cleaning, and many require occasional adjustment, such as pre-load adjustment, to minimise the effects of wear.

Bearing life is often much better when the bearing is kept clean and well lubricated. However, many applications make good maintenance difficult. For example, bearings in the conveyor of a rock crusher are exposed continually to hard abrasive particles. Cleaning is of little use, because cleaning is expensive yet the bearing is contaminated again as soon as the conveyor resumes operation. Thus, a good maintenance program might lubricate the bearings frequently but not include any disassembly for cleaning. The frequent lubrication, by its nature, provides a limited kind of cleaning action, by displacing older (grit-filled) oil or grease with a fresh charge, which itself collects grit before being displaced by the next cycle.

Rolling-element Bearing Outer Race Fault Detection

Rolling-element bearings are widely used in the industries today, and hence maintenance of these bearings becomes an important task for the maintenance professionals. The rolling-element bearings wear out easily due to metal-to-metal contact, which creates faults in the outer race, inner race and ball. It is also the most vulnerable component of a machine because it is often under high load and high running speed conditions. Regular diagnostics of rolling-element bearing faults is critical for industrial safety and operations of the machines along with reducing the maintenance costs or avoiding shutdown time. Among the outer race, inner race and ball, the outer race tends to be more vulnerable to faults and defects.

There is still a room for discussion whether the rolling element excites the natural frequencies of bearing component when it passes the fault on the outer race. Hence we need to identify the bearing outer race natural frequency and its harmonics. The bearing faults create impulses and results in strong harmonics of the fault frequencies in the spectrum of vibration signals. These fault frequencies are sometimes masked by adjacent frequencies in the spectra due to their little energy. Hence, a very high spectral resolution is often needed to identify these frequencies during a FFT analysis. The natural frequencies of a rolling element bearing with the free boundary conditions are 3 kHz. Therefore, in order to use the bearing component resonance bandwidth method to detect the bearing fault at an initial stage a high frequency range accelerometer should be adopted, and data obtained from a long duration needs to be acquired. A fault characteristic frequency can only be identified when the fault extent is severe, such as that of a presence of a hole in the outer race. The harmonics of fault frequency is a more sensitive indicator of a bearing outer race fault. For a more serious detection of defected bearing faults waveform, spectrum and envelope techniques will help reveal these faults. However, if a high frequency demodulation is used in the envelope analysis in order to detect bearing fault characteristic frequencies, the maintenance professionals have to be more careful in the analysis because of resonance, as it may or may not contain fault frequency components.

Using spectral analysis as a tool to identify the faults in the bearings faces challenges due to issues like low energy, signal smearing, cyclostationarity etc. High resolution is often desired to differentiate the fault frequency components from the other high-amplitude adjacent frequencies. Hence, when the signal is sampled for FFT analysis, the sample length should be large enough to give adequate frequency resolution in the spectrum. Also, keeping the computation time and memory within limits and avoiding unwanted aliasing may be demanding. However, a minimal frequency resolution required can be obtained by estimating the bearing fault frequencies and other vibration frequency components and its harmonics due to shaft speed, misalignment, line frequency, gearbox etc.

Packing

Some bearings use a thick grease for lubrication, which is pushed into the gaps between the bearing surfaces, also known as *packing*. The grease is held in place by a plastic, leather, or rubber gasket (also called a *gland*) that covers the inside and outside edges of the bearing race to keep the grease from escaping.

Bearings may also be packed with other materials. Historically, the wheels on railroad cars used sleeve bearings packed with *waste* or loose scraps of cotton or wool fiber soaked in oil, then later used solid pads of cotton.

Ring Oiler

Bearings can be lubricated by a metal ring that rides loosely on the central rotating shaft of the bearing. The ring hangs down into a chamber containing lubricating oil. As the bearing rotates, viscous adhesion draws oil up the ring and onto the shaft, where the oil migrates into the bearing to lubricate it. Excess oil is flung off and collects in the pool again.

Splash Lubrication

Some machines contain a pool of lubricant in the bottom, with gears partially immersed in the liquid, or crank rods that can swing down into the pool as the device operates. The spinning wheels fling oil into the air around them, while the crank rods slap at the surface of the oil, splashing it randomly on the interior surfaces of the engine. Some small internal combustion engines specifically contain special plastic *flinger wheels* which randomly scatter oil around the interior of the mechanism.

Pressure Lubrication

For high speed and high power machines, a loss of lubricant can result in rapid bearing heating and damage due to friction. Also in dirty environments the oil can become contaminated with dust or debris that increases friction. In these applications, a fresh supply of lubricant can be continuously supplied to the bearing and all other contact surfaces, and the excess can be collected for filtration, cooling, and possibly reuse. Pressure oiling is commonly used in large and complex internal combustion engines in parts of the engine where directly splashed oil cannot reach, such as up into overhead valve assemblies. High speed turbochargers also typically require a pressurized oil system to cool the bearings and keep them from burning up due to the heat from the turbine.

Composite Bearings

Composite bearings are designed with a self-lubricating polytetrafluroethylene (PTFE) liner with a laminated metal backing. The PTFE liner offers consistent, controlled friction as well as durability whilst the metal backing ensures the composite bearing is robust and capable of withstanding high loads and stresses throughout its long life. Its design also makes it lightweight-one tenth the weight of a traditional rolling element bearing.

Types

There are many different types of bearings. Newer versions of more enabling designs are in development being tested, in which will reduce friction, increase bearing load, increase momentum build-up, and speed.

Type	Description	Friction	Stiff-ness[†]	Speed	Life	Notes
Plain bearing	Rubbing surfaces, usually with lubricant; some bearings use pumped lubrication and behave similarly to fluid bearings.	Depends on materials and construction, PTFE has coefficient of friction ~0.05-0.35, depending upon fillers added	Good, provided wear is low, but some slack is normally present	Low to very high	Low to very high - depends upon application and lubrication	Widely used, relatively high friction, suffers from stiction in some applications. Depending upon the application, lifetime can be higher or lower than rolling element bearings.
Rolling element bearing	Ball or rollers are used to prevent or minimise rubbing	Rolling coefficient of friction with steel can be ~0.005 (adding resistance due to seals, packed grease, preload and misalignment can increase friction to as much as 0.125)	Good, but some slack is usually present	Moderate to high (often requires cooling)	Moderate to high (depends on lubrication, often requires maintenance)	Used for higher moment loads than plain bearings with lower friction
Jewel bearing	Off-center bearing rolls in seating	Low	Low due to flexing	Low	Adequate (requires maintenance)	Mainly used in low-load, high precision work such as clocks. Jewel bearings may be very small.
Fluid bearing	Fluid is forced between two faces and held in by edge seal	Zero friction at zero speed, low	Very high	Very high (usually limited to a few hundred feet per second at/by seal)	Virtually infinite in some applications, may wear at startup/shutdown in some cases. Often negligible maintenance.	Can fail quickly due to grit or dust or other contaminants. Maintenance free in continuous use. Can handle very large loads with low friction.

Magnetic bearings	Faces of bearing are kept separate by magnets (electromagnets or eddy currents)	Zero friction at zero speed, but constant power for levitation, eddy currents are often induced when movement occurs, but may be negligible if magnetic field is quasi-static	Low	No practical limit	Indefinite. Maintenance free. (with electromagnets)	Active magnetic bearings (AMB) need considerable power. Electrodynamic bearings (EDB) do not require external power.
Flexure bearing	Material flexes to give and constrain movement	Very low	Low	Very high.	Very high or low depending on materials and strain in application. Usually maintenance free.	Limited range of movement, no backlash, extremely smooth motion
Composite bearing	Plain bearing shape with PTFE liner on the interface between bearing and shaft with a laminated metal backing. PTFE acts as a lubricant.	PTFE and use of filters to dial in friction as necessary for friction control.	Good depending on laminated metal backing	Low to very high	Very high; PTFE and fillers ensure wear and corrosion resistance	Widely used, controls friction, reduces stick slip, PTFE reduces static friction

†Stiffness is the amount that the gap varies when the load on the bearing changes, it is distinct from the friction of the bearing.

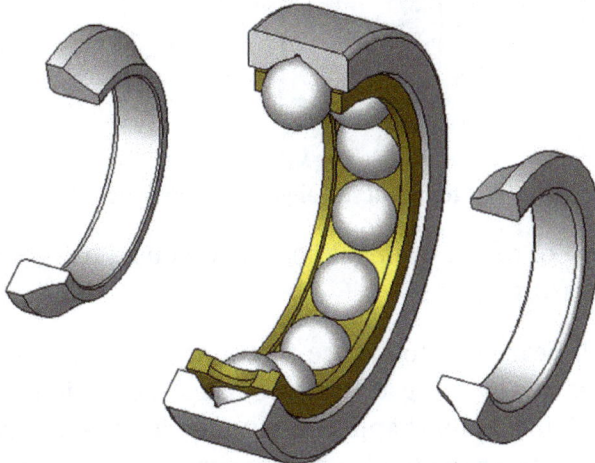

Ball Bearing

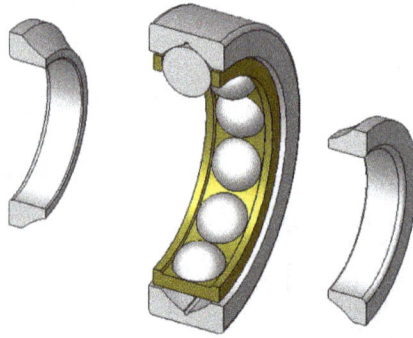

A 4-point angular contact ball bearing

A ball bearing for skateboard wheels with a plastic cage

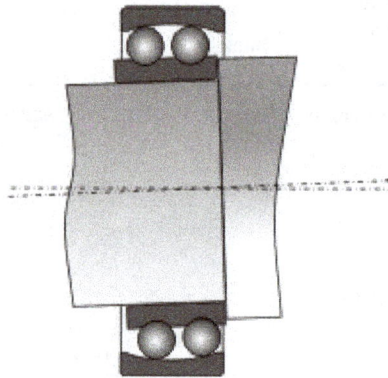

Wingquist's a self-aligning ball bearing

A ball bearing is a type of rolling-element bearing that uses balls to maintain the separation between the bearing races.

The purpose of a ball bearing is to reduce rotational friction and support radial and axial loads. It achieves this by using at least two races to contain the balls and transmit the loads through the balls. In most applications, one race is stationary and the other is attached to the rotating assembly (e.g., a hub or shaft). As one of the bearing races rotates it causes the balls to rotate as well. Because the balls are rolling they have a much lower coefficient of friction than if two flat surfaces were sliding against each other.

Ball bearings tend to have lower load capacity for their size than other kinds of rolling-element bearings due to the smaller contact area between the balls and races. However, they can tolerate some misalignment of the inner and outer races.

History

Although bearings had been developed since ancient times, the first modern recorded patent on ball bearings was awarded to Philip Vaughan, a Welsh inventor and ironmaster who created the first design for a ball bearing in Carmarthen in 1794. His was the first modern ball-bearing design, with the ball running along a groove in the axle assembly.

Jules Suriray, a Parisian bicycle mechanic, designed the first radial style ball bearing in 1869, which was then fitted to the winning bicycle ridden by James Moore in the world's first bicycle road race, Paris-Rouen, in November 1869.

Common Designs

There are several common designs of ball bearing, each offering various trade-offs. They can be made from many different materials, including: stainless steel, chrome steel, and ceramic (silicon nitride (Si_3N_4)). A hybrid ball bearing is a bearing with ceramic balls and races of metal.

Angular Contact

An *angular contact* ball bearing uses axially asymmetric races. An axial load passes in a straight line through the bearing, whereas a radial load takes an oblique path that tends to want to separate the races axially. So the angle of contact on the inner race is the same as that on the outer race. Angular contact bearings better support "combined loads" (loading in both the radial and axial directions) and the contact angle of the bearing should be matched to the relative proportions of each. The larger the contact angle (typically in the range 10 to 45 degrees), the higher the axial load supported, but the lower the radial load. In high speed applications, such as turbines, jet engines, and dentistry equipment, the centrifugal forces generated by the balls changes the contact angle at the inner and outer race. Ceramics such as silicon nitride are now regularly used in such applications due to their low density (40% of steel). These materials significantly reduce centrifugal force and function well in high temperature environments. They also tend to wear in a similar way to bearing steel—rather than cracking or shattering like glass or porcelain.

Most bicycles use angular-contact bearings in the headsets because the forces on these bearings are in both the radial and axial direction.

Axial

An *axial* ball bearing uses side-by-side races. An axial load is transmitted directly through the bearing, while a radial load is poorly supported and tends to separate the races,so that a larger radial load is likely to damage the bearing.

Deep-groove

In a *deep-groove* radial bearing, the race dimensions are close to the dimensions of the balls that run in it. Deep-groove bearings can support higher loads.

Construction Types

Conrad

The *Conrad*-style ball bearing is named after its inventor, Robert Conrad, who was awarded British patent 12,206 in 1903 and U.S. patent 822,723 in 1906. These bearings are assembled by placing the inner ring into an eccentric position relative to the outer ring, with the two rings in contact at one point, resulting in a large gap opposite the point of contact. The balls are inserted through the gap and then evenly distributed around the bearing assembly, causing the rings to become concentric. Assembly is completed by fitting a cage to the balls to maintain their positions relative to each other. Without the cage, the balls would eventually drift out of position during operation, causing the bearing to fail. The cage carries no load and serves only to maintain ball position.

Conrad bearings have the advantage that they are able to withstand both radial and axial loads, but have the disadvantage of lower load capacity due to the limited number of balls that can be loaded into the bearing assembly. Probably the most familiar industrial ball bearing is the deep-groove Conrad style. The bearing is used in most of the mechanical industries.

Slot-fill

In a *slot-fill* radial bearing, the inner and outer races are notched on one face so that when the notches are aligned, balls can be slipped in the resulting slot to assemble the bearing. A slot-fill bearing has the advantage that more balls can be assembled (even allowing a *full complement* design), resulting in a higher radial load capacity than a Conrad bearing of the same dimensions and material type. However, a slot-fill bearing cannot carry a significant axial load, and the slots cause a discontinuity in the races that can have a small but adverse effect on strength.

Relieved Race

Relieved race ball bearings are 'relieved' as the name suggests by basically have either the OD of the inner ring reduced on one side, or the ID of the outer ring increased on one side. This allows a greater number of balls to be assembled into either the inner or outer race, and then press fit over the relief. Sometimes the outer ring will be heated to facilitate assembly. Like the slot-fill construction, relieved race construction allows a greater number of balls than Conrad construction, up to and including full complement, and the extra ball count gives extra load capacity. However, a relieved race bearing can only support significant axial loads in one direction ('away from' the relieved race).

Fractured Race

Another way of fitting more balls into a radial ball bearing is by radially 'fracturing' (slicing) one of the rings all the way through, loading the balls in, re-assembling the fractured portion, and then using a pair of steel bands to hold the fractured ring sections together in alignment. Again, this allows more balls, including full ball complement, however unlike with either slot fill or relieved race constructions, it can support significant axial loading in either direction.

Rows

There are two *row* designs: *single-row* bearings and *double-row* bearings. Most ball bearings are a single-row design, which means there is one row of bearing balls. This design works with radial and thrust loads.

A *double-row* design has two rows of bearing balls. Their disadvantage is they need better alignment than single-row bearings.

Flanged

Bearings with a flange on the outer ring simplify axial location. The housing for such bearings can consist of a through-hole of uniform diameter, but the entry face of the housing (which may be either the outer or inner face) must be machined truly normal to the hole axis. However such flanges are very expensive to manufacture. A more cost effective arrangement of the bearing outer ring, with similar benefits, is a snap ring groove at either or both ends of the outside diameter. The snap ring assumes the function of a flange.

Caged

Cages are typically used to secure the balls in a Conrad-style ball bearing. In other construction types they may decrease the number of balls depending on the specific cage shape, and thus reduce the load capacity. Without cages the tangential position is stabilized by sliding of two convex surfaces on each other. With a cage the tangential position is stabilized by a sliding of a convex surface in a matched concave surface, which avoids dents in the balls and has lower friction. Caged roller bearings were invented by John Harrison in the mid-18th century as part of his work on chronographs. Caged bearings were used more frequently during wartime steel shortages for bicycle wheel bearings married to replaceable cups.

Hybrid Ball Bearings Using Ceramic Balls

Ceramic bearing balls can weigh up to 40% less than steel ones, depending on size and material. This reduces centrifugal loading and skidding, so hybrid ceramic bearings can operate 20% to 40% faster than conventional bearings. This means that the outer race groove exerts less force inward against the ball as the bearing spins. This reduction in force reduces the friction and rolling resistance. The lighter balls allow the bearing to spin faster, and uses less energy to maintain its speed.

While ceramic hybrid bearings use ceramic balls in place of steel ones, they are constructed with steel inner and outer rings; hence the *hybrid* designation. While the ceramic material itself is stronger than steel, it is also stiffer, which results in increased stresses on the rings, and hence decreased load capacity. Ceramic balls are electrically insulating, which can prevent 'arcing' failures if current should be passed through the bearing. Ceramic balls can also be effective in environments where lubrication may not be available (such as in space applications).

Self-aligning

Wingquist developed self-aligning ball bearing

Self-aligning ball bearings, such as the Wingquist bearing shown in the picture, are constructed with the inner ring and ball assembly contained within an outer ring that has a spherical raceway. This construction allows the bearing to tolerate a small angular misalignment resulting from shaft or housing deflections or improper mounting. The bearing was used mainly in bearing arrangements with very long shafts, such as transmission shafts in textile factories. One drawback of the self-aligning ball bearings is a limited load rating, as the outer raceway has very low osculation (radius is much larger than ball radius). This led to the invention of the spherical roller bearing, which has a similar design, but use rollers instead of balls. Also the spherical roller thrust bearing is an invention that derives from the findings by Wingquist.

Operating Conditions

Lifespan

The calculated life for a bearing is based on the load it carries and its operating speed. The industry standard usable bearing lifespan is inversely proportional to the bearing load cubed. Nominal maximum load of a bearing, is for a lifespan of 1 million rotations, which at 50 Hz (i.e., 3000 RPM) is a lifespan of 5.5 working hours. 90% of bearings of that type have at least that lifespan, and 50% of bearings have a lifespan at least 5 times as long.

The industry standard life calculation is based upon the work of Lundberg and Palmgren performed in 1947. The formula assumes the life to be limited by metal fatigue and that the life distribution can be described by a Weibull distribution. Many variations of the formula exist that include factors for material properties, lubrication, and loading. Factoring for loading may be viewed as a tacit admission that modern materials demonstrate a different relationship between load and life than Lundberg and Palmgren determined .

Failure Modes

If a bearing is not rotating, maximum load is determined by force that causes plastic deformation of elements or raceways. The indentations caused by the elements can concentrate stresses and generate cracks at the components. Maximum load for not or very slowly rotating bearings is called "static" maximum load.

Also if a bearing is not rotating, oscillating forces on the bearing can cause impact damage to the bearing race or the rolling elements, known as brinelling. A second lesser form called false brinelling occurs if the bearing only rotates across a short arc and pushes lubricant out away from the rolling elements.

For a rotating bearing, the dynamic load capacity indicates the load to which the bearing endures 1,000,000 cycles.

If a bearing is rotating, but experiences heavy load that lasts shorter than one revolution, static max load must be used in computations, since the bearing does not rotate during the maximum load.

If a sideways torque is applied to a deep groove radial bearing, an uneven force in the shape of an ellipse is applied on the outer ring by the rolling elements, concentrating in two regions on opposite sides of the outer ring. If the outer ring is not strong enough, or if it is not sufficiently braced by the supporting structure, the outer ring will deform into an oval shape from the sideways torque stress, until the gap is large enough for the rolling elements to escape. The inner ring then pops out and the bearing structurally collapses.

A sideways torque on a radial bearing also applies pressure to the cage that holds the rolling elements at equal distances, due to the rolling elements trying to all slide together at the location of highest sideways torque. If the cage collapses or breaks apart, the rolling elements group together, the inner ring loses support, and may pop out of the center.

Maximum Load

In general, maximum load on a ball bearing is proportional to outer diameter of the bearing times width of bearing (where width is measured in direction of axle).

Bearings have static load ratings. These are based on not exceeding a certain amount of plastic deformation in the raceway. These ratings may be exceeded by a large amount for certain applications.

Lubrication

For a bearing to operate properly, it needs to be lubricated. In most cases the lubricant is based on elastohydrodynamic effect (by oil or grease) but working at extreme temperatures dry lubricated bearings are also available.

For a bearing to have its nominal lifespan at its nominal maximum load, it must be lubricated with a lubricant (oil or grease) that has at least the minimum dynamic viscosity recommended for that bearing.

The recommended dynamic viscosity is inversely proportional to diameter of bearing.

The recommended dynamic viscosity decreases with rotating frequency. As a rough indication: for less than 3000 RPM, recommended viscosity increases with factor 6 for a factor 10 decrease in speed, and for more than 3000 RPM, recommended viscosity decreases with factor 3 for a factor 10 increase in speed.

For a bearing where average of outer diameter of bearing and diameter of axle hole is 50 mm, and that is rotating at 3000 RPM, recommended dynamic viscosity is 12 mm²/s.

Note that dynamic viscosity of oil varies strongly with temperature: a temperature increase of 50–70 °C causes the viscosity to decrease by factor 10.

If the viscosity of lubricant is higher than recommended, lifespan of bearing increases, roughly proportional to square root of viscosity. If the viscosity of the lubricant is lower than recommended, the lifespan of the bearing decreases, and by how much depends on which type of oil being used. For oils with EP ('extreme pressure') additives, the lifespan is proportional to the square root of dynamic viscosity, just as it was for too high viscosity, while for ordinary oil's lifespan is proportional to the square of the viscosity if a lower-than-recommended viscosity is used.

Lubrication can be done with a grease, which has advantages that grease is normally held within the bearing releasing the lubricant oil as it is compressed by the balls. It provides a protective barrier for the bearing metal from the environment, but has disadvantages that this grease must be replaced periodically, and maximum load of bearing decreases (because if bearing gets too warm, grease melts and runs out of bearing). Time between grease replacements decreases very strongly with diameter of bearing: for a 40 mm bearing, grease should be replaced every 5000 working hours, while for a 100 mm bearing it should be replaced every 500 working hours.

Lubrication can also be done with an oil, which has advantage of higher maximum load, but needs some way to keep oil in bearing, as it normally tends to run out of it. For oil lubrication it is recommended that for applications where oil does not become warmer than 50 °C, oil should be replaced once a year, while for applications where oil does not become warmer than 100 °C, oil should be replaced 4 times per year. For car engines, oil becomes 100 °C but the engine has an oil filter to maintain oil quality; therefore, the oil is usually changed less frequently than the oil in bearings.

Direction of Load

Most bearings are meant for supporting loads perpendicular to axle ("radial loads"). Whether they can also bear axial loads, and if so, how much, depends on the type of bearing. Thrust bearings (commonly found on lazy susans) are specifically designed for axial loads.

For single-row deep-groove ball bearings, SKF's documentation says that maximum axial load is circa 50% of maximum radial load, but it also says that "light" and/or "small" bearings can take axial loads that are 25% of maximum radial load.

For single-row edge-contact ball bearings, axial load can be circa 2 times max radial load, and for cone-bearings maximum axial load is between 1 and 2 times maximum radial load.

Often Conrad style ball bearings will exhibit contact ellipse truncation under axial load. What that means is that either the ID of the outer ring is large enough, or the OD of the inner ring is small enough, so as to reduce the area of contact between the balls and raceway. When this is the case, it can significantly increase the stresses in the bearing, often invalidating general rules of thumb regarding relationships between radial and axial load capacity. With construction types other than Conrad, one can further decrease the outer ring ID and increase the inner ring OD to guard against this.

If both axial and radial loads are present, they can be added vectorially, to result in total load on bearing, which in combination with nominal maximum load can be used to predict lifespan. However, in order to correctly predict the rating life of ball bearings the ISO/TS 16281 should be used with the help of a calculation software.

Avoiding Undesirable Axial Load

The part of a bearing that rotates (either axle hole or outer circumference) must be fixed, while for a part that does not rotate this is not necessary (so it can be allowed to slide). If a bearing is loaded axially, both sides must be fixed.

If an axle has two bearings, and temperature varies, axle shrinks or expands, therefore it is not admissible for both bearings to be fixed on both their sides, since expansion of axle would exert axial forces that would destroy these bearings. Therefore, at least one of bearings must be able to slide.

A 'freely sliding fit' is one where there is at least a 4 μm clearance, presumably because surface-roughness of a surface made on a lathe is normally between 1.6 and 3.2 μm.

Fit

Bearings can withstand their maximum load only if the mating parts are properly sized. Bearing manufacturers supply tolerances for the fit of the shaft and the housing so that this can be achieved. The material and hardness may also be specified.

Fittings that are not allowed to slip are made to diameters that prevent slipping and consequently the mating surfaces cannot be brought into position without force. For small bearings this is best done with a press because tapping with a hammer damages both bearing and shaft, while for large bearings the necessary forces are so great that there is no alternative to heating one part before fitting, so that thermal expansion allows a temporary sliding fit.

Avoiding Torsional Loads

If a shaft is supported by two bearings, and the center-lines of rotation of these bearings are not the same, then large forces are exerted on the bearing that may destroy it. Some very small amount of misalignment is acceptable, and how much depends on type of bearing. For bearings that are specifically made to be 'self-aligning', acceptable misalignment is between 1.5 and 3 degrees of arc. Bearings that are not designed to be self-aligning can accept misalignment of only 2–10 minutes of arc.

Applications

In general, ball bearings are used in most applications that involve moving parts. Some of these applications have specific features and requirements:

- Hard drive bearings used to be highly spherical, and were said to be the best spherical manufactured shapes, but this is no longer true, and more and more are being replaced with fluid bearings.

- German ball bearing factories were often a target of allied aerial bombings during World War II; such was the importance of the ball bearing to the German war industry.

- In horology, the company Jean Lassale designed a watch movement that used ball bearings to reduce the thickness of the movement. Using 0.20 mm balls, the Calibre 1200 was only 1.2 mm thick, which still is the thinnest mechanical watch movement.

- Aerospace bearings are used in many applications on commercial, private and military aircraft including pulleys, gearboxes and jet engine shafts. Materials include M50 tool steel (AMS6491), Carbon chrome steel (AMS6444), the corrosion resistant AMS5930, 440C stainless steel, silicon nitride (ceramic) and titanium carbide-coated 440C.

- A skateboard wheel contains two bearings, which are subject to both axial and radial time-varying loads. Most commonly bearing 608-2Z is used (a deep groove ball bearing from series 60 with 8 mm bore diameter)

- Yo-Yos, there are ball bearings in the center of many new, ranging from beginner to professional or competition grade, Yo-Yos.

Designation

The ball size increases as the series increases, for any given inner diameter or outer diameter (not both). The larger the ball the greater the load carrying capacity. Series 200 and 300 are the most common.

Composite Bearing

A composite bearing is used to maintain separation and control friction between two moving parts. The distinguishing characteristic of a composite bearing is that the bearing is made from a combination of materials such as a resin reinforced with fibre and this may also include friction reducing lubricants and ingredients. A composite bearing is not simply a PTFE bearing in a carrier of another material, this is a PTFE bearing in a carrier. The plain composite bearing can be lighter than a rolling element bearing but this is not always a feature as some composites are extremely dense which results in lower porosity. Another distinctive feature of the composite bearing is its lightweight design - it can be one-tenth the weight of the traditional rolling element bearing. No heavy metals are used in its manufacture.

Composite bearings can be customized to meet the individual requirements of many applications, such as wear- or high-temperature resistance. The weight of the composite bearing can vary depending on its backing. The PTFE liner can be applied on steel or aluminum backing. Through filler compounds, various properties of the composite bearing, such as resistance to creep, wear and electrical conductivity, can be optimized.

Technology

A composite bearing is a bearing made from a composite material, such as fibre reinforced resin, or plastic. These will often contain friction reducing ingredients such as PTFE but this is not the only material for reducing friction and wear and lubricating when the bearing is running dry (without external lubricants). PTFE in itself is not a good bearing material as it is inherently soft and deforms under pressure, so the use of PTFE as a liner is not as beneficial as having PTFE combined into a solid and strong matrix of resin and fibre. A composite bearing is a bearing with a liner of PTFE compound and a metal backing. PTFE is a fluorocarbon solid, as it is a high-molecular-weight compound consisting wholly of carbon and fluorine. PTFE is hydrophobic: neither water nor water-containing substances wet PTFE. Components engineered with PTFE offer consistent, controlled friction over their lifetime.

PTFE is often used as a non-stick coating for pans and other cookware. It is very non-reactive, partly because of the strength of carbon-fluorine bonds, so it is often used in containers and pipe work for reactive and corrosive chemicals. PTFE can also be used as a machinery lubricant to reduce friction, wear and energy consumption.

PTFE is self-lubricating, so wet lubrication and replenishment of lubrication is not necessary but can be added if a reduction of the coefficient of friction or wear is demanded depending on the application.

Application and Uses

Automotive

Automotive manufacturers are striving to meet growing demand for lighter and more fuel-efficient vehicles and evaluate components and materials used in various car applications. Composite bearings are useful components in automotive design and are used in a range of applications throughout the car, from the powertrain to the car interior. In addition to enhancing automotive performance, composite bearings' split ring design allows them to be press fitted, without the need for adhesive or excessive assembly force.

Steering Yoke

The steering rack has a bearing situated at the yoke - the interface between the steering rack and the steering column. The yoke is designed to prevent the separation of the steering rack from the steering column, whilst allowing the steering rack to move freely the in transverse direction. The yoke determines a motorist's ability to feel the road surface and the vehicle's maneuverability. With their PTFE liner, composite bearings reduce friction in the steering yoke.

Belt Tensioner

A belt tensioner is a device designed to maintain tension in the engine's timing belt. There is a spring device in the belt tensioner where the bearing is located. The spring device oscillates back and forth at 2° for about 60 cycles per minute.

Composite bearings ensure an appropriate and consistent level of torque and damping to maintain the correct tension in the drive belt while the engine is in operation.

Door Hinges

In door hinges, bearings sit between the hinge pin and housing to ensure smooth movement of the door when it is opened and closed by passengers. Composite bearings are used in a number of automotive hinge systems due to their durability under high loads and corrosion resistance.

Bearings also play a role in obtaining a quality paint finish on the car. Composite bearings' PTFE compound liners are conductive and can transfer electricity to the hinges to facilitate the electrostatic painting process.

They are also hydrophobic and repel paint, minimizing the risk of excess droplets impairing the paintwork finish.

Seat Mechanism

In adjustable seats, seat mechanisms facilitate movement. A bearing fits between the linkage and the pin and is designed to provide correct levels of torque. Composite bearings can be used in the pivot points in seat components to maintain torque, allowing passengers to adjust their seat easily and smoothly for a comfortable experience.

Bicycle

Bicycle designers strive to reduce weight and increase the performance of their bicycles without sacrificing quality and strength. The bicycle industry is demanding lightweight, high-performance products with maintenance-free components.

Fork

The bicycle fork enhances rider comfort in rough terrain by enabling the shaft attached to the bicycle frame to slide within the housing attached to the wheels. The bearing sits between the shaft and housing. The liner of PTFE within composite bearings enhances shock absorption as it acts as a cushion, while the lightweight metal backing helps to reduce bicycle weight.

Shock Absorbers

Front shocks, key components in suspension bicycles, are designed to reduce the impact of bumps and jolts for a smoother ride. The bearing sits between the inner shaft and outer housing to facilitate smooth movement in the mechanism for optimum performance. Composite bearings with a PTFE liner act as a cushion, absorbing excess vibrations to further enhance the movement.

Headsets

Composite bearings can be used in the headset. Composite bearing can be lightweight to support weight reduction efforts across the overall bicycle.

Pedal and Brake Pivots

The low friction that bearings can provide helps to reduce "stick slip", avoiding any undesirable jerking motion. Pedals rotate on bearings that connect the spindle to the end of the crank and the body of the pedal.

Derailleurs

The derailleur is used in the bicycle's gear system. The change in cable tension to switch gears moves a chain from side to side, "derailing" the chain onto different sprockets and, therefore, different gears.

Solar

Solar has become a viable source of energy. According to the International Energy Agency, Concentrated Solar Power (CSP) could be responsible for up to 11.5% of global electricity production by 2050. The life expectancy of the CSP plant could be up to 40 years (13)and energy companies are looking for components that will last the lifetime of the CSP plant.

Solar Tracking System

Concentrating Solar Power (CSP) plants use concentrated solar radiation as a high-temperature energy source to produce electrical power. A solar tracker is a device for concentrating solar reflectors toward the sun. Bearings are used in the pivot points to both support the structure in a parabolic trough and to rotate the mirrors on heliostats (solar tower).

Composite bearings can be used in the parabolic trough and solar power tower to rotate the mirrors. They can withstand the loads in CSP applications, are weather and corrosion resistant and also offer low and constant friction (no stick-slip effect) over the mechanism's life cycle.

Piston Pumps in Off-highway Construction Equipment

Off-highway construction equipment from excavators to simple loaders employs hydraulic transmission systems as a primary source of motion. Hydraulic piston pumps, mechanical devices used to convert mechanical energy into hydraulic energy, are typically used in off-highway construction equipment and driven by an electric motor or a combustion engine.

Bearings are mounted on the piston pump shaft, which transfers drive torque to the cylinder block. The bearing's role is to ensure smooth movement and reduce energy use. Composite bearings' self-lubricating PTFE layer allows consistent, low friction in pump mechanisms for minimal energy use and reduced maintenance requirements.

Wetting

Water beads on a fabric that has been made nonwetting by chemical treatment

Wetting is the ability of a liquid to maintain contact with a solid surface, resulting from intermolecular interactions when the two are brought together. The degree of wetting (wettability) is determined by a force balance between adhesive and cohesive forces. Wetting deals with the three phases of materials: gas, liquid, and solid. It is now a center of attention in nanotechnology and nanoscience studies due to the advent of many nanomaterials in the past two decades (e.g. graphene, carbon nanotube).

Figure: Contact angle for a liquid droplet on a solid surface

Wetting is important in the bonding or adherence of two materials. Wetting and the surface forces that control wetting are also responsible for other related effects, including capillary effects.

There are two types of wetting: non-reactive wetting and active wetting.

Explanation

Adhesive forces between a liquid and solid cause a liquid drop to spread across the surface. Cohesive forces within the liquid cause the drop to ball up and avoid contact with the surface.

Contact angle	Degree of wetting	Strength of:	
		Solid/liquid interactions	Liquid/liquid interactions
$\theta = 0$	Perfect wetting	strong	weak
$0 < \theta < 90°$	high wettability	strong	strong
		weak	weak
$90° \le \theta < 180°$	low wettability	weak	strong
$\theta = 180°$	perfectly non-wetting	weak	strong

The contact angle (θ), as seen in Figure, is the angle at which the liquid–vapor interface meets the solid–liquid interface. The contact angle is determined by the result between adhesive and cohesive forces. As the tendency of a drop to spread out over a flat, solid surface increases, the contact angle decreases. Thus, the contact angle provides an inverse measure of wettability.

A contact angle less than 90° (low contact angle) usually indicates that wetting of the surface is very favorable, and the fluid will spread over a large area of the surface. Contact angles greater than 90° (high contact angle) generally means that wetting of the surface is unfavorable, so the fluid will minimize contact with the surface and form a compact liquid droplet.

For water, a wettable surface may also be termed hydrophilic and a nonwettable surface hydrophobic. Superhydrophobic surfaces have contact angles greater than 150°, showing almost no contact between the liquid drop and the surface. This is sometimes referred to as the "Lotus effect". The table describes varying contact angles and their corresponding solid/liquid and liquid/liquid interactions. For nonwater liquids, the term lyophilic is used for low contact angle conditions and lyophobic is used when higher contact angles result. Similarly, the terms omniphobic and omniphilic apply to both polar and apolar liquids.

High-energy vs. Low-energy Surfaces

Liquids can interact with two main types of solid surfaces. Traditionally, solid surfaces have been divided into high-energy solids and low-energy types. The relative energy of a solid has to do with the bulk nature of the solid itself. Solids such as metals, glasses, and ceramics are known as 'hard solids' because the chemical bonds that hold them together (e.g., covalent, ionic, or metallic) are very strong. Thus, it takes a large input of energy to break these solids (alternatively large amount of energy is required to cut the bulk and make two separate surfaces so high surface energy), so they are termed "high energy". Most molecular liquids achieve complete wetting with high-energy surfaces.

The other type of solids is weak molecular crystals (e.g., fluorocarbons, hydrocarbons, etc.) where the molecules are held together essentially by physical forces (e.g., van der Waals and hydrogen bonds). Since these solids are held together by weak forces, a very low input of energy is required to break them, thus they are termed "low energy". Depending on the type of liquid chosen, low-energy surfaces can permit either complete or partial wetting.

Dynamic surfaces have been reported that undergo changes in surface energy upon the application of an appropriate stimuli. For example, a surface presenting photon-driven molecular motors was shown to undergo changes in water contact angle when switched between bistable conformations of differing surface energies.

Wetting of Low-energy Surfaces

Low-energy surfaces primarily interact with liquids through dispersion (van der Waals) forces. William Zisman had several key findings in the work that he did:

Zisman observed that $\cos \theta$ increases linearly as the surface tension (γ_{LV}) of the liquid decreased. Thus, he was able to establish a linear function between $\cos \theta$ and the surface tension (γ_{LV}) for various organic liquids.

A surface is more wettable when γ_{LV} and θ is low. Zisman termed the intercept of these lines when $\cos \theta = 1$, as the critical surface tension (γ_c) of that surface. This critical surface tension is an important parameter because it is a characteristic of only the solid.

Knowing the critical surface tension of a solid, it is possible to predict the wettability of the surface. The wettability of a surface is determined by the outermost chemical groups of the solid. Differences in wettability between surfaces that are similar in structure are due to differences in packing of the atoms. For instance, if a surface has branched chains, it will have poorer packing than a surface with straight chains.

Ideal Solid Surfaces

An ideal surface is flat, rigid, perfectly smooth, and chemically homogeneous, and has zero contact angle hysteresis. Zero hysteresis implies the advancing and receding contact angles are equal. In other words, only one thermodynamically stable contact angle exists. When a drop of liquid is placed on such a surface, the characteristic contact angle is formed as depicted in Fig. Furthermore, on an ideal surface, the drop will return to its original shape if it is disturbed. The following derivations apply only to ideal solid surfaces; they are only valid for the state in which the interfaces are not moving and the phase boundary line exists in equilibrium.

Minimization of Energy, Three Phases

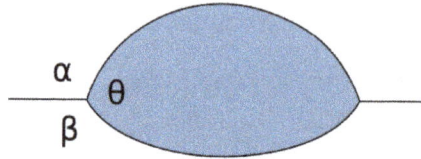

Figure: Coexistence of three fluid phases in mutual contact; here, α, β, and θ each indicate both a phase and its contact angle.

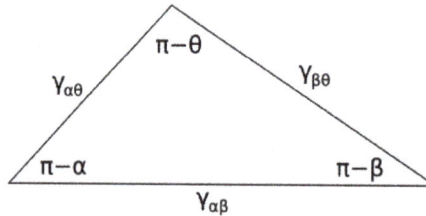

Figure: Neumann's triangle relating the surface energies and contact angles of three fluid phases coexisting in static equilibrium, as depicted in Figure.

Figure shows the line of contact where three phases meet. In equilibrium, the net force per unit length acting along the boundary line between the three phases must be zero. The components of net force in the direction along each of the interfaces are given by:

$$\gamma_{\alpha\theta} + \gamma_{\theta\beta}\cos\theta + \gamma_{\alpha\beta}\cos\alpha = 0$$

$$\gamma_{\alpha\theta}\cos\theta + \gamma_{\theta\beta} + \gamma_{\alpha\beta}\cos\beta = 0$$

$$\gamma_{\alpha\theta}\cos\alpha + \gamma_{\theta\beta}\cos\beta + \gamma_{\alpha\beta} = 0$$

where α, β, and θ are the angles shown and γ_{ij} is the surface energy between the two indicated phases. These relations can also be expressed by an analog to a triangle known as Neumann's triangle, shown in Figure 4. Neumann's triangle is consistent with the geometrical restriction that $\alpha + \beta + \theta = 2\pi$, and applying the law of sines and law of cosines to it produce relations that describe how the interfacial angles depend on the ratios of surface energies.

Because these three surface energies form the sides of a triangle, they are constrained by the triangle inequalities, $\gamma_{ij} < \gamma_{jk} + \gamma_{ik}$ meaning that no one of the surface tensions can exceed the sum of the other two. If three fluids with surface energies that do not follow these inequalities are brought into contact, no equilibrium configuration consistent with Figure will exist.

Simplification to Planar Geometry, Young's Relation

If the β phase is replaced by a flat rigid surface, as shown in Figure, then $\beta = \pi$, and the second net force equation simplifies to the Young equation,

$$\gamma_{SG} = \gamma_{SL} + \gamma_{LG} \cos \theta$$

which relates the surface tensions between the three phases: solid, liquid and gas. Subsequently, this predicts the contact angle of a liquid droplet on a solid surface from knowledge of the three surface energies involved. This equation also applies if the "gas" phase is another liquid, immiscible with the droplet of the first "liquid" phase.

Real Smooth Surfaces and the Young Contact Angle

The Young equation assumes a perfectly flat and rigid surface often referred to as an ideal surface. In many cases, surfaces are far from this ideal situation, and two are considered here: the case of rough surfaces and the case of smooth surfaces that are still real (finitely rigid). Even in a perfectly smooth surface, a drop will assume a wide spectrum of contact angles ranging from the so-called advancing contact angle, θ_A, to the so-called receding contact angle, θ_R. The equilibrium contact angle (θ_c) can be calculated from θ_A and θ_R as was shown by Tadmor as,

$$\theta_c = \arccos\left(\frac{r_A \cos\theta_A + r_R \cos\theta_R}{r_A + r_R}\right)$$

where

$$r_A = \left(\frac{\sin^3\theta_A}{2 - 3\cos\theta_A + \cos^3\theta_A}\right)^{1/3} \; ; \; r_R = \left(\frac{\sin^3\theta_R}{2 - 3\cos\theta_R + \cos^3\theta_R}\right)^{1/3}$$

The Young–Dupré Equation and Spreading Coefficient

The Young–Dupré equation (Thomas Young 1805; Anthanase Dupré and Paul Dupré 1869) dictates that neither γ_{SG} nor γ_{SL} can be larger than the sum of the other two surface energies. The consequence of this restriction is the prediction of complete wetting when $\gamma_{SG} > \gamma_{SL} + \gamma_{LG}$ and zero wetting when $\gamma_{SL} > \gamma_{SG} + \gamma_{LG}$. The lack of a solution to the Young–Dupré equation is an indicator that there is no equilibrium configuration with a contact angle between 0 and 180° for those situations.

A useful parameter for gauging wetting is the *spreading parameter S*,

$$S = \gamma_{SG} - (\gamma_{SL} + \gamma_{LG})$$

When $S > 0$, the liquid wets the surface completely (complete wetting). When $S < 0$, partial wetting occurs.

Combining the spreading parameter definition with the Young relation yields the Young–Dupré equation:

$$S = \gamma_{LG}(\cos \theta - 1)$$

which only has physical solutions for θ when $S < 0$.

Nonideal Rough Solid Surfaces

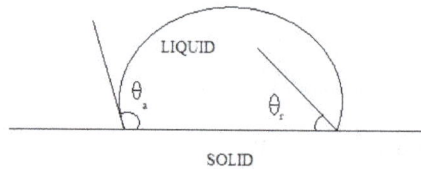

Figure: Schematic of advancing and receding contact angles

Unlike ideal surfaces, real surfaces do not have perfect smoothness, rigidity, or chemical homogeneity. Such deviations from ideality result in phenomenon called contact-angle hysteresis, which is defined as the difference between the advancing (θ_a) and receding (θ_r) contact angles

$$H = \theta_a - \theta_r$$

In simpler terms, contact angle hysteresis is essentially the displacement of a contact line such as the one in Figure, by either expansion or retraction of the droplet. Figure depicts the advancing and receding contact angles. The advancing contact angle is the maximum stable angle, whereas the receding contact angle is the minimum stable angle. Contact-angle hysteresis occurs because many different thermodynamically stable contact angles are found on a nonideal solid. These varying thermodynamically stable contact angles are known as metastable states.

Such motion of a phase boundary, involving advancing and receding contact angles, is known as dynamic wetting. When a contact line advances, covering more of the surface with liquid, the contact angle is increased and generally is related to the velocity of the contact line. If the velocity of a contact line is increased without bound, the contact angle increases, and as it approaches 180°, the gas phase will become entrained in a thin layer between the liquid and solid. This is a kinetic nonequilibrium effect which results from the contact line moving at such a high speed that complete wetting cannot occur.

A well-known departure from ideality is when the surface of interest has a rough texture. The rough texture of a surface can fall into one of two categories: homogeneous or heterogeneous. A homogeneous wetting regime is where the liquid fills in the roughness grooves of a surface. A heterogeneous wetting regime, though, is where the surface is a composite of two types of patches. An important example of such a composite surface is one composed of patches of both air and solid. Such surfaces have varied effects on the contact angles of wetting liquids. Cassie–Baxter and Wenzel are the two main models that attempt to describe the wetting of textured surfaces. However, these equations only apply when the drop size is sufficiently large compared with the surface roughness scale. When the droplet size is comparable to that of the underlying pillars, the effect of line tension should be considered. .

Wenzel's Model

Figure: Wenzel model

The Wenzel model (Robert N. Wenzel 1936) describes the homogeneous wetting regime, as seen in Figure, and is defined by the following equation for the contact angle on a rough surface:

$$\cos\theta^* = r\cos\theta$$

where θ^* is the apparent contact angle which corresponds to the stable equilibrium state (i.e. minimum free energy state for the system). The roughness ratio, r, is a measure of how surface roughness affects a homogeneous surface. The roughness ratio is defined as the ratio of true area of the solid surface to the apparent area.

θ is the Young contact angle as defined for an ideal surface. Although Wenzel's equation demonstrates the contact angle of a rough surface is different from the intrinsic contact angle, it does not describe contact angle hysteresis.

Cassie–Baxter Model

Figure: Cassie–Baxter model

When dealing with a heterogeneous surface, the Wenzel model is not sufficient. A more complex model is needed to measure how the apparent contact angle changes when various materials are involved. This heterogeneous surface, like that seen in Figure, is explained using the Cassie–Baxter equation (Cassie's law):

$$\cos\theta^* = r_f\, f \cos\theta_Y + f - 1$$

Here the r_f is the roughness ratio of the wet surface area and f is the fraction of solid surface area wet by the liquid. It is important to realize that when $f = 1$ and $r_f = r$, the Cassie–Baxter equations becomes the Wenzel equation. On the other hand, when there are many different fractions of surface roughness, each fraction of the total surface area is denoted by f_i.

A summation of all f_i equals 1 or the total surface. Cassie–Baxter can also be recast in the following equation:

$$\gamma \cos\theta^* = \sum_{n=1}^{N} f_i(\gamma_{i,sv} - \gamma_{i,sl})$$

Here γ is the Cassie–Baxter surface tension between liquid and vapor, the $\gamma_{i,sv}$ is the solid vapor surface tension of every component and $\gamma_{i,sl}$ is the solid liquid surface tension of every component. A case that is worth mentioning is when the liquid drop is placed on the substrate and creates small air pockets underneath it. This case for a two-component system is denoted by:

$$\gamma \cos\theta^* = f_1(\gamma_{1,sv} - \gamma_{1,sl}) - (1 - f_1)\gamma$$

Here the key difference to notice is that there is no surface tension between the solid and the vapor for the second surface tension component. This is because of the assumption that the surface of air that is exposed is under the droplet and is the only other substrate in the system. Subsequently the equation is then expressed as $(1 - f)$. Therefore, the Cassie equation can be easily derived from the Cassie–Baxter equation. Experimental results regarding the surface properties of Wenzel versus Cassie–Baxter systems showed the effect of pinning for a Young angle of 180 to 90°, a region classified under the Cassie–Baxter model. This liquid/air composite system is largely hydrophobic. After that point, a sharp transition to the Wenzel regime was found where the drop wets the surface, but no further than edges of the drop.

Precursor Film

With the advent of high resolution imaging, researchers have started to obtain experimental data which have led them to question the assumptions of the Cassie–Baxter equation when calculating the apparent contact angle. These groups believe the apparent contact angle is largely dependent on the triple line. The triple line, which is in contact with the heterogeneous surface, cannot rest on the heterogeneous surface like the rest of the drop. In theory, it should follow the surface imperfection. This bending in triple line is unfavorable and is not seen in real-world situations. A theory that preserves the Cassie–Baxter equation while at the same time explaining the presence of minimized energy state of the triple line hinges on the idea of a precursor film. This film of submicrometer thickness advances ahead of the motion of the droplet and is found around the triple line. Furthermore, this precursor film allows the triple line to bend and take different conformations that were originally considered unfavorable. This precursor fluid has been observed using environmental scanning electron microscopy (ESEM) in surfaces with pores formed in the bulk. With the introduction of the precursor film concept, the triple line can follow energetically feasible conformations and thereby correctly explaining the Cassie–Baxter model.

"Petal Effect" vs. "Lotus Effect"

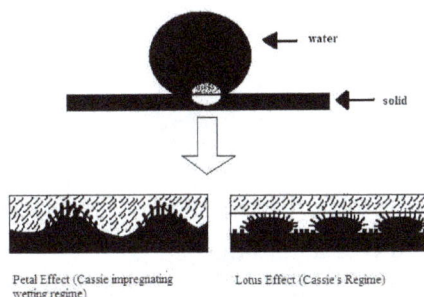

Figure: "Petal effect" vs. "lotus effect"

The intrinsic hydrophobicity of a surface can be enhanced by being textured with different length scales of roughness. The red rose takes advantage of this by using a hierarchy of micro- and nanostructures on each petal to provide sufficient roughness for superhydrophobicity. More specifically, each rose petal has a collection of micropapillae on the surface and each papilla, in turn, has many nanofolds. The term "petal effect" describes the fact that a water droplet on the surface of a rose petal is spherical in shape, but cannot roll off even if the petal is turned upside down. The water drops maintain their spherical shape due to the superhydrophobicity of the petal (contact angle of about 152.4°), but do not roll off because the petal surface has a high adhesive force with water.

When comparing the "petal effect" to the "lotus effect", it is important to note some striking differences. The surface structure of the lotus leaf and the rose petal, as seen in Figure 9, can be used to explain the two different effects. The lotus petal has a randomly rough surface and low contact angle hysteresis, which means the water droplet is not able to wet the microstructure spaces between the spikes. This allows air to remain inside the texture, causing a heterogeneous surface composed of both air and solid. As a result, the adhesive force between the water and the solid surface is extremely low, allowing the water to roll off easily (i.e. "self-cleaning" phenomenon).

The rose petal's micro- and nanostructures are larger in scale than those of the lotus leaf, which allows the liquid film to impregnate the texture. However, as seen in Figure, the liquid can enter the larger-scale grooves, but it cannot enter into the smaller grooves. This is known as the Cassie impregnating wetting regime. Since the liquid can wet the larger-scale grooves, the adhesive force between the water and solid is very high. This explains why the water droplet will not fall off even if the petal is tilted at an angle or turned upside down. This effect will fail if the droplet has a volume larger than 10 µl because the balance between weight and surface tension is surpassed.

Cassie–Baxter to Wenzel Transition

In the Cassie–Baxter model, the drop sits on top of the textured surface with trapped air underneath. During the wetting transition from the Cassie state to the Wenzel state, the air pockets are no longer thermodynamically stable and liquid begins to nucleate from the middle of the drop, creating a "mushroom state" as seen in Figure 10. The penetration condition is given by:

$$\cos\theta_C = \frac{\phi-1}{r-\phi}$$

where

- θ_C is the critical contact angle

- Φ is the fraction of solid/liquid interface where drop is in contact with surface

- r is solid roughness (for flat surface, r = 1)

The penetration front propagates to minimize the surface energy until it reaches the edges of the drop, thus arriving at the Wenzel state. Since the solid can be considered an absorptive material due to its surface roughness, this phenomenon of spreading and imbibition is called hemiwicking. The contact angles at which spreading/imbibition occurs are between 0 and $\pi/2$.

The Wenzel model is valid between θ_C and $\pi/2$. If the contact angle is less than Θ_C, the penetration front spreads beyond the drop and a liquid film forms over the surface. Figure 11 depicts the transition from the Wenzel state to the surface film state. The film smoothes the surface roughness and the Wenzel model no longer applies. In this state, the equilibrium condition and Young's relation yields:

$$\cos\theta^* = \phi\cos\theta_C + (1-\phi)^[$$

By fine-tuning the surface roughness, it is possible to achieve a transition between both superhydrophobic and superhydrophilic regions. Generally, the rougher the surface, the more hydrophobic it is.

Spreading Dynamics

If a drop is placed on a smooth, horizontal surface, it is generally not in the equilibrium state. Hence, it spreads until an equilibrium contact radius is reached (partial wetting). While taking into account capillary, gravitational, and viscous contributions, the drop radius as a function of time can be expressed as

$$r(t) = r_e \left[1 - \exp\left(-\left(\frac{2\gamma_{LG}}{r_e^{12}} + \frac{\rho g}{9 r_e^{10}} \right) \frac{24 \lambda V^4 (t + t_0)}{\pi^2 \eta} \right) \right]^{\frac{1}{6}}$$

For the complete wetting situation, the drop radius at any time during the spreading process is given by

$$r(t) = \left[\left(\gamma_{LG} \frac{96 \lambda V^4}{\pi^2 \eta} (t + t_0) \right)^{\frac{1}{2}} + \left(\frac{\lambda(t + t_0)}{\eta} \right)^{\frac{2}{3}} \frac{24 \rho g V^{\frac{3}{8}}}{7 \cdot 96^{\frac{1}{3}} \pi^{\frac{4}{3}} \gamma_{LG}^{\frac{1}{3}}} \right]^{\frac{1}{6}}$$

where

- γ_{LG} = Surface tension of the fluid
- V = Drop volume
- η = Viscosity of the fluid
- ρ = Density of the fluid
- g = Gravitational constant
- λ = Shape factor λ = 37.1 m^{-1}
- t_0 = Experimental delay time
- r_e = Drop radius in the equilibrium

Modifying wetting properties

Surfactants

Many technological processes require control of liquid spreading over solid surfaces. When a drop is placed on a surface, it can completely wet, partially wet, or not wet the surface. By reducing the surface tension with surfactants, a nonwetting material can be made to become partially or completely wetting. The excess free energy (σ) of a drop on a solid surface is:

$$\sigma = \gamma S + PV + \pi R^2 (\gamma_{SL} - \gamma_{SV})$$

- γ is the liquid–vapor interfacial tension

- γ_{SL} is the solid–liquid interfacial tension

- γ_{SV} is the solid–vapor interfacial tension

- S is the area of liquid–vapor interface

- P is the excess pressure inside liquid

- R is the radius of droplet base

Based on this equation, the excess free energy is minimized when γ decreases, γ_{SL} decreases, or γ_{SV} increases. Surfactants are absorbed onto the liquid–vapor, solid–liquid, and solid–vapor interfaces, which modify the wetting behavior of hydrophobic materials to reduce the free energy. When surfactants are absorbed onto a hydrophobic surface, the polar head groups face into the solution with the tail pointing outward. In more hydrophobic surfaces, surfactants may form a bilayer on the solid, causing it to become more hydrophilic. The dynamic drop radius can be characterized as the drop begins to spread. Thus, the contact angle changes based on the following equation:

$$\cos\theta(t) = \cos\theta_0 + (\cos\theta_\infty - \cos\theta_0)(1 - e^{\frac{-t}{\tau}})$$

- θ_o is initial contact angle

- θ_∞ is final contact angle

- τ is the surfactant transfer time scale

As the surfactants are absorbed, the solid–vapor surface tension increases and the edges of the drop become hydrophilic. As a result, the drop spreads.

Surface Changes

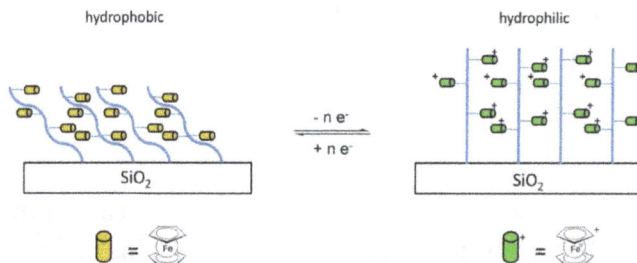

Strands of an uncharged ferrocene-substituted polymer are tethered to a hydrophobic silica surface. Oxidation of the ferrocenyl groups produces a hydrophilic surface due to electrostatic attractions between the resulting charges and the polar solvent.

Ferrocene is a redox-active organometallic compound which can be incorporated into various monomers and used to make polymers which can be tethered onto a surface.

Vinylferrocene (ferroceneylethene) can be prepared by a Wittig reaction and then polymerised to form polyvinylferrocene (PVFc), an analogue of polystyrene. Another polymer which can be formed is poly(2-(methacryloyloxy)ethyl ferrocenecarboxylate), PFcMA. Both PVFc and PFcMA have been tethered onto silica wafers and the wettability measured when the polymer chains are uncharged and when the ferrocene moieties are oxidised to produce positively charged groups, as illustrated at right. The contact angle with water on the PFcMA-coated wafers was 70° smaller following oxidation, while in the case of PVFc the decrease was 30°, and the switching of wettability has been shown to be reversible. In the PFcMA case, the effect of longer chains with more ferrocene groups (and also greater molar mass) has been investigated, and it was found that longer chains produce significantly larger contact angle reductions.

Surface Finishing

Surface finishing is a broad range of industrial processes that alter the surface of a manufactured item to achieve a certain property. Finishing processes may be employed to: improve appearance, adhesion or wettability, solderability, corrosion resistance, tarnish resistance, chemical resistance, wear resistance, hardness, modify electrical conductivity, remove burrs and other surface flaws, and control the surface friction. In limited cases some of these techniques can be used to restore original dimensions to salvage or repair an item. An unfinished surface is often called *mill finish*.

Surface finishing processes can be categorized by how they affect the workpiece:

- Removing or reshaping finishing

- Adding or altering finishing

Mechanical processes may also be categorized together because of similarities the final surface finish.

Adding and Altering

- Blanching

- Burnishing

- Case hardening

- Ceramic glaze

- Cladding

- Corona treatment

- Diffusion processes:
    - Carburizing
    - Nitriding
- Electroplating
- Galvanizing
- Gilding
- Glazing
- Knurling
- Painting
- Passivation/Conversion coating
    - Anodizing
    - Bluing
    - Chromate conversion coating
    - Phosphate conversion coating
        - Parkerizing
    - Plasma electrolytic oxidation
- Peening
    - Shot peening
    - Laser peening
- Pickling
- Plasma spraying
- Powder coating
- Thin-film deposition
    - Chemical vapor deposition (CVD)
    - Electroplating
    - Electrophoretic deposition (EPD)

- o Mechanical plating
- o Sputter deposition
- o Physical vapor deposition (PVD)
- o Vacuum plating
- Vitreous enamel

Removing and Reshaping

- Abrasive blasting
  - o Sandblasting
- Burnishing
- Chemical-mechanical planarization (CMP)
- Electropolishing
- Flame polishing
- Gas cluster ion beam
- Grinding
- Industrial etching
- Laser ablation
- Laser engraving
- Linishing
- Magnetic field-assisted finishing
- Mass finishing processes
  - o Tumble finishing
  - o Vibratory finishing
- Peening
  - o Shot peening
  - o Laser peening
- Pickling

- Polishing

  - Buffing

  - Lapping

- Superfinishing

Mechanical Finishing

Mechanical finishing processes include:

- Abrasive blasting

  - Sandblasting

- Burnishing

- Grinding

- Mass finishing processes

  - Tumble finishing

  - Vibratory finishing

- Polishing

  - Buffing

  - Lapping

The use of abrasives in metal polishing results in what is considered a "mechanical finish".

Metal Finish Designations

#3 Finish

Also known as grinding, roughing or rough grinding. These finishes are coarse in nature and usually are a preliminary finish applied before manufacturing. An example would be grinding gates off of castings, deburring or removing excess weld material. It is coarse in appearance and applied by using 36–100 grit abrasive.

When the finish is specified as #3, the material is polished to a uniform 60–80 grit.

#4 Architectural finish

Also known as brushed, directional or satin finish. A #4 architectural finish is characterized by fine polishing grit lines that are uniform and directional in appearance. It is produced by polishing the metal with a 120–180 grit belt or wheel finish and then softened with an 80–120 grit greaseless compound or a medium non woven abrasive belt or pad.

#4 Dairy or sanitary finish

This finish is commonly used for the medical and food industry and almost exclusively used on stainless steel. This finish is much finer than a #4 architectural finish. This finish enhances the physical appearance of the metal as well as increases the sanitary benefits. One takes great care to remove any surface defects in the metal, like pits, that could allow bacteria to grow. A #4 dairy or sanitary finish is produced by polishing with a 180–240 grit belt or wheel finish softened with 120–240 grit greaseless compound or a fine non woven abrasive belt or pad.

#6 Finish

Also known as a fine satin finish. This finish is produced by polishing with a 220–280 grit belt or wheel softened with a 220–230 greaseless compound or very fine non woven abrasive belt or pad. Polishing lines will be soft and less reflective than a #4 architectural finish.

#7 Finish

A #7 finish is produced by polishing with a 280–320 belt or wheel and sisal buffing with a cut and color compound. This is a semi-bright finish that will still have some polishing lines but they will be very dull. Carbon steel and iron are commonly polished to a #7 finish before chrome plating. A #7 finish can be made bright by color buffing with coloring compound and a cotton buff. This is commonly applied to keep polishing costs down when a part needs to be shiny but not flawless.

#8 Finish

Also known as a mirror finish. This finish is produced by polishing with at least a 320 grit belt or wheel finish. Care will be taken in making sure all surface defects are removed. The part is sisal buffed and then color buffed to achieve a mirror finish. The quality of this finish is dependent on the quality of the metal being polished. Some alloys of steel and aluminum cannot be brought to a mirror finish. Castings that have slag or pits will also be difficult, if not impossible, to polish to a #8.

Surfactant

Surfactants are compounds that lower the surface tension (or interfacial tension) between two liquids or between a liquid and a solid. Surfactants may act as detergents, wetting agents, emulsifiers, foaming agents, and dispersants.

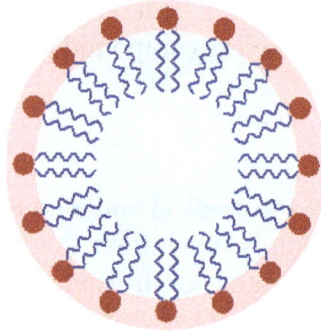

Schematic diagram of a micelle of oil in aqueous suspension, such as might occur in an emulsion of oil in water. In this example the surfactant molecules' oil-soluble tails project into the oil, while the water-soluble ends remain in contact with the water phase

Etymology and Definition

The term *surfactant* is a blend of *surface active agent*.

In the United States National Library of Medicine's Medical Subject Headings (MeSH) vocabulary, *surfactant* is reserved for the meaning pulmonary surfactant. For the more general meaning, *surface active agent/s* is the heading.

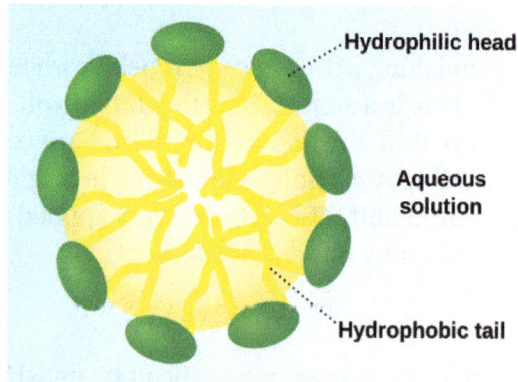

Schematic diagram of a micelle – the lipophilic tails of the surfactant ions remain inside the oil because they interact more strongly with oil than with water. The polar "heads" of the surfactant molecules coating the micelle interact more strongly with water, so they form a hydrophilic outer layer that forms a barrier between micelles. This inhibits the oil droplets, the hydrophobic cores of micelles, from merging into fewer, larger droplets ("emulsion breaking") of the micelle. The compounds that coat a micelle are typically amphiphilic in nature, meaning that micelles may be stable either as droplets of aprotic solvents such as oil in water, or as protic solvents such as water in oil. When the droplet is aprotic it sometimes is known as a reverse micelle.

Composition and Structure

Surfactants are usually organic compounds that are amphiphilic, meaning they contain both hydrophobic groups (their *tails*) and hydrophilic groups (their *heads*). Therefore, a surfactant contains both a water-insoluble (or oil-soluble) component and a water-soluble component. Surfactants will diffuse in water and adsorb at interfaces between air and water or at the interface between oil and water, in the case where water is mixed with oil. The water-insoluble hydrophobic group may extend out of the bulk water phase, into the air or into the oil phase, while the water-soluble head group remains in the water phase.

World production of surfactants is estimated at 15 Mton/y, of which about half are soaps. Other surfactants produced on a particularly large scale are linear alkylbenzenesulfonates (1700 kton/y), lignin sulfonates (600 kton/y), fatty alcohol ethoxylates (700 ktons/y), and alkylphenol ethoxylates (500 kton/y).

Sodium stearate, the most common component of most soap, which comprises about 50% of commercial surfactants.

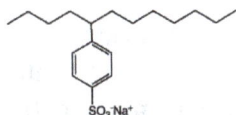

4-(5-Dodecyl) benzenesulfonate, a linear dodecylbenzenesulfonate, one of the most common surfactants.

Structure of Surfactant Phases in Water

In the bulk aqueous phase, surfactants form aggregates, such as micelles, where the hydrophobic tails form the core of the aggregate and the hydrophilic heads are in contact with the surrounding liquid. Other types of aggregates can also be formed, such as spherical or cylindrical micelles or lipid bilayers. The shape of the aggregates depends on the chemical structure of the surfactants, namely the balance in size between hydrophilic head and hydrophobic tail. A measure of this is the HLB, Hydrophilic-lipophilic balance. Surfactants reduce the surface tension of water by adsorbing at the liquid-air interface. The relation that links the surface tension and the surface excess is known as the Gibbs isotherm.

Dynamics of Surfactants at Interfaces

The dynamics of surfactant adsorption is of great importance for practical applications such as in foaming, emulsifying or coating processes, where bubbles or drops are rapidly generated and need to be stabilized. The dynamics of adsorption depend on the diffusion coefficient of the surfactant. As the interface is created, the adsorption is limited by the diffusion of the surfactant to the interface. In some cases, there can exist an energetic barrier to adsorption or desorption of the surfactant. If such a barrier limits the adsorption rate, the dynamics are said to be 'kinetically limited'. Such energy barriers can be due to steric or electrostatic repulsions. The surface rheology of surfactant layers, including the elasticity and viscosity of the layer, play an important role in the stability of foams and emulsions.

Characterization of Interfaces and Surfactant Layers

Interfacial and surface tension can be characterized by classical methods such as the -pendant or spinning drop method. Dynamic surface tensions, i.e. surface tension as a function of time, can be obtained by the maximum bubble pressure apparatus

The structure of surfactant layers can be studied by ellipsometry or X-Ray reflectivity.

Surface rheology can be characterized by the oscillating drop method or shear surface rheometers such as double-cone, double-ring or magnetic rod shear surface rheometer.

Detergents in Biochemistry and Biotechnology

In solution, detergents help solubilize a variety of chemical species by dissociating aggregates and unfolding proteins. Popular surfactants in the biochemistry laboratory are SDS and CTAB. Detergents are key reagents to extract protein by lysis of the cells and tissues: They disorganize the membrane's lipidic bilayer (SDS, Triton X-100, X-114, CHAPS, DOC, and NP-40), and solubilize proteins. Milder detergents such as octyl thioglucoside, octyl glucoside or dodecyl maltoside are used to solubilize membrane proteins such as enzymes and receptors without denaturing them. Non-solubilized material is harvested by centrifugation or other means. For electrophoresis, for example, proteins are classically treated with SDS to denature the native tertiary and quaternary structures, allowing the separation of proteins according to their molecular weight.

Detergents have also been used to decellularise organs. This process maintains a matrix of proteins that preserves the structure of the organ and often the microvascular network. The process has been successfully used to prepare organs such as the liver and heart for transplant in rats. Pulmonary surfactants are also naturally secreted by type II cells of the lung alveoli in mammals.

Classification of Surfactants

The "tail" of most surfactants are fairly similar, consisting of a hydrocarbon chain, which can be branched, linear, or aromatic. Fluorosurfactants have fluorocarbon chains. Siloxane surfactants have siloxane chains.

Many important surfactants include a polyether chain terminating in a highly polar anionic group. The polyether groups often comprise ethoxylated (polyethylene oxide-like) sequences inserted to increase the hydrophilic character of a surfactant. Polypropylene oxides conversely, may be inserted to increase the lipophilic character of a surfactant.

Surfactant molecules have either one tail or two; those with two tails are said to be *double-chained*.

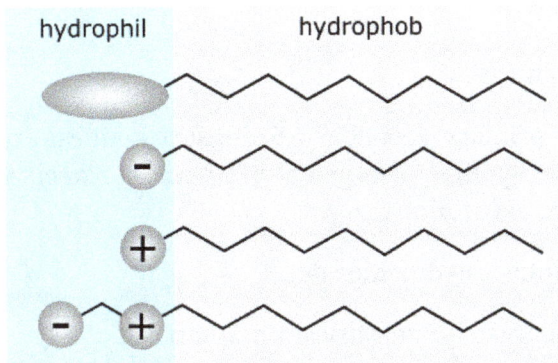

Surfactant classification according to the composition of their head: nonionic, anionic, cationic, amphoteric.

Most commonly, surfactants are classified according to polar head group. A non-ionic surfactant has no charged groups in its head. The head of an ionic surfactant carries a net positive, or negative charge. If the charge is negative, the surfactant is more specifically called anionic; if the charge is positive, it is called cationic. If a surfactant contains a head with two oppositely charged groups, it is termed zwitterionic. Commonly encountered surfactants of each type include:

Anionic

Sulfate, Sulfonate, and Phosphate Esters

Anionic surfactants contain anionic functional groups at their head, such as sulfate, sulfonate, phosphate, and carboxylates. Prominent alkyl sulfates include ammonium lauryl sulfate, sodium lauryl sulfate (sodium dodecyl sulfate, SLS, or SDS), and the related alkyl-ether sulfates sodium laureth sulfate (sodium lauryl ether sulfate or SLES), and sodium myreth sulfate.

Others include:

- Docusate (dioctyl sodium sulfosuccinate)
- Perfluorooctanesulfonate (PFOS)
- Perfluorobutanesulfonate
- Alkyl-aryl ether phosphates
- Alkyl ether phosphates

Carboxylates

These are the most common surfactants and comprise the alkyl carboxylates (soaps), such as sodium stearate. More specialized species include sodium lauroyl sarcosinate and carboxylate-based fluorosurfactants such as perfluorononanoate, perfluoroocta-noate (PFOA or PFO).

Cationic Head Groups

- pH-dependent primary, secondary, or tertiary amines: Primary and secondary amines become positively charged at pH < 10:"Bordwell pKa Table (Acidity in DMSO)". Retrieved 11 May 2014.

    o Octenidine dihydrochloride;

- Permanently charged quaternary ammonium salts:

    o Cetrimonium bromide (CTAB)

    o Cetylpyridinium chloride (CPC)

    o Benzalkonium chloride (BAC)

    o Benzethonium chloride (BZT)

    o Dimethyldioctadecylammonium chloride

    o Dioctadecyldimethylammonium bromide (DODAB)

Zwitterionic Surfactants

Zwitterionic (amphoteric) surfactants have both cationic and anionic centers attached to the same molecule. The cationic part is based on primary, secondary, or tertiary amines or quaternary ammonium cations. The anionic part can be more variable and include sulfonates, as in the sultaines CHAPS (3-[(3-Cholamidopropyl)dimethylam-monio]-1-propanesulfonate) and cocamidopropyl hydroxysultaine. Betaines such as cocamidopropyl betaine have a carboxylate with the ammonium. The most common biological zwitterionic surfactants have a phosphate anion with an amine or ammoni-um, such as the phospholipids phosphatidylserine, phosphatidylethanolamine, phos-phatidylcholine, and sphingomyelins.

Nonionic Surfactant

Many long chain alcohols exhibit some surfactant properties. Prominent among these are the fatty alcohols, cetyl alcohol, stearyl alcohol, and cetostearyl alcohol (consisting predominantly of cetyl and stearyl alcohols), and oleyl alcohol.

- Polyethylene glycol alkyl ethers (Brij): $CH_3-(CH_2)_{10-16}-(O-C_2H_4)_{1-25}-OH$:

    o Octaethylene glycol monododecyl ether

    o Pentaethylene glycol monododecyl ether

- Polypropylene glycol alkyl ethers: $CH_3-(CH_2)_{10-16}-(O-C_3H_6)_{1-25}-OH$
- Glucoside alkyl ethers: $CH_3-(CH_2)_{10-16}-(O-Glucoside)_{1-3}-OH$:
  - Decyl glucoside,
  - Lauryl glucoside
  - Octyl glucoside
- Polyethylene glycol octylphenyl ethers: $C_8H_{17}-(C_6H_4)-(O-C_2H_4)_{1-25}-OH$:
  - Triton X-100
- Polyethylene glycol alkylphenyl ethers: $C_9H_{19}-(C_6H_4)-(O-C_2H_4)_{1-25}-OH$:
  - Nonoxynol-9
- Glycerol alkyl esters:
  - Glyceryl laurate
- Polyoxyethylene glycol sorbitan alkyl esters: Polysorbate
- Sorbitan alkyl esters: Spans
- Cocamide MEA, cocamide DEA
- Dodecyldimethylamine oxide
- Block copolymers of polyethylene glycol and polypropylene glycol: Poloxamers
- Polyethoxylated tallow amine (POEA).

According to the Composition of their Counter-ion

In the case of ionic surfactants, the counter-ion can be:

- Monatomic / Inorganic:
  - Cations: metals : alkali metal, alkaline earth metal, transition metal
  - Anions: halides: chloride (Cl^-), bromide (Br^-), iodide (I^-)
- Polyatomic / Organic:
  - Cations: ammonium, pyridinium, triethanolamine (TEA)
  - Anions: tosyls, trifluoromethanesulfonates, methyl sulfate

In Pharmacy

A wetting agent is a surfactant that, when dissolved in water, lowers the advancing contact angle, aids in displacing an air phase at the surface, and replaces it with a liquid phase. Examples of application of wetting to pharmacy and medicine include the displacement of air from the surface of sulfur, charcoal, and other powders for the purpose of dispersing these drugs in liquid vehicles; the displacement of air from the matrix of cotton pads and bandages so that medicinal solutions can be absorbed for application to various body areas; the displacement of dirt and debris by the use of detergents in the washing of wounds; and the application of medicinal lotions and sprays to surface of skin and mucous membranes.

Current Market and Forecast

The annual global production of surfactants was 13 million tonnes in 2008. In 2014, the world market for surfactants reached a volume of more than 33 billion US-dollars. Market researchers expect annual revenues to increase by 2.5% per year to around 40.4 billion US-dollars until 2022. The commercially most significant type of surfactants is currently the anionic surfactant alkyl benzene sulfonate (LAS), which is widely used in cleaners and detergents.

Health and Environmental Controversy

Surfactants are routinely deposited in numerous ways on land and into water systems, whether as part of an intended process or as industrial and household waste. Some of them are known to be toxic to animals, ecosystems, and humans, and can increase the diffusion of other environmental contaminants. As a result, there are proposed or voluntary restrictions on the use of some surfactants. For example, PFOS is a persistent organic pollutant as judged by the Stockholm Convention. Additionally, PFOA has been subject to a voluntary agreement by the U.S. Environmental Protection Agency and eight chemical companies to reduce and eliminate emissions of the chemical and its precursors.

The two major surfactants used in the year 2000 were linear alkylbenzene sulfonates (LAS) and the alkyl phenol ethoxylates (APE). They break down in the aerobic conditions found in sewage treatment plants and in soil to the metabolite nonylphenol, which is thought to be an endocrine disruptor.

Ordinary dishwashing detergent, for example, will promote water penetration in soil, but the effect would last only a few days (many standard laundry detergent powders contain levels of chemicals such as alkali and chelating agents that can be damaging to plants and should not be applied to soils). Commercial soil wetting agents will continue to work for a considerable period, but they will eventually be degraded by soil micro-organisms. Some can, however, interfere with the life-cycles of some aquatic organisms, so care should be taken to prevent run-off of these products into streams, and excess product should not be washed down.

Anionic surfactants can be found in soils as the result of sludge application, wastewater irrigation, and remediation processes. Relatively high concentrations of surfactants together with multimetals can represent an environmental risk. At low concentrations, surfactant application is unlikely to have a significant effect on trace metal mobility.

Biosurfactants

Biosurfactants are surface-active substances synthesised by living cells. Interest in microbial surfactants has been steadily increasing in recent years due to their diversity, environmentally friendly nature, possibility of large-scale production, selectivity, performance under extreme conditions, and potential applications in environmental protection. A few of the popular examples of microbial biosurfactants includes Emulsan produced by *Acinetobacter calcoaceticus*, Sophorolipids produced by several yeasts belonging to *candida* and the *starmerella* clade, and Rhamnolipid produced by *Pseudomonas aeruginosa* etc.

Biosurfactants enhance the emulsification of hydrocarbons, have the potential to solubilise hydrocarbon contaminants and increase their availability for microbial degradation. The use of chemicals for the treatment of a hydrocarbon polluted site may contaminate the environment with their by-products, whereas biological treatment may efficiently destroy pollutants, while being biodegradable themselves. Hence, biosurfactant-producing microorganisms may play an important role in the accelerated bioremediation of hydrocarbon-contaminated sites. These compounds can also be used in enhanced oil recovery and may be considered for other potential applications in environmental protection. Other applications include herbicides and pesticides formulations, detergents, healthcare and cosmetics, pulp and paper, coal, textiles, ceramic processing and food industries, uranium ore-processing, and mechanical dewatering of peat.

Several microorganisms are known to synthesise surface-active agents; most of them are bacteria and yeasts. When grown on hydrocarbon substrate as the carbon source, these microorganisms synthesise a wide range of chemicals with surface activity, such as glycolipid, phospholipid, and others. These chemicals are synthesised to emulsify the hydrocarbon substrate and facilitate its transport into the cells. In some bacterial species such as *Pseudomonas aeruginosa*, biosurfactants are also involved in a group motility behavior called swarming motility.

Safety and Environmental Risks

Most anionic and nonionic surfactants are nontoxic, having LD50 comparable to sodium chloride. The situation for cationic surfactants is more diverse. Dialkyldimethylammonium chlorides have very low LD50's (5 g/kg) but alkylbenzyldimethylammonium chloride has an LD50 of 0.35 g/kg. Prolonged exposure of skin to surfactants can cause chafing because surfactants (e.g., soap) disrupts the lipid coating that protects skin (and other) cells.

Biosurfactants and Deepwater Horizon

The use of biosurfactants as a way to remove petroleum from contaminated sites has been studied and found to be safe and effective in the removal of petroleum products from soil. Other studies found that surfactants are often more toxic than the oil that is being dispersed, and the combination of the oil and the surfactant can be more toxic than either alone. Biosurfactants were not used by BP after the Deepwater Horizon oil spill. However, unprecedented amounts of Corexit (active ingredient: Tween-80), were sprayed directly into the ocean at the leak and on the sea-water's surface, the theory being that the surfactants isolate droplets of oil, making it easier for petroleum-consuming microbes to digest the oil.

Biosurfactants produced by microbe or bacteria can be used to enhance oil production by microbial enhanced oil recovery method (MEOR).

Applications

Surfactants play an important role as cleaning, wetting, dispersing, emulsifying, foaming and anti-foaming agents in many practical applications and products, including:

- Detergents

- Fabric softeners

- Emulsions

- Soaps

- Paints

- Adhesives

- Inks

- Anti-fogs

- Ski waxes, snowboard wax

- Deinking of recycled papers, in flotation, washing and enzymatic processes

- Laxatives

- Agrochemical formulations

    o   Herbicides (some)

    o   Insecticides

- Quantum dot in order to manipulate growth and assembly of the dots, reactions on their surface, electrical properties, etc., it is important to understand how surfactants arrange on the surface of the quantum dots

- Biocides (sanitizers)

- Cosmetics:

    o Shampoos

    o Hair conditioners (after shampoo)

    o Toothpastes

- Spermicides (nonoxynol-9)

- Firefighting

- Pipelines, liquid drag reducing agent

- Alkali Surfactant Polymers (used to mobilize oil in oil wells)

- Ferrofluids

- Leak Detectors

Brinelling

Brinelling is the permanent indentation of a hard surface. It is named after the Brinell scale of hardness, in which a small ball is pushed against a hard surface at a preset level of force, and the depth and diameter of the mark indicates the Brinell hardness of the surface. Brinelling is a process of wear in which similar marks are pressed into the surface of a moving part, such as bearings or hydraulic pistons. The brinelling is usually undesirable, as the parts often mate with other parts in very close proximity. The very small indentations can quickly lead to improper operation, like chattering or excess vibration, which in turn can accelerate other forms of wear, such as spalling and galling.

Introduction

Brinelling is a material surface failure caused by Hertz contact stress that exceeds the material limit. It usually occurs in situations where a significant load force is distributed over a relatively small surface area. Brinelling typically results from a heavy or repeated impact load, either while stopped or during rotation, though it can also be caused by just one application of a force greater than the material limit.

Brinelling can be caused by a heavy load resting on a stationary bearing for an extended length of time. The result is a permanent dent or "brinell mark". The brinell marks will often appear in evenly spaced patterns along the bearing races, resembling the primary elements of the bearing, such as rows of indented lines for needle or roller bearings or rounded indentations in ball bearings. It is a common cause of roller bearing failures, and loss of preload in bolted joints when a hardened washer is not used. For example, brinelling occurs in casters when the ball bearings within the swivel head produce grooves in the hard cap, thus degrading performance by increasing the required swivel force.

Avoiding Brinelling Damage

Engineers can use the Brinell hardness of materials in their calculations to avoid this mode of failure. A rolling element bearing's static load rating is defined to avoid this failure type. Increasing the number of elements can provide better distribution of the load, so bearings intended for a large load may have many balls, or use needles instead. This decreases the chances of brinelling, but increases friction and other factors. However, although roller and ball bearings work well for radial and thrust loading, they are often prone to brinelling when very high impact loading, lateral loading, or vibration are experienced. Babbitt bearings or bronze bushings are often used instead of roller bearings in applications where such loads exist, such as in automotive crankshafts or pulley sheaves, to decrease the possibility of brinelling by distributing the force over a very large surface area.

A common cause of brinelling is the use of improper installation procedures. Brinelling often occurs when pressing bearings into holes or onto shafts. Care must usually be taken to ensure that pressure is applied to the proper bearing race to avoid transferring the pressure from one race to the other through the balls or rollers. If pressing force is applied to the wrong race, brinelling can occur to either or both of the races. The act of pressing or clamping can also leave brinell marks, especially if the vise or press has serrated jaws or roughened surfaces. Flat pressing plates are often used in the pressing of bearings, while soft copper, brass, or aluminum jaw covers are often used in vises to help avoid brinell marks from being forced into the workpiece.

False Brinelling

A similar-looking kind of damage is called false brinelling and is caused by fretting wear. This occurs when contacting bodies vibrate against each other in the presence of very small loads, which pushes lubricant out of the contact surface area, and the bearing assembly cannot move far enough to redistribute the displaced lubricant. The result is a finely polished surface that resembles a brinell mark, but has not permanently deformed either contacting surface. This type of false brinelling usually occurs in bearings during transportation, between the time of manufacture and installation. The polished surfaces are often mistaken for brinelling, although no actual damage to the bearing exists. The false brinelling will disappear after a short break-in period of operation.

Fretting wear can also occur during operation, causing deeper indentations. This occurs when small vibrations form in the rotating shaft and become harmonically in sync with the speed of rotation, causing circular oscillations in the shaft. The oscillation causes the shaft to move in precession, and the timing of the rotation speed causes the balls or rollers to contact the races only when they are in similar positions. This forms wear marks caused by contact with the bearings and the races in specific areas, but not in others, leaving an uneven wear-pattern that resembles brinelling. However, the marks are usually too wide and do not exactly match the shape of the bearing, and therefore this type of wear can be differentiated from true brinelling.

False Brinelling

False-Brinelling of a bearing

False brinelling is damage caused by fretting, with or without corrosion, that causes imprints that look similar to brinelling, but are caused by a different mechanism. Brinell damage is characterized by permanent material deformation (without loss of material) and occurs during one load event, whereas false brinelling is characterized by material wear or removal and occurs over an extended time from vibration and light loads.

The basic cause of false brinelling is that the design of the bearing does not have a method for redistribution of lubricant without large rotational movement of all bearing surfaces in the raceway. Lubricant is pushed out of a loaded region during small oscillatory movements and vibration where the bearings surfaces repeatedly do not move very far. Without lubricant, wear is increased when the small oscillatory movements occur again. It is possible for the resulting wear debris to oxidize and form an abrasive compound which further accelerates wear.

Mechanism of Action

In normal operation, a rolling-element bearing has the rollers and races separated by a thin layer of lubricant such as grease or oil. Although these lubricants normally appear liquid (not solid), under high pressure they act as solids and keep the bearing and race from touching.

If the lubricant is removed, the bearings and races can touch directly. While bearings and races appear smooth to the eye, they are microscopically rough. Thus, high points of each surface can touch, but "valleys" do not. The bearing load is thus spread over much less area increasing the contact stress, causing pieces of each surface to break off or to become pressure-welded then break off when the bearing rolls on.

The broken-off pieces are also called *wear debris*. Wear debris is bad because it is relatively large compared to the surrounding surface finish and thus creates more regions of high contact stress. Worse, the steel in ordinary bearings can oxidize (rust), producing a more abrasive compound which accelerates wear.

Examples

The discovery of false brinelling is unclear, but one story describes how, in the 1930s, new automobiles were loaded onto trains for delivery; when they were unloaded, some would show severe wheel bearing damage. On further inspection, it turned out that many wheel bearings were slightly damaged. The damage was eventually traced to rocking of the autos and the regular impact every time a railroad car wheel passed a track joint. These conditions led to false brinelling.

Although the auto-delivery problem has been solved, there are many modern examples. For example, generators or pumps may fail or need service, so it is common to have a nearby spare unit which is left off most of the time but brought into service when needed. Surprisingly, however, vibration from the operating unit can cause bearing failure in the unit which is switched off. When that unit is turned on, the bearings may be noisy due to damage, and may fail completely within a few days or weeks even though the unit and its bearings are otherwise new. Common solutions include: keeping the spare unit at a distance from the one which is on and vibrating; manually rotating shafts of the spare units on a regular (for example, weekly) basis; or regularly switching between the units so that both are in regular (for example, weekly) operation.

Until recently, bicycle headsets tended to suffer from false brinelling in the "straight ahead" steering position, due to small movements caused by flexing of the fork. Good modern headsets incorporate a plain bearing to accommodate this flexing, leaving the ball race to provide pure rotational movement.

Gear

A gear or cogwheel is a rotating machine part having cut teeth, or cogs, which mesh with another toothed part to transmit torque. Geared devices can change the speed, torque, and direction of a power source. Gears almost always produce a change in torque, creating a mechanical advantage, through their gear ratio, and thus may be considered a simple machine. The teeth on the two meshing gears all have the same shape. Two or more meshing gears, working in a sequence, are called a gear train or a *transmission*. A gear can mesh with a linear toothed part, called a rack, thereby producing translation instead of rotation.

Two meshing gears transmitting rotational motion. Note that the smaller gear is rotating faster. Although the larger gear is rotating less quickly, its torque is proportionally greater. One subtlety of this particular arrangement is that the linear speed at the pitch diameter is the same on both gears.

Multiple reducer gears in microwave oven (ruler for scale)

The gears in a transmission are analogous to the wheels in a crossed, belt pulley system. An advantage of gears is that the teeth of a gear prevent slippage.

When two gears mesh, if one gear is bigger than the other, a mechanical advantage is produced, with the rotational speeds, and the torques, of the two gears differing in proportion to their diameters.

In transmissions with multiple gear ratios—such as bicycles, motorcycles, and cars—the term "gear" as in "first gear" refers to a gear ratio rather than an actual physical gear. The term describes similar devices, even when the gear ratio is continuous rather than discrete, or when the device does not actually contain gears, as in a continuously variable transmission.

History

Early examples of gears date from the 4th century BCE in China (Zhan Guo times - Late East Zhou dynasty), which have been preserved at the Luoyang Museum of Henan Province, China. The earliest gears in Europe were circa CE 50 by Hero of Alexandria, but they can be traced back to the Greek mechanics of the Alexandrian school in the 3rd century BCE and were greatly developed by the Greek polymath Archimedes (287–212 BCE).

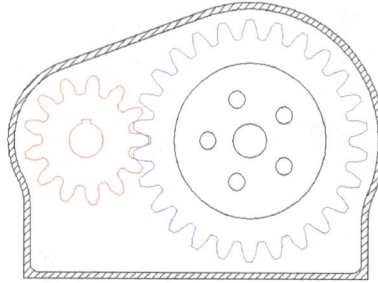

Single stage gear reducer.

Examples of further development include:

- Ma Jun (c. 200–265 CE) used gears as part of a south-pointing chariot.

- The Antikythera mechanism is an example of a very early and intricate geared device, designed to calculate astronomical positions. Its time of construction is now estimated between 150 and 100 BCE.

- The water-powered grain-mill, the water-powered saw mill, fulling mill, and other applications of watermill often used gears.

- The first mechanical clocks were built in CE 725.

  - The 1386 Salisbury cathedral clock may be the world's oldest working mechanical clock.

Comparison With Drive Mechanisms

The definite velocity ratio that teeth give gears provides an advantage over other drives (such as traction drives and V-belts) in precision machines such as watches that depend upon an exact velocity ratio. In cases where driver and follower are proximal, gears also have an advantage over other drives in the reduced number of parts required; the downside is that gears are more expensive to manufacture and their lubrication requirements may impose a higher operating cost per hour.

Types

External vs Internal Gears

An *external gear* is one with the teeth formed on the outer surface of a cylinder or cone.

Conversely, an *internal gear* is one with the teeth formed on the inner surface of a cylinder or cone. For bevel gears, an internal gear is one with the pitch angle exceeding 90 degrees. Internal gears do not cause output shaft direction reversal.

Internal gear

Spur

Spur gear

Spur gears or *straight-cut gears* are the simplest type of gear. They consist of a cylinder or disk with teeth projecting radially. Though the teeth are not straight-sided (but usually of special form to achieve a constant drive ratio, mainly involute but less commonly cycloidal), the edge of each tooth is straight and aligned parallel to the axis of rotation. These gears mesh together correctly only if fitted to parallel shafts. No axial thrust is created by the tooth loads. Spur gears are excellent at moderate speeds but tend to be noisy at high speeds.

Helical

Helical gears Top: parallel configuration Bottom: crossed configuration

Helical or "dry fixed" gears offer a refinement over spur gears. The leading edges of the teeth are not parallel to the axis of rotation, but are set at an angle. Since the gear is curved, this angling makes the tooth shape a segment of a helix. Helical gears can be meshed in *parallel* or *crossed* orientations. The former refers to when the shafts are parallel to each other; this is the most common orientation. In the latter, the shafts are non-parallel, and in this configuration the gears are sometimes known as "skew gears".

The angled teeth engage more gradually than do spur gear teeth, causing them to run more smoothly and quietly. With parallel helical gears, each pair of teeth first make contact at a single point at one side of the gear wheel; a moving curve of contact then grows gradually across the tooth face to a maximum then recedes until the teeth break contact at a single point on the opposite side. In spur gears, teeth suddenly meet at a line contact across their entire width causing stress and noise. Spur gears make a characteristic whine at high speeds. For this reason spur gears are used in low speed applications and in situations where noise control is not a problem, and helical gears are used in high speed applications, large power transmission, or where noise abatement is important. The speed is considered high when the pitch line velocity exceeds 25 m/s.

A disadvantage of helical gears is a resultant thrust along the axis of the gear, which must be accommodated by appropriate thrust bearings, and a greater degree of sliding friction between the meshing teeth—often addressed with additives in the lubricant.

Skew Gears

For a 'crossed' or 'skew' configuration, the gears must have the same pressure angle and normal pitch; however, the helix angle and handedness can be different. The relationship between the two shafts is actually defined by the helix angle(s) of the two shafts and the handedness, as defined:

$E = \beta_1 + \beta_2$ for gears of the same handedness

$E = \beta_1 - \beta_2$ for gears of opposite handedness

Where β is the helix angle for the gear. The crossed configuration is less mechanically sound because there is only a point contact between the gears, whereas in the parallel configuration there is a line contact.

Quite commonly, helical gears are used with the helix angle of one having the negative of the helix angle of the other; such a pair might also be referred to as having a right-handed helix and a left-handed helix of equal angles. The two equal but opposite angles add to zero: the angle between shafts is zero—that is, the shafts are *parallel*. Where the sum or the difference (as described in the equations above) is not zero the shafts are *crossed*. For shafts *crossed* at right angles, the helix angles are of the same hand because they must add to 90 degrees. (This is the case with the gears in the illustration above: they mesh correctly in the crossed configuration: for the parallel config-uration, one of the helix angles should be reversed. The gears illustrated cannot mesh with the shafts parallel.)

- 3D Animation of helical gears (parallel axis)

- 3D Animation of helical gears (crossed axis)

Double Helical

Herringbone Gears

Double helical gears and herringbone gears are similar but the difference is that her-ringbone gears don't have a groove in the middle like double helical gears do. Double helical gears overcome the problem of axial thrust presented by single helical gears by using two sets of teeth that are set in a V shape. A double helical gear can be thought of as two mirrored helical gears joined together. This arrangement cancels out the net axial thrust, since each half of the gear thrusts in the opposite direction resulting in a net axial force of zero. This arrangement can remove the need for thrust bearings. However, double helical gears are more difficult to manufacture due to their more com-plicated shape.

For both possible rotational directions, there exist two possible arrangements for the oppositely-oriented helical gears or gear faces. One arrangement is stable, and the oth-er is unstable. In a stable orientation, the helical gear faces are oriented so that each axial force is directed toward the center of the gear. In an unstable orientation, both axial forces are directed away from the center of the gear. In both arrangements, the total (or *net*) axial force on each gear is zero when the gears are aligned correctly. If the gears become misaligned in the axial direction, the unstable arrangement generates a net force that may lead to disassembly of the gear train, while the stable arrangement generates a net corrective force. If the direction of rotation is reversed, the direction of the axial thrusts is also reversed, so a stable configuration becomes unstable, and *vice versa*.

Stable double helical gears can be directly interchanged with spur gears without any need for different bearings.

Bevel

Bevel Gear

A bevel gear is shaped like a right circular cone with most of its tip cut off. When two bevel gears mesh, their imaginary vertices must occupy the same point. Their shaft axes also intersect at this point, forming an arbitrary non-straight angle between the shafts. The angle between the shafts can be anything except zero or 180 degrees. Bevel gears with equal numbers of teeth and shaft axes at 90 degrees are called *miter gears*.

Spiral Bevels

Spiral bevel gears

Spiral bevel gears can be manufactured as Gleason types (circular arc with non-constant tooth depth), Oerlikon and Curvex types (circular arc with constant tooth depth), Klingelnberg Cyclo-Palloid (Epicycloide with constant tooth depth) or Klingelnberg Palloid. Spiral bevel gears have the same advantages and disadvantages relative to their straight-cut cousins as helical gears do to spur gears. Straight bevel gears are generally used only at speeds below 5 m/s (1000 ft/min), or, for small gears, 1000 r.p.m.

Note: The cylindrical gear tooth profile corresponds to an involute, but the bevel gear tooth profile to an octoid. All traditional bevel gear generators (like Gleason, Klingelnberg, Heidenreich & Harbeck, WMW Modul) manufacture bevel gears with an octoidal tooth profile. IMPORTANT: For 5-axis milled bevel gear sets it is important to choose the same calculation / layout like the conventional manufacturing method. Simplified calculated bevel gears on the basis of an equivalent cylindrical gear in normal section with an involute tooth form show a deviant tooth form with reduced tooth strength by 10-28% without offset and 45% with offset [Diss. Hünecke, TU Dresden]. Furthermore, the "involute bevel gear sets" cause more noise.

Hypoid

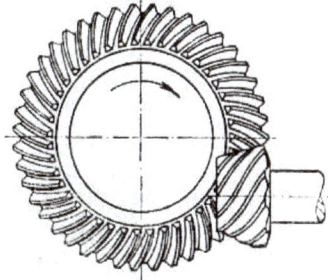

Hypoid gear

Hypoid gears resemble spiral bevel gears except the shaft axes do not intersect. The pitch surfaces appear conical but, to compensate for the offset shaft, are in fact hyperboloids of revolution. Hypoid gears are almost always designed to operate with shafts at 90 degrees. Depending on which side the shaft is offset to, relative to the angling of the teeth, contact between hypoid gear teeth may be even smoother and more gradual than with spiral bevel gear teeth, but also have a sliding action along the meshing teeth as it rotates and therefore usually require some of the most viscous types of gear oil to avoid it being extruded from the mating tooth faces, the oil is normally designated HP (for hypoid) followed by a number denoting the viscosity. Also, the pinion can be designed with fewer teeth than a spiral bevel pinion, with the result that gear ratios of 60:1 and higher are feasible using a single set of hypoid gears. This style of gear is most common in motor vehicle drive trains, in concert with a differential. Whereas a regular (nonhypoid) ring-and-pinion gear set is suitable for many applications, it is not ideal for vehicle drive trains because it generates more noise and vibration than a hypoid does. Bringing hypoid gears to market for mass-production applications was an engineering improvement of the 1920s.

Crown

Crown gear

Crown gears or *contrate gears* are a particular form of bevel gear whose teeth project at right angles to the plane of the wheel; in their orientation the teeth resemble the points on a crown. A crown gear can only mesh accurately with another bevel gear, although crown gears are sometimes seen meshing with spur gears. A crown gear is also sometimes meshed with an escapement such as found in mechanical clocks.

Worm

Worm gear

Worms resemble screws. A worm is meshed with a *worm wheel*, which looks similar to a spur gear.

Worm-and-gear sets are a simple and compact way to achieve a high torque, low speed gear ratio. For example, helical gears are normally limited to gear ratios of less than 10:1 while worm-and-gear sets vary from 10:1 to 500:1. A disadvantage is the potential for considerable sliding action, leading to low efficiency.

4-start worm and wheel

A worm gear is a species of helical gear, but its helix angle is usually somewhat large (close to 90 degrees) and its body is usually fairly long in the axial direction. These attributes give it screw like qualities. The distinction between a worm and a helical gear is that at least one tooth persists for a full rotation around the helix. If this occurs, it is a 'worm'; if not, it is a 'helical gear'. A worm may have as few as one tooth. If that tooth persists for several turns around the helix, the worm appears, superficially, to have more than one tooth, but what one in fact sees is the same tooth reappearing at intervals along the length of the worm. The usual screw nomenclature applies: a one-toothed worm is called *single thread* or *single start*; a worm with more than one tooth is called *multiple thread* or *multiple start*. The helix angle of a worm is not usually specified. Instead, the lead angle, which is equal to 90 degrees minus the helix angle, is given.

In a worm-and-gear set, the worm can always drive the gear. However, if the gear attempts to drive the worm, it may or may not succeed. Particularly if the lead angle is small, the gear's teeth may simply lock against the worm's teeth, because the force component circumferential to the worm is not sufficient to overcome friction.

Worm-and-gear sets that do lock are called self locking, which can be used to advantage, as for instance when it is desired to set the position of a mechanism by turning the worm and then have the mechanism hold that position. An example is the machine head found on some types of stringed instruments.

If the gear in a worm-and-gear set is an ordinary helical gear only a single point of contact is achieved. If medium to high power transmission is desired, the tooth shape of the gear is modified to achieve more intimate contact by making both gears partially envelop each other. This is done by making both concave and joining them at a saddle point; this is called a **cone-drive** or "Double enveloping".

Worm gears can be right or left-handed, following the long-established practice for screw threads.

- 3D Animation of a worm-gear set

Non-circular

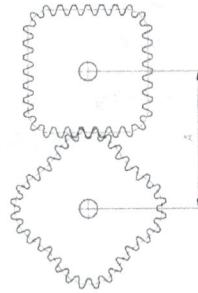

Non-circular gears

Non-circular gears are designed for special purposes. While a regular gear is optimized to transmit torque to another engaged member with minimum noise and wear and maximum efficiency, a non-circular gear's main objective might be ratio variations, axle displacement oscillations and more. Common applications include textile machines, potentiometers and continuously variable transmissions.

Rack and Pinion

A rack is a toothed bar or rod that can be thought of as a sector gear with an infinitely large radius of curvature. Torque can be converted to linear force by meshing a rack with a pinion: the pinion turns; the rack moves in a straight line. Such a mechanism is used in automobiles to convert the rotation of the steering wheel into the left-to-right motion of the tie rod(s). Racks also feature in the theory of gear geometry, where, for instance, the tooth shape of an interchangeable set of gears may be specified for the rack, (infinite radius), and the tooth shapes for gears of particular actual radii are then derived from that. The rack and pinion gear type is employed in a rack railway.

Rack and pinion gearing

Epicyclic

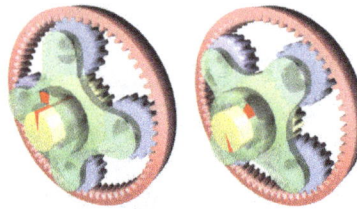

Epicyclic gearing

In epicyclic gearing one or more of the gear axes moves. Examples are sun and planet gearing, cycloidal drive, and mechanical differentials.

Sun and Planet

Sun (yellow) and planet (red) gearing

Sun and planet gearing is a method of converting reciprocating motion into rotary motion that was used in steam engines. James Watt used it on his early steam engines to get around the patent on the crank, but it also provided the advantage of increasing the flywheel speed so Watt could use a lighter flywheel.

In the illustration, the sun is yellow, the planet red, the reciprocating arm is blue, the flywheel is green and the driveshaft is gray.

Harmonic Gear

Harmonic gearing

A *harmonic gear* is a specialized gearing mechanism often used in industrial motion control, robotics and aerospace for its advantages over traditional gearing systems, including lack of backlash, compactness and high gear ratios.

Cage Gear

Cage gear in Pantigo Windmill, Long Island (with the driving gearwheel disengaged)

A *cage gear*, also called a *lantern gear* or *lantern pinion* has cylindrical rods for teeth, parallel to the axle and arranged in a circle around it, much as the bars on a round bird cage or lantern. The assembly is held together by disks at each end, into which the tooth rods and axle are set. Cage gears are more efficient than solid pinions, and dirt can fall through the rods rather than becoming trapped and increasing wear. They can be constructed with very simple tools as the teeth are not formed by cutting or milling, but rather by drilling holes and inserting rods.

Sometimes used in clocks, the *cage gear* should always be driven by a gearwheel, not used as the driver. The *cage gear* was not initially favoured by conservative clock makers. It became popular in turret clocks where dirty working conditions were most commonplace. Domestic American clock movements often used them.

Magnetic Gear

All cogs of each gear component of magnetic gears act as a constant magnet with periodic alternation of opposite magnetic poles on mating surfaces. Gear components are mounted with a backlash capability similar to other mechanical gearings. Although they cannot exert as much force as a traditional gear, such gears work without touching and so are immune to wear, have very low noise and can slip without damage making them very reliable. They can be used in configurations that are not possible for gears that must be physically touching and can operate with a non-metallic barrier completely separating the driving force from the load. The magnetic coupling can transmit force into a hermetically sealed enclosure without using a radial shaft seal, which may leak.

Nomenclature

General Nomenclature

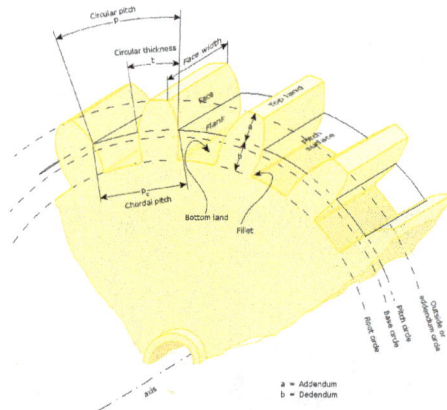

Rotational frequency, n

Measured in rotation over time, such as RPM.

Angular frequency, ω

Measured in radians/second. rad/second

Number of teeth, N

How many teeth a gear has, an integer. In the case of worms, it is the number of thread starts that the worm has.

Gear, wheel

The larger of two interacting gears or a gear on its own.

Pinion

The smaller of two interacting gears.

Path of contact

Path followed by the point of contact between two meshing gear teeth.

Line of action, pressure line

Line along which the force between two meshing gear teeth is directed. It has the same direction as the force vector. In general, the line of action changes from moment to moment during the period of engagement of a pair of teeth. For involute gears, however, the tooth-to-tooth force is always directed along the same line—that is, the line of action is constant. This implies that for involute gears the path of contact is also a straight line, coincident with the line of action—as is indeed the case.

Axis

Axis of revolution of the gear; center line of the shaft.

Pitch point

Point where the line of action crosses a line joining the two gear axes.

Pitch circle, pitch line

Circle centered on and perpendicular to the axis, and passing through the pitch point. A predefined diametral position on the gear where the circular tooth thickness, pressure angle and helix angles are defined.

Pitch diameter, d

A predefined diametral position on the gear where the circular tooth thickness, pressure angle and helix angles are defined. The standard pitch diameter is a basic dimension and cannot be measured, but is a location where other measurements are made. Its value is based on the number of teeth, the normal module (or normal diametral pitch), and the helix angle. It is calculated as:

in metric units or in imperial units.

Module or modulus, m

Since it is impractical to calculate circular pitch with irrational numbers, mechanical engineers usually use a scaling factor that replaces it with a regular value instead. This is known as the *module* or *modulus* of the wheel and is simply defined as

$$m = p \, / \, \pi$$

where m is the module and p the circular pitch. The units of module are customarily millimeters; an *English Module* is sometimes used with the units of inches. When the diametral pitch, DP, is in English units,

in conventional metric units.

The distance between the two axis becomes

$$m = 25.4 >$$

where a is the axis distance, z_1 and z_2 are the number of cogs (teeth) for each of the two wheels (gears). These numbers (or at least one of them) is often chosen among primes to create an even contact between every cog of both wheels, and thereby avoid unnecessary wear and damage. An even uniform gear wear is achieved by ensuring the tooth counts of the two gears meshing together are relatively prime to each other; this occurs when the greatest common divisor (GCD) of each gear tooth count equals 1, e.g. GCD(16,25)=1; if a 1:1 gear ratio is desired a relatively prime gear may be inserted in between the two gears; this maintains the 1:1 ratio but reverses the gear direction; a second relatively prime gear could also be inserted to restore the original rotational direction while maintaining uniform wear with all 4 gears in this case. Mechanical engineers, at least in continental Europe, usually use the module instead of circular pitch. The module, just like the circular pitch, can be used for all types of cogs, not just evolvent based straight cogs.

Operating pitch diameters

Diameters determined from the number of teeth and the center distance at which gears operate. Example for pinion:

$$d_w = \frac{2a}{u+1} = \frac{2a}{\frac{z_2}{z_1}+1}.$$

Pitch surface

In cylindrical gears, cylinder formed by projecting a pitch circle in the axial direction. More generally, the surface formed by the sum of all the pitch circles as one moves along the axis. For bevel gears it is a cone.

Angle of action

Angle with vertex at the gear center, one leg on the point where mating teeth first make contact, the other leg on the point where they disengage.

Arc of action

Segment of a pitch circle subtended by the angle of action.

Pressure angle,

> The complement of the angle between the direction that the teeth exert force on each other, and the line joining the centers of the two gears. For involute gears, the teeth always exert force along the line of action, which, for involute gears, is a straight line; and thus, for involute gears, the pressure angle is constant.

Outside diameter,

> Diameter of the gear, measured from the tops of the teeth.

Root diameter

> Diameter of the gear, measured at the base of the tooth.

Addendum, a

> Radial distance from the pitch surface to the outermost point of the tooth.

Dedendum, b

> Radial distance from the depth of the tooth trough to the pitch surface.

Whole depth,

> The distance from the top of the tooth to the root; it is equal to addendum plus dedendum or to working depth plus clearance.

Clearance

> Distance between the root circle of a gear and the addendum circle of its mate.

Working depth

> Depth of engagement of two gears, that is, the sum of their operating addendums.

Circular pitch, p

> Distance from one face of a tooth to the corresponding face of an adjacent tooth on the same gear, measured along the pitch circle.

Diametral pitch, DP

> $DP = N >$

> Ratio of the number of teeth to the pitch diameter. Could be measured in teeth per inch or teeth per centimeter, but conventionally has units of per inch of diameter. Where the module, m, is in metric units

> $DP = 25.4 >$ in English units

Base circle

> In involute gears, the tooth profile is generated by the involute of the base circle. The radius of the base circle is somewhat smaller than that of the pitch circle

Base pitch, normal pitch,

> In involute gears, distance from one face of a tooth to the corresponding face of an adjacent tooth on the same gear, measured along the base circle

Interference

> Contact between teeth other than at the intended parts of their surfaces

Interchangeable set

> A set of gears, any of which mates properly with any other

Helical Gear Nomenclature

Helix angle, ψ

> Angle between a tangent to the helix and the gear axis. It is zero in the limiting case of a spur gear, albeit it can considered as the hypotenuse angle as well.

Normal circular pitch, p_n

> Circular pitch in the plane normal to the teeth.

Transverse circular pitch, p

> Circular pitch in the plane of rotation of the gear. Sometimes just called "circular pitch". $p_n - p\cos(\psi)$

Several other helix parameters can be viewed either in the normal or transverse planes. The subscript n usually indicates the normal.

Worm Gear Nomenclature

Lead

> Distance from any point on a thread to the corresponding point on the next turn of the same thread, measured parallel to the axis.

Linear pitch, p

> Distance from any point on a thread to the corresponding point on the adjacent thread, measured parallel to the axis. For a single-thread worm, lead and linear pitch are the same.

Lead angle,

> Angle between a tangent to the helix and a plane perpendicular to the axis. Note that the complement of the helix angle is usually given for helical gears.

Pitch diameter,

> Same as described earlier in this list. Note that for a worm it is still measured in a plane perpendicular to the gear axis, not a tilted plane.

Subscript w denotes the worm, subscript g denotes the gear.

Tooth Contact Nomenclature

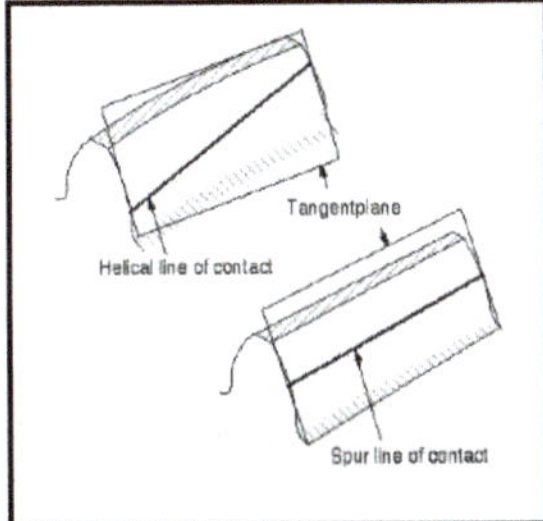

Line of contact	Path of action	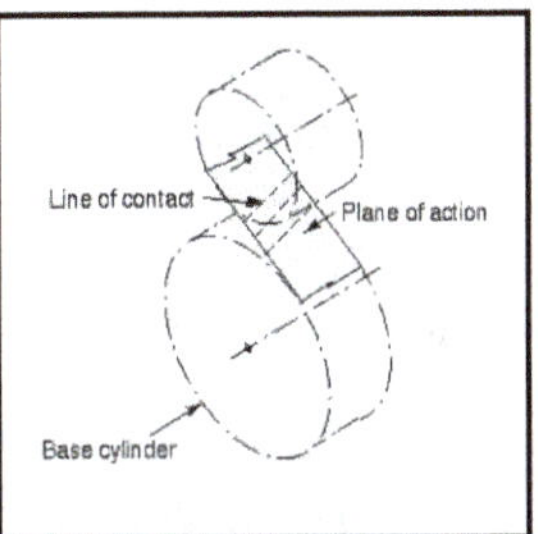 Line of action	Plane of action

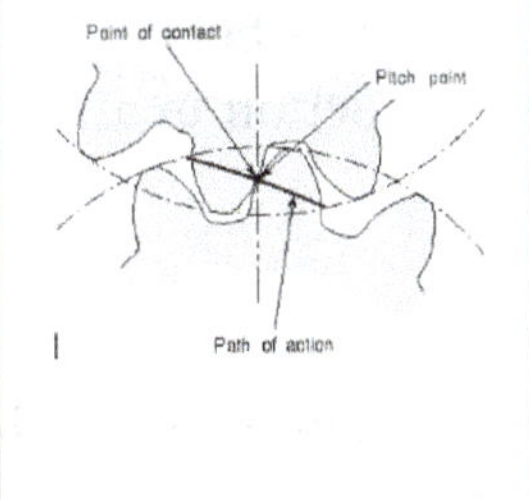

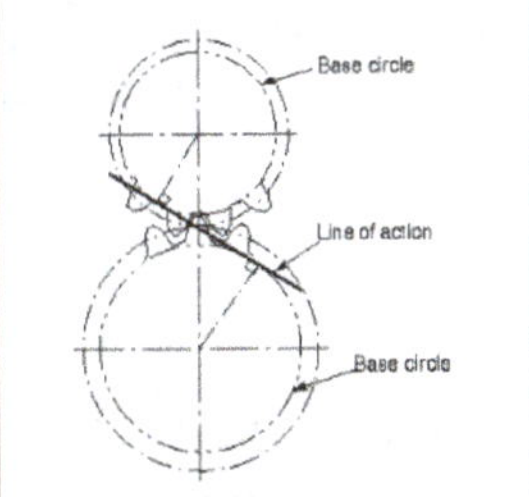

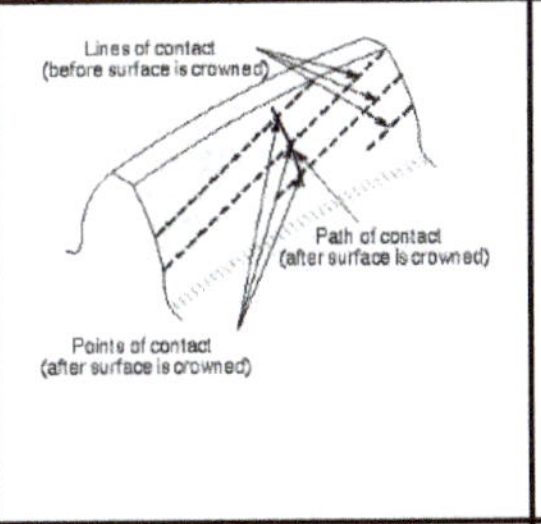

Lines of contact (helical gear)	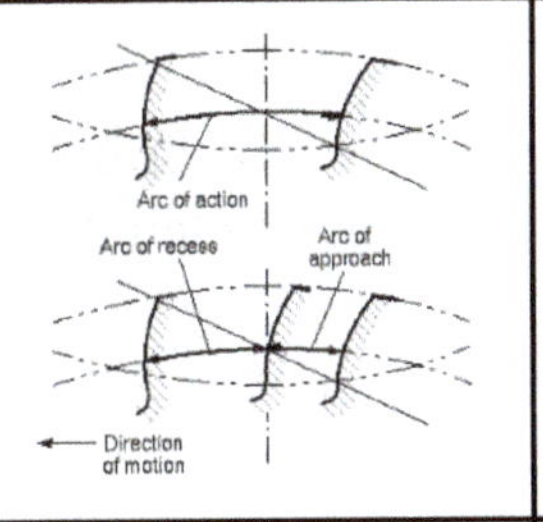 Arc of action	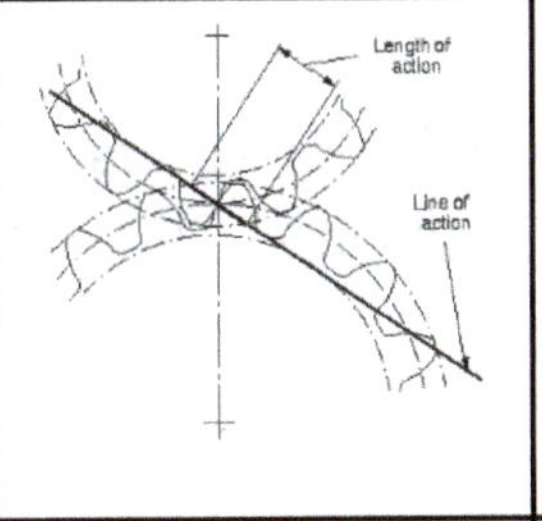 Length of action	Limit diameter

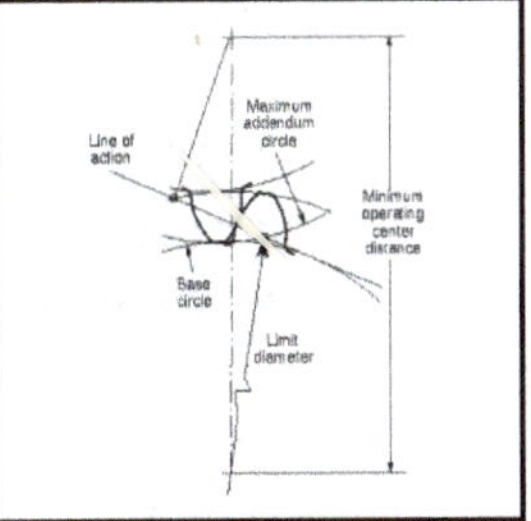

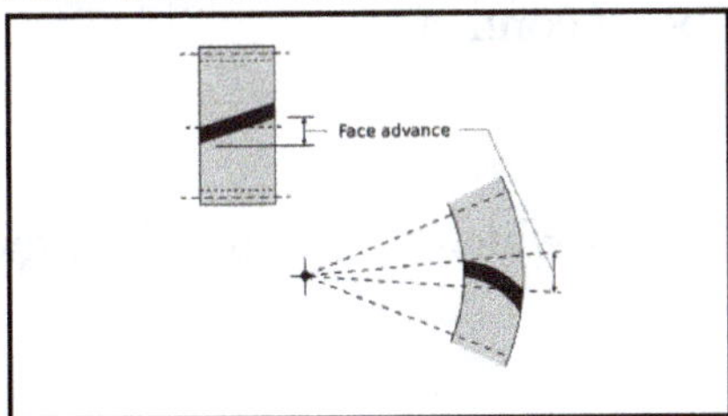

Face advance	Zone of action

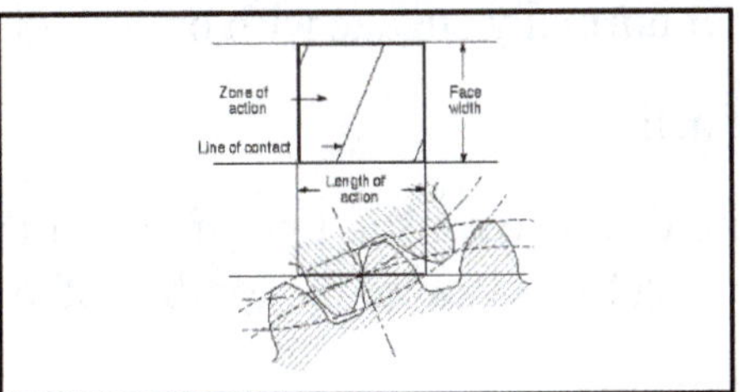

Point of contact

> Any point at which two tooth profiles touch each other.

Line of contact

> A line or curve along which two tooth surfaces are tangent to each other.

Path of action

> The locus of successive contact points between a pair of gear teeth, during the phase of engagement. For conjugate gear teeth, the path of action passes through the pitch point. It is the trace of the surface of action in the plane of rotation.

Line of action

> The path of action for involute gears. It is the straight line passing through the pitch point and tangent to both base circles.

Surface of action

> The imaginary surface in which contact occurs between two engaging tooth surfaces. It is the summation of the paths of action in all sections of the engaging teeth.

Plane of action

> The surface of action for involute, parallel axis gears with either spur or helical teeth. It is tangent to the base cylinders.

Zone of action (contact zone)

> For involute, parallel-axis gears with either spur or helical teeth, is the rectangular area in the plane of action bounded by the length of action and the effective face width.

Path of contact

> The curve on either tooth surface along which theoretical single point contact occurs during the engagement of gears with crowned tooth surfaces or gears that normally engage with only single point contact.

Length of action

> The distance on the line of action through which the point of contact moves during the action of the tooth profile.

Arc of action, Q_t

> The arc of the pitch circle through which a tooth profile moves from the beginning to the end of contact with a mating profile.

Arc of approach, Q_a

> The arc of the pitch circle through which a tooth profile moves from its beginning of contact until the point of contact arrives at the pitch point.

Arc of recess, Q_r

> The arc of the pitch circle through which a tooth profile moves from contact at the pitch point until contact ends.

Contact ratio, m_c, ε

> The number of angular pitches through which a tooth surface rotates from the beginning to the end of contact. In a simple way, it can be defined as a measure of the average number of teeth in contact during the period during which a tooth comes and goes out of contact with the mating gear.

Transverse contact ratio, m_p, ε_α

> The contact ratio in a transverse plane. It is the ratio of the angle of action to the angular pitch. For involute gears it is most directly obtained as the ratio of the length of action to the base pitch.

Face contact ratio, m_F, ε_β

> The contact ratio in an axial plane, or the ratio of the face width to the axial pitch. For bevel and hypoid gears it is the ratio of face advance to circular pitch.

Total contact ratio, m_t, ε_γ

> The sum of the transverse contact ratio and the face contact ratio.

$$\epsilon_\gamma = \epsilon_\alpha + \epsilon_\beta$$

$$m_t = m_p + m_F$$

Modified contact ratio, m_o

> For bevel gears, the square root of the sum of the squares of the transverse and face contact ratios.

$$m_o = \sqrt{m_p^2 + m_F^2}$$

Limit diameter

> Diameter on a gear at which the line of action intersects the maximum (or minimum for internal pinion) addendum circle of the mating gear. This is also referred to as the start of active profile, the start of contact, the end of contact, or the end of active profile.

Start of active profile (SAP)

> Intersection of the limit diameter and the involute profile.

Face advance

>   Distance on a pitch circle through which a helical or spiral tooth moves from
>   the position at which contact begins at one end of the tooth trace on the pitch
>   surface to the position where contact ceases at the other end.

Tooth Thickness Nomenclature

Tooth thickness	Thickness relationships	Chordal thickness
Tooth thickness measurement over pins	Span measurement	Long and short addendum teeth

Circular thickness

>   Length of arc between the two sides of a gear tooth, on the specified datum
>   circle.

Transverse circular thickness

>   Circular thickness in the transverse plane.

Normal circular thickness

>   Circular thickness in the normal plane. In a helical gear it may be considered as
>   the length of arc along a normal helix.

Axial thickness

>   In helical gears and worms, tooth thickness in an axial cross section at the stan-
>   dard pitch diameter.

Base circular thickness

In involute teeth, length of arc on the base circle between the two involute curves forming the profile of a tooth.

Normal chordal thickness

Length of the chord that subtends a circular thickness arc in the plane normal to the pitch helix. Any convenient measuring diameter may be selected, not necessarily the standard pitch diameter.

Chordal addendum (chordal height)

Height from the top of the tooth to the chord subtending the circular thickness arc. Any convenient measuring diameter may be selected, not necessarily the standard pitch diameter.

Profile shift

Displacement of the basic rack datum line from the reference cylinder, made non-dimensional by dividing by the normal module. It is used to specify the tooth thickness, often for zero backlash.

Rack shift

Displacement of the tool datum line from the reference cylinder, made non-dimensional by dividing by the normal module. It is used to specify the tooth thickness.

Measurement over pins

Measurement of the distance taken over a pin positioned in a tooth space and a reference surface. The reference surface may be the reference axis of the gear, a datum surface or either one or two pins positioned in the tooth space or spaces opposite the first. This measurement is used to determine tooth thickness.

Span measurement

Measurement of the distance across several teeth in a normal plane. As long as the measuring device has parallel measuring surfaces that contact on an unmodified portion of the involute, the measurement wis along a line tangent to the base cylinder. It is used to determine tooth thickness.

Modified addendum teeth

Teeth of engaging gears, one or both of which have non-standard addendum.

Full-depth teeth

Teeth in which the working depth equals 2.000 divided by the normal diametral pitch.

Stub teeth

Teeth in which the working depth is less than 2.000 divided by the normal diametral pitch.

Equal addendum teeth

Teeth in which two engaging gears have equal addendums.

Long and short-addendum teeth

Teeth in which the addendums of two engaging gears are unequal.

Pitch Nomenclature

Pitch is the distance between a point on one tooth and the corresponding point on an adjacent tooth. It is a dimension measured along a line or curve in the transverse, normal, or axial directions. The use of the single word *pitch* without qualification may be ambiguous, and for this reason it is preferable to use specific designations such as transverse circular pitch, normal base pitch, axial pitch.

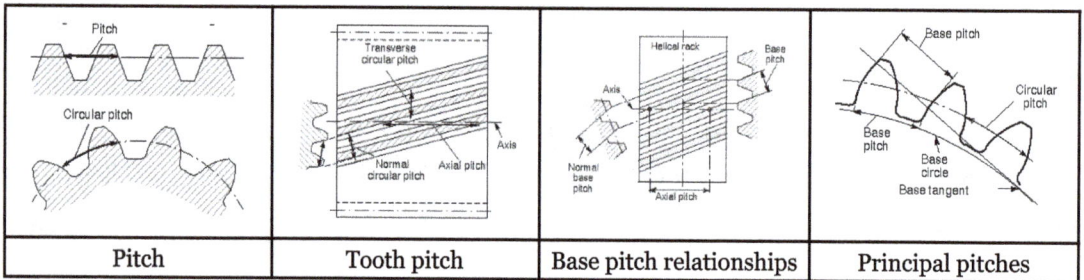

Pitch	Tooth pitch	Base pitch relationships	Principal pitches

Circular pitch, p

Arc distance along a specified pitch circle or pitch line between corresponding profiles of adjacent teeth.

Transverse circular pitch, p_t

Circular pitch in the transverse plane.

Normal circular pitch, p_n, p_e

Circular pitch in the normal plane, and also the length of the arc along the normal pitch helix between helical teeth or threads.

Axial pitch, p_x

Linear pitch in an axial plane and in a pitch surface. In helical gears and worms, axial pitch has the same value at all diameters. In gearing of other types, axial pitch may be confined to the pitch surface and may be a circular measurement.

The term axial pitch is preferred to the term linear pitch. The axial pitch of a helical worm and the circular pitch of its worm gear are the same.

Normal base pitch, p_N, p_{bn}

An involute helical gear is the base pitch in the normal plane. It is the normal distance between parallel helical involute surfaces on the plane of action in the normal plane, or is the length of arc on the normal base helix. It is a constant distance in any helical involute gear.

Transverse base pitch, p_b, p_{bt}

In an involute gear, the pitch is on the base circle or along the line of action. Corresponding sides of involute gear teeth are parallel curves, and the base pitch is the constant and fundamental distance between them along a common normal in a transverse plane.

Diametral pitch (transverse), P_d

Ratio of the number of teeth to the standard pitch diameter in inches.

$$P_d = \frac{N}{d} = \frac{25.4}{m} = \frac{\pi}{p}$$

Normal diametral pitch, P_{nd}

Value of diametral pitch in a normal plane of a helical gear or worm.

$$P_{nd} = \frac{P_d}{\cos \psi}$$

Angular pitch, θ_N, τ

Angle subtended by the circular pitch, usually expressed in radians.

$$\tau = \frac{360}{z} \text{ degrees or } \frac{2\pi}{z} \text{ radians}$$

Backlash

Backlash is the error in motion that occurs when gears change direction. It exists because there is always some gap between the trailing face of the driving tooth and the leading face of the tooth behind it on the driven gear, and that gap must be closed before force can be transferred in the new direction. The term "backlash" can also be used to refer to the size of the gap, not just the phenomenon it causes; thus, one could speak of a pair of gears as having, for example, "0.1 mm of backlash." A pair of gears could be designed to have zero backlash, but this would presuppose perfection in manufacturing, uniform thermal expansion characteristics throughout the system, and no lubricant. Therefore, gear pairs are designed to have some backlash. It is usually provided by reducing the tooth thickness of each gear by half the desired gap distance. In the case of a large gear and a small pinion, however, the backlash is usually taken entirely off the gear and the pinion is given full sized teeth. Backlash can also be provided by moving the gears further apart. The backlash of a gear train equals the sum of the backlash of each pair of gears, so in long trains backlash can become a problem.

For situations that require precision, such as instrumentation and control, backlash can be minimised through one of several techniques. For instance, the gear can be split along a plane perpendicular to the axis, one half fixed to the shaft in the usual manner, the other half placed alongside it, free to rotate about the shaft, but with springs between the two halves providing relative torque between them, so that one achieves, in effect, a single gear with expanding teeth. Another method involves tapering the teeth in the axial direction and letting the gear slide in the axial direction to take up slack.

Shifting of Gears

In some machines (e.g., automobiles) it is necessary to alter the gear ratio to suit the task, a process known as gear shifting or changing gear. There are several ways of shifting gears, for example:

- Manual transmission

- Automatic transmission

- Derailleur gears, which are actually sprockets in combination with a roller chain

- Hub gears (also called epicyclic gearing or sun-and-planet gears)

There are several outcomes of gear shifting in motor vehicles. In the case of vehicle noise emissions, there are higher sound levels emitted when the vehicle is engaged in lower gears. The design life of the lower ratio gears is shorter, so cheaper gears may be used, which tend to generate more noise due to smaller overlap ratio and a lower mesh stiffness etc. than the helical gears used for the high ratios. This fact has been used to analyze vehicle-generated sound since the late 1960s, and has been incorporated into the simulation of urban roadway noise and corresponding design of urban noise barriers along roadways.

Tooth Profile

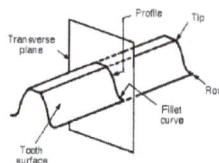

Profile of a spur gear

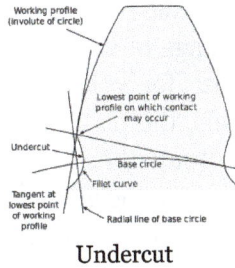

Undercut

A profile is one side of a tooth in a cross section between the outside circle and the root circle. Usually a profile is the curve of intersection of a tooth surface and a plane or surface normal to the pitch surface, such as the transverse, normal, or axial plane.

The fillet curve (root fillet) is the concave portion of the tooth profile where it joins the bottom of the tooth space.[2]

As mentioned near the beginning of the article, the attainment of a nonfluctuating velocity ratio is dependent on the profile of the teeth. Friction and wear between two gears is also dependent on the tooth profile. There are a great many tooth profiles that provides a constant velocity ratio. In many cases, given an arbitrary tooth shape, it is possible to develop a tooth profile for the mating gear that provides a constant velocity ratio. However, two constant velocity tooth profiles are the most commonly used in modern times: the *cycloid* and the *involute*. The cycloid was more common until the late 1800s. Since then, the involute has largely superseded it, particularly in drive train applications. The cycloid is in some ways the more interesting and flexible shape; however the involute has two advantages: it is easier to manufacture, and it permits the center-to-center spacing of the gears to vary over some range without ruining the constancy of the velocity ratio. Cycloidal gears only work properly if the center spacing is exactly right. Cycloidal gears are still used in mechanical clocks.

An undercut is a condition in generated gear teeth when any part of the fillet curve lies inside of a line drawn tangent to the working profile at its point of juncture with the fillet. Undercut may be deliberately introduced to facilitate finishing operations. With undercut the fillet curve intersects the working profile. Without undercut the fillet curve and the working profile have a common tangent.

Gear Materials

Numerous nonferrous alloys, cast irons, powder-metallurgy and plastics are used in the manufacture of gears. However, steels are most commonly used because of their high strength-to-weight ratio and low cost. Plastic is commonly used where cost or weight is a concern. A properly designed plastic gear can replace steel in many cases because it has many desirable properties, including dirt tolerance, low speed meshing, the ability to "skip" quite well and the ability to be made with materials that don't need additional lubrication. Manufacturers have used plastic gears to reduce costs in consumer items including copy machines, optical storage devices, cheap dynamos, consumer audio equipment, servo motors, and printers. Another advantage of the use of plastics, formerly (such as in the 1980s), was the reduction of repair costs for certain expensive machines. In cases of severe jamming (as of the paper in a printer), the plastic gear teeth would be torn free of their substrate, allowing the drive mechanism to then spin freely (instead of damaging itself by straining against the jam). This use of "sacrificial" gear teeth avoided destroying the much more expensive motor and related parts. This method has been superseded, in more recent designs, by the use of clutches and torque- or current-limited motors.

Wooden gears of a historic windmill

Standard Pitches and the Module System

Although gears can be made with any pitch, for convenience and interchangeability standard pitches are frequently used. Pitch is a property associated with linear dimensions and so differs whether the standard values are in the Imperial (inch) or Metric systems. Using *inch* measurements, standard diametral pitch values with units of "per inch" are chosen; the *diametral pitch* is the number of teeth on a gear of one inch pitch diameter. Common standard values for spur gears are 3, 4, 5, 6, 8, 10, 12, 16, 20, 24, 32, 48, 64, 72, 80, 96, 100, 120, and 200. Certain standard pitches such as *1/10* and *1/20* in inch measurements, which mesh with linear rack, are actually (linear) *circular pitch* values with units of "inches"

When gear dimensions are in the metric system the pitch specification is generally in terms of *module* or *modulus*, which is effectively a length measurement across the *pitch diameter*. The term module is understood to mean the pitch diameter in millimeters divided by the number of teeth. When the module is based upon inch measurements, it is known as the *English module* to avoid confusion with the metric module. Module is a direct dimension, unlike diametral pitch, which is an inverse dimension ("threads per inch"). Thus, if the pitch diameter of a gear is 40 mm and the number of teeth 20, the module is 2, which means that there are 2 mm of pitch diameter for each tooth. The preferred standard module values are 0.1, 0.2, 0.3, 0.4, 0.5, 0.6, 0.8, 1.0, 1.25, 1.5, 2.0, 2.5, 3, 4, 5, 6, 8, 10, 12, 16, 20, 25, 32, 40 and 50.

Manufacture

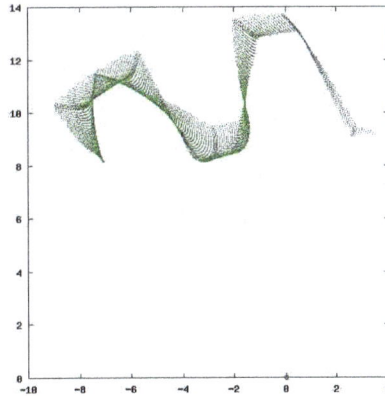

Gear Cutting simulation (playing length 1m35s).

As of 2014, an estimated 80% of all gearing produced worldwide is produced by net shape molding. Molded gearing is usually either powder metallurgy or plastic. Many gears are done when they leave the mold (including injection molded plastic and die cast metal gears), but powdered metal gears require sintering and sand castings or investment castings require gear cutting or other machining to finish them. The most common form of gear cutting is hobbing, but gear shaping, milling, and broaching also exist. 3D printing as a production method is expanding rapidly. For metal gears in the transmissions of cars and trucks, the teeth are heat treated to make them hard and more wear resistant while leaving the core soft and tough. For large gears that are prone to warp, a quench press is used.

Gear Box Designing

The essential information required for designing a Gear box are as follows. 1. The lowest output rpm, n min 2. The highest output rpm, n max 3. The number of steps z into which the range between n max and n min is divided and 4. The number of stages in which the required number of speed steps are to be achieved.

PROCEDURE FOR STEPWISE SOLUTION

Let's n1, n2, n3... n z be the rpm values available on machine tool be in geometric progression.

Rn= nz/(n1)= Ø z-1 Where Rn= Range ratio, Ø= Progression Ratio Ø= Rn (1/z-1) Z= log Rn. Ø/ log Ø Suppose a speed on one shaft yields two speed values on next shaft i.e. No. of speed steps of particular transmission group is p=2. The transmission ratio that provide two speed values must lie in following range:

I max= 2 & I min= ¼

Maximum reduction of speed is limited to four times to keep the radial dimensions of

Gear box within reasonable limits, Maximum increase of speed is limited to two times due to limitations of the pitch line velocity. I max/I min= 8 structural formula z= p1(x1) p2(x2) p3(x3) here x1= 1, x2= characteristics of x1= p1 & x3=characteristics of x1x2= p1p2 Let us understand this by an e.g. n min = 16 rpm n max =770 rpm Ø= 1.41 Rn= 770/(16)= 1.4112-1 Z= log (770/16). 1.41/ log 1.41 Z=12 Z= 2X2X3 Z= 2(1) X 2(2) X 3(4) This is called structural formula. The possible structural formula for above Z are 2(1) X 2(2) X 3(4) 2(1) X 2(2) X 3(4) 2(2) X 2(1) X 3(4) 2(6) X 2(1) X 3(2) 2(6) X 2(3) X 3(1) 2(3) X 2(6) X 3(1) 3(1) X 2(3) X 2(6) 3(1) X 2(6) X 2(3) 3(2) X 2(1) X 2(6) 3(4) X 2(1) X 2(2) 3(4) X 2(2) X 2(1) 3(2) X 2(6) X 2(1) 2(1) X 3(2) X 2(6) 2(1) X 3(4) X 2(2) 2(3) X 3(1) X 2(6) 2(6) X 3(1) X 2(3) 2(6) X 3(2) X 2(1) 2(2) X 3(4) X 2(1) In z= p1(x1) p2(x2) p3(x3)

The best version which ensures that it will be one in which the n min values of the intermediate shafts are maximum and n max values of the intermediate shafts re minimum P1>p2>p3 & x1<x2<x3 Of the above 18 structural formula3(1) X 2(3) X 2(6) is the most appropriate structural diagram let us draw structural diagram for the generated structural formula. u = No. of stages Here in this case u=3 transmission group 1)Draw u+1(Here 4) vertical line at convenient distance where first vertical line represents the transmission from motor shaft and the rest represent the transmission groups of speed box. 2) Draw array of horizontal lines (Here 12) is equal to the number of speed steps z of speed intersecting the vertical lines distance of log Ø from each other. 3) Select the best structure diagram from the above structure diagrams. The number of gears on the last shaft (spindle) should be minimum possible. The transmission ratio between spindle and the shaft preceding it should be the maximum possible i.e. speed reduction should be the maximum possible. The structure diagram shod be narrow towards the starting point (on input shaft) i.e. parabolic in nature. Speed chart structural diagram only depicts the range ratio of transmission groups but gives no information about transmission ratios. In order to determine the transmission ratios of all transmissions and the rpm values of gearbox shafts, need to draw speed chart. The line is horizontal, it corresponds to transmission ratio I =1, i.e. no speed change. The line is inclined upward, it depicts I >1, i.e., speed increase. The line is inclined downward, it depicts I <1, i.e., speed reduction. Draw z+1 No. of vertical lines and u+1 horizontal lines intersecting the vertical lines at convenient distance. (here 13 horizontal lines and 4 vertical lines.) Draw the rays depicting transmission between shaft and the shaft preceding it. The rays are drawn from the lowest rpm of last shaft keeping in mind the transmission ratio restriction condition I max≤ 2 and I min>=1/4. here. I max= 2 = ɸ 2 [(1.41) 2 = 1.9881

I min = 1/4 = 1/ ɸ 4 [(1.41) 4 = 3.9525],

Mark speeds of shaft in front of each horizontal line, starting with n min* ɸ on first line n min* ɸ2 on second line, n min* ɸ3 on third line and so on. Select the best speed chart i.e. It should have concave shape. Gear box diagram No. of teeth on the smallest gear should be such that there is no undercutting of gear teeth Z min≥17. If gears are on parallel shafts sum of no. of teeth of mating gear pair should be same. Spacing between two adjacent gears should be such that one should be such that one should disengage before other mate. Center distance between two shafts, $A=m (Z_1+Z_2)/2$ In machine tools with large inertia of driven member, friction clutch and brake should be provided on input shaft. Reversing devices should be provided so that tool can be returned to its initial position after completion of the cutting process. Reversal speed = 1.3-1.5 times greater than the cutting speed. If spindle head traverse, electric motor should be mounted on the speed box and the transmission from motor shaft to input shaft of the speed box obtained through a clutch or gear pair. Spindle is kinematic ally linked to feed mechanism, from the spindle to feed train must be shown on gearing diagram. The largest of gears on the spindle should be mounted closest to front spindle bearing.

$$Z \min = 2 k (1+\sqrt{(1 \pm i(2 \pm i)} \sin [^2 \alpha]))/((2 \pm i)\sin [^2 \alpha])$$

K=tooth addendum i= transmission ratio α= pressure angle Z min=17 Let's return to our question: n min = 16 rpm, n max =770 rpm, Ø= 1.41, n motor =1440 rpm, input shaft=630 rpm. from gearbox diagram z7 is a link gear z7/z14= 250/250

z10/z11= 63/250 z10 +z11= 86, z10=23, z11=63

z12/z13= 63/250 z12+ z13= 86, z12=17, z13=69

z15/z16= 125/63 z15=60, z18=30

z17/z18= 16/63 z17=18, z18= 72

Inspection

Overall gear geometry can be inspected and verified using various methods such as industrial CT scanning, coordinate-measuring machines, white light scanner or laser scanning. Particularly useful for plastic gears, industrial CT scanning can inspect internal geometry and imperfections such as porosity.

Important dimensional variations of gears result from variations in the combinations of the dimensions of the tools used to manufacture them. An important parameter for meshing qualities such as backlash and noise generation is the variation of the actual contact point as the gear rotates, or the instantaneous pitch radius. Precision gears were frequently inspected by a method that produced a paper "gear tape" record showing variations with a resolution of .0001 inches as the gear was rotated.

The American Gear Manufacturers Association was organized in 1916 to formulate quality standards for gear inspection to reduce noise from automotive timing gears; in 1993 AGMA assumed leadership of the ISO committee governing international standards for gearing. The *ANSI/AGMA 2000 A88 Gear Classi ication and Inspection Handbook* specifies quality numbers from Q3 to Q15 to represent the accuracy of tooth geometry; the higher the number the better the tolerance. Some dimensions can be measured to millionths of an inch in controlled-environment rooms.

Gear Model in Modern Physics

Modern physics adopted the gear model in different ways. In the nineteenth century, James Clerk Maxwell developed a model of electromagnetism in which magnetic field lines were rotating tubes of incompressible fluid. Maxwell used a gear wheel and called it an "idle wheel" to explain the electric current as a rotation of particles in opposite directions to that of the rotating field lines.

More recently, quantum physics uses "quantum gears" in their model. A group of gears can serve as a model for several different systems, such as an artificially constructed nanomechanical device or a group of ring molecules.

The Three Wave Hypothesis compares the wave–particle duality to a bevel gear.

Gear Mechanism in Natural World

Issus coleoptratus

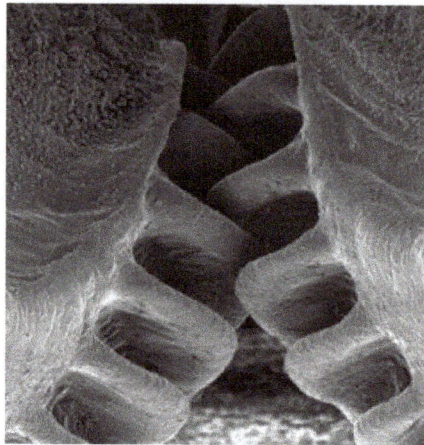

A functioning gear mechanism discovered in *Issus coleoptratus*, a planthopper species common in Europe

The gear mechanism was previously considered exclusively artificial—but in 2013, scientists from the University of Cambridge announced their discovery that the juvenile form of a common insect *Issus* (species Issus coleoptratus), found in many European gardens, has a gear-like mechanism in its hind legs. Each leg has joints that form two 180-degree, helix-shaped strips with twelve fully interlocking spur type gear teeth. The joint rotates like mechanical gears and synchronizes Issus's legs when it jumps.

Triboelectric Effect

The triboelectric effect (also known as triboelectric charging) is a type of contact electrification in which certain materials become electrically charged after they come into frictional contact with a different material. Rubbing glass with fur, or a comb through the hair, can build up triboelectricity. Most everyday static electricity is triboelectric. The polarity and strength of the charges produced differ according to the materials, surface roughness, temperature, strain, and other properties.

The triboelectric effect is not very predictable, and only broad generalizations can be made. Amber, for example, can acquire an electric charge by contact and separation (or friction) with a material like wool. This property was first recorded by Thales of Miletus. The word "electricity" is derived from William Gilbert's initial coinage, "electra", which originates in the Greek word for amber, *ēlektron*. Other examples of materials that can acquire a significant charge when rubbed together include glass rubbed with silk, and hard rubber rubbed with fur.

The triboelectric effect is now considered to be very close to the phenomenon of adhesion, where two materials composed of different molecules tend to stick together on contact due to a form of chemical reaction. This is very close to a chemical bond; the adjacent dissimilar molecules exchange electrons. And when one material is physically moved away from the other, the bonding forces we experience are regarded by us as 'friction'. The result is that excess electrons are left behind in one material, while a deficit occurs in the other.

Triboelectric Series

John Carl Wilcke published the first triboelectric series in a 1757 paper on static charges. Materials are often listed in order of the polarity of charge separation when they are touched with another object. A material towards the bottom of the series, when touched to a material near the top of the series, will acquire a more negative charge. The farther away two materials are from each other on the series, the greater the charge transferred. Materials near to each other on the series may not exchange any charge, or may even exchange the opposite of what is implied by the list. This can be caused by rubbing, by contaminants or oxides, or other variables. Lists vary somewhat as to the exact order of some materials, since the relative charge varies for nearby materials. From actual tests, there is little or no measurable difference in charge affinity between metals, probably because the rapid motion of conduction electrons cancels such differences.

Cause

Although the part 'tribo-' the two materials only need to come into contact for electrons to be exchanged. After coming into contact, a chemical bond is formed between parts of the two surfaces, called adhesion, and charges move from one material to the other to equalize their electrochemical potential. This is what creates the net charge imbalance between the objects. When separated, some of the bonded atoms have a tendency to keep extra electrons, and some a tendency to give them away, though the imbalance will be partially destroyed by tunneling or electrical breakdown (usually corona discharge). In addition, some materials may exchange ions of differing mobility, or exchange charged fragments of larger molecules.

The triboelectric effect is related to friction only because they both involve adhesion. However, the effect is greatly enhanced by rubbing the materials together, as they touch and separate many times. For surfaces with differing geometry, rubbing may also lead to heating of protrusions, causing pyroelectric charge separation which may add to the existing contact electrification, or which may oppose the existing polarity. Surface nano-effects are not well understood, and the atomic force microscope has enabled rapid progress in this field of physics.

Sparks

Because the surface of the material is now electrically charged, either negatively or positively, any contact with an uncharged conductive object or with an object having substantially different charge may cause an electrical discharge of the built-up static electricity: a spark. A person simply walking across a carpet may build up a potential of many thousands of volts, enough to cause a spark one centimeter long or more. Low humidity in the ambient air increases the voltage at which electrical discharge occurs by increasing the ability of the insulating material to hold charge by decreasing the conductivity of the air, making it difficult for the charge build-up to dissipate gradually. Simply removing a nylon shirt or corset can also create sparks. Car travel can lead to a build-up of charge on the driver and passengers due to friction between the driver's clothes and the leather or plastic furnishings inside the vehicle. This charge can then be relaxed as a spark to the metal car body, fuel dispensers, or nearby door handles, etc. When the vehicle's body itself builds up a static charge (acting as a Faraday cage) it can relax through the carbon in the tires. If it remains charged when parked, sparks may jump from the door frame to occupants as they make contact with the ground.

This type of discharge is often harmless because the energy ($1/2V^2C$) of the spark is very small, being typically several tens of micro joules in cold dry weather, and much less than that in humid conditions. However, such sparks can ignite flammable vapours.

In Aircraft and Spacecraft

Aircraft flying in weather will develop a static charge from air friction on the airframe. The static can be discharged with static dischargers or static wicks.

NASA follows what they call the Triboelectrification Rule whereby they will cancel a launch if the launch vehicle is predicted to pass through certain types of clouds. Flying through high-level clouds can generate "P-static" (P for precipitation), which can create static around the launch vehicle that will interfere with radio signals sent by or to the vehicle. This may prevent transmitting of telemetry to the ground or, if the need arises, sending a signal to the vehicle, particularly critical signals for the flight termination system. When a hold is put in place due to the triboelectrification rule, it remains until Space Wing and observer personnel such as those in reconnaissance aircraft indicate that the skies are clear.

Risks and Counter-measures

Ignition

The effect is of considerable industrial importance in terms of both safety and potential damage to manufactured goods. Static discharge is a particular hazard in grain elevators owing to the danger of a dust explosion. The spark produced is fully able to ignite flammable vapours, for example, petrol, ether fumes as well as methane gas. For bulk fuel deliveries and aircraft fueling a grounding connection is made between the vehicle and the receiving tank prior to opening the tanks. When fueling vehicles at a retail station touching metal on the car before opening the gas tank or touching the nozzle may decrease one's risk of static ignition of fuel vapors.

In the Workplace

Means have to be provided to discharge carts which may carry such volatile liquids, flammable gasses, or oxygen in hospitals. Even where only a small charge is produced, it can result in dust particles being attracted to the rubbed surface. In the case of textile manufacture this can lead to a permanent grimy mark where the cloth comes in contact with dust accumulations held by a static charge. Dust attraction may be reduced by treating insulating surfaces with an antistatic cleaning agent.

Damage to Electronics

Some electronic devices, most notably CMOS integrated circuits and MOSFET transistors, can be accidentally destroyed by high-voltage static discharge. Such components are usually stored in a conductive foam for protection. Grounding oneself by touching the workbench, or using a special bracelet or anklet is standard practice while handling unconnected integrated circuits. Another way of dissipating charge is by using conducting materials such as carbon black loaded rubber mats in operating theatres, for example.

Devices containing sensitive components must be protected during normal use, installation, and disconnection, accomplished by designed-in protection at external connections where needed. Protection may be through the use of more robust devices or protective countermeasures at the device's external interfaces. These may be opto-isolators, less sensitive types of transistors, and static bypass devices such as metal oxide varistors.

Tribometer

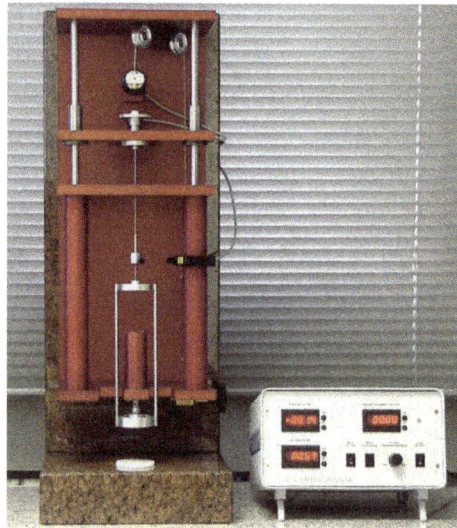

Static Friction Tribometer

A tribometer is an instrument that measures tribological quantities, such as coefficient of friction, friction force, and wear volume, between two surfaces in contact. It was invented by the 18th century Dutch scientist Musschenbroek

A tribotester is the general name given to a machine or device used to perform tests and simulations of wear, friction and lubrication which are the subject of the study of tribology. Often tribotesters are extremely specific in their function and are fabricated by manufacturers who desire to test and analyze the long-term performance of their products. An example is that of orthopedic implant manufactures who have spent considerable sums of money to develop tribotesters that accurately reproduce the motions and forces that occur in human hip joints so that they can perform accelerated wear tests of their products.

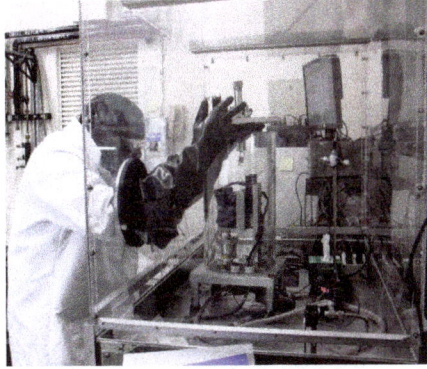

Hydrogen Tribometer

Theory

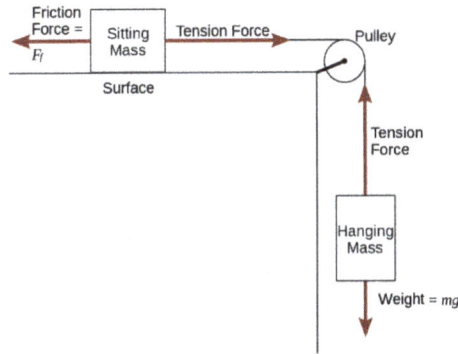

A simple tribometer is described by a hanging mass and a mass resting on a horizontal surface, connected to each other via a string and pulley. The coefficient of friction, μ, when the system is stationary, is determined by increasing the hanging mass until the moment that the resting mass begins to slide. Then using the general equation for friction force:

$$F = \mu \mathbb{N}$$

Where N, the normal force, is equal to the weight (mass x gravity) of the sitting mass (m_T) and F, the loading force, is equal to the weight (mass x gravity) of the hanging mass (m_H).

To determine the kinetic coefficient of friction the hanging mass is increased or decreased until the mass system moves at a constant speed.

In both cases, the coefficient of friction is simplified to the ratio of the two masses:

$$\mu = m_H \, / \, m_T$$

In most test applications using tribometers, wear is measured by comparing the mass or surfaces of test specimens before and after testing. Equipment and methods used to examine the worn surfaces include optical microscopes, scanning electron microscopes, optical interferometry and mechanical roughness testers.

Types

Tribometers are often referred to by the specific contact arrangement they simulate or by the original equipment developer. Several arrangements are:

- Four ball
- Pin on disc
- Block on ring
- Bouncing ball
- Fretting test machine
- Twin disc

Bouncing Ball

A *bouncing ball* tribometer consists of a ball which is impacted at an angle against a surface. During a typical test, a ball is slid on an angle along a track until it impacts a surface and then bounces off of the surface. The friction produced in the contact between the ball and the surface results in a horizontal force on the surface and a rotational force on the ball. Frictional force is determined by finding the rotational speed of the ball using high speed photography or by measuring the force on the horizontal surface. Pressure in the contact is very high due to the large instantaneous force caused by the impact with the ball.

Bouncing ball tribometers have been used to determine the shear characteristics of lubricants under high pressures such as is found in ball bearings or gears.

Pin on Disc

A *pin on disc tribometer* consists of a stationary "pin" under an applied load in contact with a rotating disc. The pin can have any shape to simulate a specific contact, but spherical tips are often used to simplify the contact geometry. Coefficient of friction is determined by the ratio of the frictional force to the loading force on the pin.

The pin on disc test has proved useful in providing a simple wear and friction test for low friction coatings such as diamond-like carbon coatings on valve train components in internal combustion engines.

Surface Finish

Surface finish, also known as surface texture or surface topography, is the nature of a surface as defined by the three characteristics of lay, surface roughness, and waviness. It comprises the small local deviations of a surface from the perfectly flat ideal (a true plane).

Surface texture is one of the important factors that control friction and transfer layer formation during sliding. Considerable efforts have been made to study the influence of surface texture on friction and wear during sliding conditions. Surface textures can be isotropic or anisotropic. Sometimes, stick-slip friction phenomena can be observed during sliding depending on surface texture.

Each manufacturing process (such as the many kinds of machining) produces a surface texture. The process is usually optimized to ensure that the resulting texture is usable. If necessary, an additional process will be added to modify the initial texture. The latter process may be grinding (abrasive cutting), polishing, lapping, abrasive blasting, honing, electrical discharge machining (EDM), milling, lithography, industrial etching/chemical milling, laser texturing, or other processes.

Lay

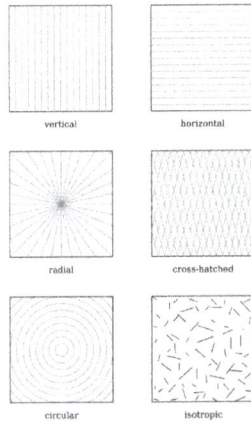

vertical   horizontal

radial   cross-hatched

circular   isotropic

Examples of various lay patterns

Lay is the direction of the predominant surface pattern ordinarily determined by the production method used.

Surface Roughness

Surface roughness commonly shortened to *roughness*, is a measure of the finely spaced surface irregularities. In engineering, this is what is usually meant by "surface finish".

Waviness

Waviness is the measure of surface irregularities with a spacing greater than that of surface roughness. These usually occur due to warping, vibrations, or deflection during machining.

Measurement

Surface finish may be measured in two ways: *contact* and *non-contact* methods. Contact methods involve dragging a measurement stylus across the surface; these instruments are called profilometers. Non-contact methods include: interferometry, confocal microscopy, focus variation, structured light, electrical capacitance, electron microscopy, and photogrammetry.

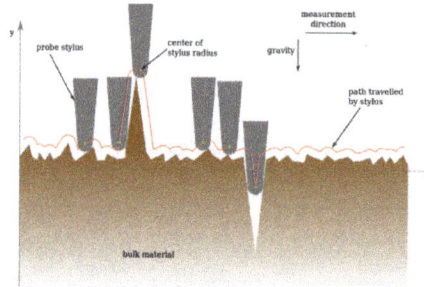

How a profilometer works

Specification

In the United States, surface finish is usually specified using the ASME Y14.36M standard. The other common standard is International Organization for Standardization (ISO) 1302.

c	d	Lay	a	Surface parameter
a	=	Parallel		D F S-L / Rz N C V
e d b	⊥	Perpendicular		
	X	Cross-hatch	D	Tolerance direction, upper (U) or lower (L)
	M	Multi-directional	F	Filter type, for example "2RC"
b Secondary surface parameter	C	Circular	S	Short filter cutoff, for removing noise
c Manufacturing method	R	Radial	L	Long filter cutoff, for removing waviness
e Minimum material removal	P	Particulate	R	Profile type, primary (P), waviness (W), or roughness (R)
			z	Parameter type, for example "a" for Ra or "3z" for R3z
			N	Assesment length; multiple of sampling length, usually 5
Material removal not allowed Material removal required			C	Comparison rule, "max" for 100%, "16%" for 116%
			V	Specified value in micrometers

Manufacturing

Many factors contribute to the surface finish in manufacturing. In forming processes, such as molding or metal forming, surface finish of the die determines the surface finish of the workpiece. In machining the interaction of the cutting edges and the microstructure of the material being cut both contribute to the final surface finish.

In general, the cost of manufacturing a surface increases as the surface finish improves. Any given manufacturing process is usually optimized enough to ensure that the resulting texture is usable for the part's intended application. If necessary, an additional process will be added to modify the initial texture. The expense of this additional process must be justified by adding value in some way—principally better function or longer lifespan. Parts that have sliding contact with others may work better or last longer if the roughness is lower. Aesthetic improvement may add value if it improves the saleability of the product.

A practical example is as follows. An aircraft maker contracts with a vendor to make parts. A certain grade of steel is specified for the part because it is strong enough and hard enough for the part's function. The steel is machinable although not free-machining. The vendor decides to mill the parts. The milling can achieve the specified roughness (for example, ≤ 3.2 µm) as long as the machinist uses premium-quality inserts in the end mill and replaces the inserts after every 20 parts (as opposed to cutting hundreds before changing the inserts). There is no need to add a second operation (such as grinding or polishing) after the milling as long as the milling is done well enough (correct inserts, frequent-enough insert changes, and clean coolant). The inserts and coolant cost money, but the costs that grinding or polishing would incur (more time and additional materials) would cost even more than that. Obviating the second operation results in a lower unit cost and thus a lower price. The competition between vendors elevates such details from minor to crucial importance. It was certainly possible to make the parts in a slightly less efficient way (two operations) for a slightly higher price; but only one vendor can get the contract, so the slight difference in efficiency is magnified by competition into the great difference between the prospering and shuttering of firms.

Just as different manufacturing processes produce parts at various tolerances, they are also capable of different roughnesses. Generally these two characteristics are linked: manufacturing processes that are dimensionally precise create surfaces with low roughness. In other words, if a process can manufacture parts to a narrow dimensional tolerance, the parts will not be very rough.

Due to the abstractness of surface finish parameters, engineers usually use a tool that has a variety of surface roughnesses created using different manufacturing methods.

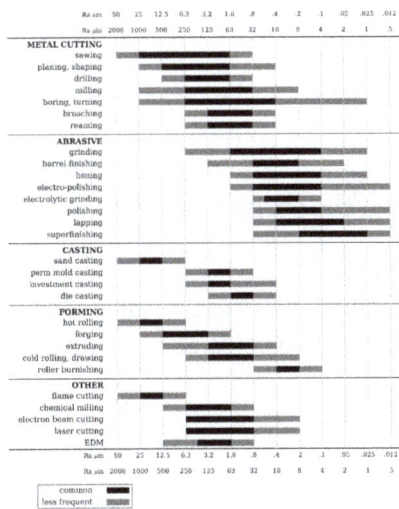

Capillary Surface

In fluid mechanics and mathematics, a capillary surface is a surface that represents the interface between two different fluids. As a consequence of being a surface, a capillary surface has no thickness in slight contrast with most real fluid interfaces.

Capillary surfaces are of interest in mathematics because the problems involved are very nonlinear and have interesting properties, such as discontinuous dependence on boundary data at isolated points. In particular, static capillary surfaces with gravity absent have constant mean curvature, so that a minimal surface is a special case of static capillary surface.

They are also of practical interest for fluid management in space (or other environments free of body forces), where both flow and static configuration are often dominated by capillary effects.

The Stress Balance Equation

The defining equation for a capillary surface is called the stress balance equation, which can be derived by considering the forces and stresses acting on a small volume that is partly bounded by a capillary surface. For a fluid meeting another fluid (the "other" fluid notated with bars) at a surface S, the equation reads

$$(\sigma_{ij} - \bar{\sigma}_{ij})\hat{\mathbf{n}} = -\gamma\hat{\mathbf{n}}(\nabla_S \cdot \hat{\mathbf{n}}) + \nabla_S\gamma \quad ; \quad \nabla_S\gamma = \nabla\gamma - \hat{\mathbf{n}}(\hat{\mathbf{n}} \cdot \nabla\gamma)$$

where $\hat{\mathbf{n}}$ is the unit normal pointing toward the "other" fluid (the one whose quantities are notated with bars), σ_{ij} is the stress tensor (note that on the left is a tensor-vector product), γ is the surface tension associated with the interface, and ∇_S is the surface gradient. Note that the quantity $-\nabla_S \cdot \hat{\mathbf{n}}$ is twice the mean curvature of the surface.

In fluid mechanics, this equation serves as a boundary condition for interfacial flows, typically complementing the Navier–Stokes equations. It describes the discontinuity in stress that is balanced by forces at the surface. As a boundary condition, it is somewhat unusual in that it introduces a new variable: the surface S that defines the interface. It's not too surprising then that the stress balance equation normally mandates its own boundary conditions.

For best use, this vector equation is normally turned into 3 scalar equations via dot product with the unit normal and two selected unit tangents:

$$((\sigma_{ij} - \bar{\sigma}_{ij})\hat{\mathbf{n}}) \cdot \hat{\mathbf{n}} = -\gamma\nabla_S \cdot \hat{\mathbf{n}}$$

$$((\sigma_{ij} - \bar{\sigma}_{ij})\hat{\mathbf{n}}) \cdot \hat{\mathbf{t}}_1 = \nabla_S\gamma \cdot \hat{\mathbf{t}}_1$$

$$((\sigma_{ij} - \bar{\sigma}_{ij})\hat{\mathbf{n}}) \cdot \hat{\mathbf{t}}_2 = \nabla_S\gamma \cdot \hat{\mathbf{t}}_2$$

Note that the products lacking dots are tensor products of tensors with vectors (resulting in vectors similar to a matrix-vector product), those with dots are dot products. The first equation is called the normal stress equation, or the normal stress boundary condition. The second two equations are called tangential stress equations.

The Stress Tensor

The stress tensor is related to velocity and pressure. Its actual form will depend on the specific fluid being dealt with, for the common case of incompressible Newtonian flow the stress tensor is given by

$$\sigma_{ij} = -\begin{pmatrix} p & 0 & 0 \\ 0 & p & 0 \\ 0 & 0 & p \end{pmatrix} + \mu \begin{pmatrix} 2\dfrac{\partial u}{\partial x} & \dfrac{\partial u}{\partial y}+\dfrac{\partial v}{\partial x} & \dfrac{\partial u}{\partial z}+\dfrac{\partial w}{\partial x} \\ \dfrac{\partial v}{\partial x}+\dfrac{\partial u}{\partial y} & 2\dfrac{\partial v}{\partial y} & \dfrac{\partial v}{\partial z}+\dfrac{\partial w}{\partial y} \\ \dfrac{\partial w}{\partial x}+\dfrac{\partial u}{\partial z} & \dfrac{\partial w}{\partial y}+\dfrac{\partial v}{\partial z} & 2\dfrac{\partial w}{\partial z} \end{pmatrix} = -pI + \mu(\nabla \mathbf{v} + (\nabla \mathbf{v})^T)$$

where p is the pressure in the fluid, $\mathbf{v}$ is the velocity, and μ is the viscosity.

Static Interfaces

In the absence of motion, the stress tensors yield only hydrostatic pressure so that $\sigma_{ij} = -pI$, regardless of fluid type or compressibility. Considering the normal and tangential equations,

$$\bar{p} - p = \gamma \nabla \cdot \hat{\mathbf{n}}$$

$$0 = \nabla \gamma \cdot \hat{\mathbf{t}}$$

The first equation establishes that curvature forces are balanced by pressure forces. The second equation implies that a static interface cannot exist in the presence of non-zero surface tension gradient.

If gravity is the only body force present, the Navier–Stokes equations simplify significantly:

$$0 = -\nabla p + \rho \mathbf{g}$$

If coordinates are chosen so that gravity is nonzero only in the z direction, this equation degrades to a particularly simple form:

$$\frac{dp}{dz} = \rho g \quad \Rightarrow \quad p = \rho g z + p_0$$

where p_0 is an integration constant that represents some reference pressure at $z = 0..$ Substituting this into the normal stress equation yields what is known as the Young-Laplace equation:

$$\bar{\rho} g z + \bar{p}_0 - (\rho g z + p_0) = \gamma \nabla \cdot \hat{\mathbf{n}} \quad \Rightarrow \quad \Delta \rho g z + \Delta p = \gamma \nabla \cdot \hat{\mathbf{n}}$$

where is the (constant) pressure difference across the interface, and is the difference in density. Note that, since this equation defines a surface, is the coordinate of the capillary surface. This nonlinear partial differential equation when supplied with the right boundary conditions will define the static interface.

The pressure difference above is a constant, but its value will change if the coordinate is shifted. The linear solution to pressure implies that, unless the gravity term is absent, it is always possible to define the coordinate so that . Nondimensionalized, the Young-Laplace equation is usually studied in the form

$$\kappa z + \lambda = \nabla \cdot \hat{\mathbf{n}}$$

where (if gravity is in the negative z direction) κ is positive if the denser fluid is "inside" the interface, negative if it is "outside", and zero if there is no gravity or if there is no difference in density between the fluids.

This nonlinear equation has some rich properties, especially in terms of existence of unique solutions. For example, the nonexistence of solution to some boundary value problem implies that, physically, the problem can't be static. If a solution does exist, normally it'll exist for very specific values of , which is representative of the pressure jump across the interface. This is interesting because there isn't another physical equation to determine the pressure difference. In a capillary tube, for example, implementing the contact angle boundary condition will yield a unique solution for exactly one value of . Solutions often aren't unique, this implies that there are multiple static interfaces possible; while they may all solve the same boundary value problem, the minimization of energy will normally favor one. Different solutions are called *configurations* of the interface.

Energy Consideration

A deep property of capillary surfaces is the surface energy that is imparted by surface tension:

$$E_S = \gamma_S A_S$$

where is the area of the surface being considered, and the total energy is the summation of all energies. Note that *every* interface imparts energy. For example, if there are two different fluids (say liquid and gas) inside a solid container with gravity and other energy potentials absent, the energy of the system is

$$E = \sum \gamma_S A_S = \gamma_{LG} A_{LG} + \gamma_{SG} A_{SG} + \gamma_{SL} A_{SL}$$

where the subscripts, and respectively indicate the liquid–gas, solid–gas, and solid–liquid interfaces. Note that inclusion of gravity would require consideration of the volume enclosed by the capillary surface and the solid walls.

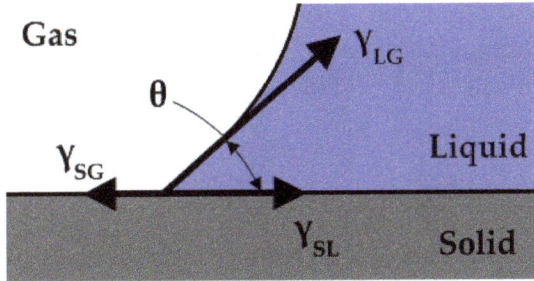

Illustration of distributed forces at the contact line, with the contact line perpendicular to the image. The vertical part of is balanced by a slight deformation of the solid (not shown and inconsequential to this context).

Typically the surface tension values between the solid–gas and solid–liquid interfaces are not known. This does not pose a problem; since only changes in energy are of primary interest. If the net solid area is a constant, and the contact angle is known, it may be shown that (again, for two different fluids in a solid container)

$$E = \gamma_{SL}(A_{SL} + A_{SG}) + \gamma_{LG}A_{LG} + \gamma_{LG}A_{SG}\cos(\theta)$$

so that

$$\frac{\Delta E}{\gamma_{LG}} = \Delta A_{LG} + \Delta A_{SG}\cos(\theta) = \Delta A_{LG} - \Delta A_{SL}\cos(\theta)$$

where θ is the contact angle and the capital delta indicates the change from one configuration to another. To obtain this result, it's necessary to sum (distributed) forces at the contact line (where solid, gas, and liquid meet) in a direction tangent to the solid interface and perpendicular to the contact line:

$$0 = \sum F_{\text{Contact line}} = \gamma_{LG}\cos(\theta) + \gamma_{SL} - \gamma_{SG}$$

where the sum is zero because of the static state. When solutions to the Young-Laplace equation aren't unique, the most physically favorable solution is the one of minimum energy, though experiments (especially low gravity) show that metastable surfaces can be surprisingly persistent, and that the most stable configuration can become metastable through mechanical jarring without too much difficulty. On the other hand, a metastable surface can sometimes spontaneously achieve lower energy without any input given enough time.

Boundary Conditions

Boundary conditions for stress balance describe the capillary surface at the contact line: the line where a solid meets the capillary interface; also, volume constraints can serve as boundary conditions (a suspended drop, for example, has no contact line but clearly must admit a unique solution).

For static surfaces, the most common contact line boundary condition is the implementation of the contact angle, which specifies the angle that one of the fluids meets the
solid wall. The contact angle condition on the surface is normally written as:

$$\hat{\mathbf{n}} \cdot \hat{\mathbf{v}} = \cos(\theta)$$

where is the contact angle. This condition is imposed on the boundary (or boundaries)
of the surface. is the unit outward normal to the solid surface, and is a unit normal to .
Choice of depends on which fluid the contact angle is specified for.

For dynamic interfaces, the boundary condition showed above works well if the contact
line velocity is low. If the velocity is high, the contact angle will change ("dynamic contact angle"), and as of 2007 the mechanics of the moving contact line (or even the validity of the contact angle as a parameter) is not known and an area of active research.

Surface Roughness

The basic symbol of surface roughness

Surface roughness often shortened to roughness, is a component of surface texture.
It is quantified by the deviations in the direction of the normal vector of a real surface from its ideal form. If these deviations are large, the surface is rough; if they are
small, the surface is smooth. Roughness is typically considered to be the high-frequen-
cy, short-wavelength component of a measured surface. However, in practice it is often
necessary to know both the amplitude and frequency to ensure that a surface is fit for
a purpose.

Roughness plays an important role in determining how a real object will interact
with its environment. Rough surfaces usually wear more quickly and have higher
friction coefficients than smooth surfaces. Roughness is often a good predictor of the
performance of a mechanical component, since irregularities in the surface may form
nucleation sites for cracks or corrosion. On the other hand, roughness may promote
adhesion.

Although a high roughness value is often undesirable, it can be difficult and expensive to control in manufacturing. Decreasing the roughness of a surface will usually increase its manufacturing costs. This often results in a trade-off between the manufacturing cost of a component and its performance in application.

Roughness can be measured by manual comparison against a "surface roughness comparator", a sample of known surface roughnesses, but more generally a surface profile measurement is made with a profilometer that can be contact (typically a diamond stylus) or optical (e.g. a white light interferometer or laser scanning confocal microscope).

However, controlled roughness can often be desirable. For example, a gloss surface can be too shiny to the eye and too slippery to the finger (a touchpad is a good example) so a controlled roughness is required. This is a case where both amplitude and frequency are very important.

Parameters

A roughness value can either be calculated on a profile (line) or on a surface (area). The profile roughness parameter (Ra, Rq,...) are more common. The area roughness parameters (Sa, Sq,...) give more significant values.

Profile Roughness Parameters

Each of the roughness parameters is calculated using a formula for describing the surface. Although these parameters are generally considered to be "well known" a standard reference describing each in detail is Surfaces and their Measurement.

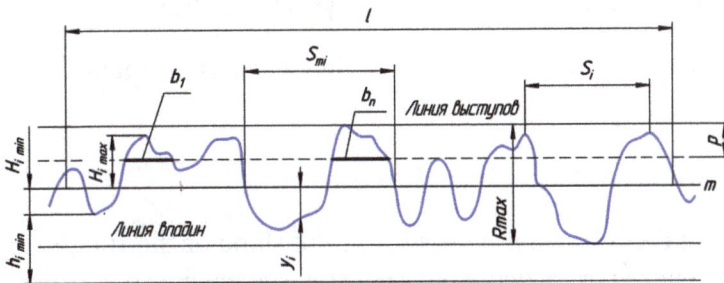

There are many different roughness parameters in use, but R_a is by far the most common, though this is often for historical reasons and not for particular merit, as the early roughness meters could *only* measure R_a. Other common parameters include R_z, R_q, and R_{sk}. Some parameters are used only in certain industries or within certain countries. For example, the R_k family of parameters is used mainly for cylinder bore linings, and the Motif parameters are used primarily within France.

Since these parameters reduce all of the information in a profile to a single number, great care must be taken in applying and interpreting them. Small changes in how the raw profile data is filtered, how the mean line is calculated, and the physics of the measurement can greatly affect the calculated parameter. With modern digital equipment it makes sense to look at the scan and make sure there aren't some obvious glitches that are skewing the values - and if there are, to re-measure.

Because it is not obvious to many users what each of the measurements really means, it is helpful to have a simulation tool that lets you "play" with key parameters and see how well (or badly) surfaces which are obviously different to the human eye are differentiated by the measures. It is clear, for example that R_a would fail to distinguish between two surfaces where one is composed of peaks on an otherwise smooth surface and the other is composed of troughs of the same amplitude. Such tools can be found in app format.

By convention every 2D roughness parameter is a capital R followed by additional characters in the subscript. The subscript identifies the formula that was used, and the R means that the formula was applied to a 2D roughness profile. Different capital letters imply that the formula was applied to a different profile. For example, Ra is the arithmetic average of the roughness profile, Pa is the arithmetic average of the unfiltered raw profile, and Sa is the arithmetic average of the 3D roughness.

Each of the formulas listed in the tables assumes that the roughness profile has been filtered from the raw profile data and the mean line has been calculated. The roughness profile contains n ordered, equally spaced points along the trace, and y_i is the vertical distance from the mean line to the ith data point. Height is assumed to be positive in the up direction, away from the bulk material.

Amplitude Parameters

Amplitude parameters characterize the surface based on the vertical deviations of the roughness profile from the mean line. Many of them are closely related to the parameters found in statistics for characterizing population samples. For example, R_a is the arithmetic average of the absolute values and R_t is the range of the collected roughness data points.

The roughness average, R_a, is the most widely used one-dimensional roughness parameter.

Slope, Spacing, and Counting Parameters

Slope parameters describe characteristics of the slope of the roughness profile. Spacing and counting parameters describe how often the profile crosses certain thresholds. These parameters are often used to describe repetitive roughness profiles, such as those produced by turning on a lathe.

Other "frequency" parameters are S_m, λ_a and λ_q. S_m is the mean spacing between peaks. Just as with real mountains it is important to define a "peak". For S_m the surface must have dipped below the mean surface before rising again to a new peak. The average wavelength λ_a and the root mean square wavelength λ_q are derived from Δ_a. When trying to understand a surface that depends on both amplitude and frequency it is not obvious which pair of metrics optimally describes the balance, so it is a good idea to do a statistical analysis of pairs of measurements (e.g. R_z and λ_a or R_a and Sm) to find the strongest correlation.

Bearing Ratio Curve Parameters

These parameters are based on the bearing ratio curve (also known as the Abbott-Firestone curve.) This includes the Rk family of parameters.

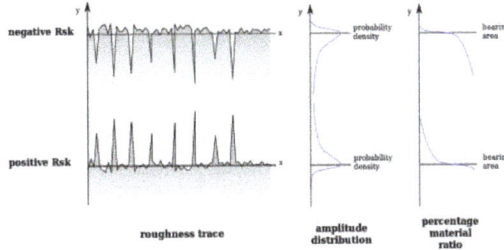

Sketches depicting surfaces with negative and positive skew. The roughness trace is on the left, the amplitude distribution curve is in the middle, and the bearing area curve (Abbott-Firestone curve) is on the right.

Fractal Theory

The mathematician Benoît Mandelbrot has pointed out the connection between surface roughness and fractal dimension.

Areal Roughness Parameters

Areal roughness parameters are defined in the ISO 25178 series. The resulting values are Sa, Sq, Sz,... . At the moment many optical measurement instruments are able to measure the surface roughness over an area. Area measurements are also possible with contact measurement systems. Multiple, closely spaced 2D scans are taken of the target area. These are then digitally stitched together using relevant software, resulting in a 3D image and accompanying areal roughness parameters.

Practical Effects

In terms of engineering surfaces, roughness is considered to be detrimental to part performance. As a consequence, most manufacturing prints establish an upper limit on roughness, but not a lower limit. An exception is in cylinder bores where oil is retained in the surface profile and a minimum roughness is required.

Roughness is often closely related to the friction and wear properties of a surface. A surface with a large R_a value, or a positive R_{sk}, will usually have high friction and wear quickly. The peaks in the roughness profile are not always the points of contact. The form and waviness (i.e. both amplitude and frequency) must also be considered.

References

- White, John H. (1985) [1978]. The American Railroad Passenger Car. 2. Baltimore, MD: Johns Hopkins University Press. p. 518. ISBN 0801827477. OCLC 11469984.

- Brumbach, Michael E.; Clade, Jeffrey A. (2003), Industrial Maintenance, Cengage Learning, pp. 112–113, ISBN 978-0-7668-2695-3.

- Brunner, Gisbert (1999). Wristwatches – Armbanduhren – Montres-bracelets. Köln, Germany: Könnemann. p. 454. ISBN 3-8290-0660-8.

- Eustathopoulos, N.; Nicholas, M.G.; Drevet B. (1999). Wettability at high temperatures. Oxford, UK: Pergamon. ISBN 0-08-042146-6.

- Schrader, M.E; Loeb, G.I. (1992). Modern Approaches to Wettability. Theory and Applications. New York: Plenum Press. ISBN 0-306-43985-9.

- Johnson, Rulon E. (1993) in Wettability Ed. Berg, John. C. New York, NY: Marcel Dekker, Inc. ISBN 0-8247-9046-4

- Rowlinson, J.S.; Widom, B. (1982). Molecular Theory of Capillarity. Oxford, UK: Clarendon Press. ISBN 0-19-855642-X.

- Rosen MJ & Kunjappu JT (2012). Surfactants and Interfacial Phenomena (4th ed.). Hoboken, New Jersey: John Wiley & Sons. p. 1. ISBN 1-118-22902-9.

- Hilbert, David; Cohn-Vossen, Stephan (1952), Geometry and the Imagination (2nd ed.), New York: Chelsea, p. 287, ISBN 978-0-8284-1087-8.

- Gunnar Dahlvig, "Construction elements and machine construction", Konstruktionselement och maskinbyggnad (in Swedish), 7, ISBN 9140115542

- Oberg, E.; Jones, F. D.; Horton, H. L.; Ryffell, H. H. (2000), Machinery's Handbook (26th ed.), Industrial Press, p. 2649, ISBN 978-0-8311-2666-7.

- N.K.Mehata,"Machine Tool Design and Numerical control", Tata McGraw- Hill Publishing Company Limited, ISBN 9780074622377, 1996.

- Siegel, Daniel M. (1991). Innovation in Maxwell's Electromagnetic Theory: Molecular Vortices, Displacement Current, and Light. University of Chicago Press. ISBN 0521353653.

- Whitehouse, David (2012). Surfaces and their Measurement. Boston: Butterworth-Heinemann. ISBN 978-0080972015.

- Degarmo, E. Paul; Black, J.; Kohser, Ronald A. (2003), Materials and Processes in Manufacturing (9th ed.), Wiley, p. 223, ISBN 0-471-65653-4.

- "Saint-Gobain Components Drive Advances in Automotive Friction Control and Manufacturing Performance" (PDF). Industrial Products Purchase. Retrieved 12 May 2016.

- "Fuel Economy and CO2 Emissions Standards, Manufacturer Pricing Strategies, and Feebates" (PDF). Oak Ridge National Laboratory. Retrieved 9 June 2016.

- Publishing, Seattle. "Bicycle Paper.com :: News :: Riding Circles Around Old Standards". www.bicyclepaper.com. Retrieved 2016-06-09.

- "Key Applications of Norglide Composite Bearings & Norslide Cable Liners in High-Performance Bicycles | Product Showcase". showcase.designnews.com. Retrieved 2016-06-09.

- Pietschnig, R. (2016). "Polymers with pendant ferrocenes". Chem. Soc. Rev. 45: 5216–5231. doi:10.1039/C6CS00196C.

7

Stress and Deformation in Solid Materials

Stress in quantum mechanics studies the properties of different materials under strain. Stress and strain by themselves also occur in particular ways in particular material. The chapter focuses on topics such as deformation, Cauchy stress tensor and residual stress. The aspects elucidated in the following chapter are of vital importance, and provides a better understanding on stress and deformation in solid materials.

Stress (Mechanics)

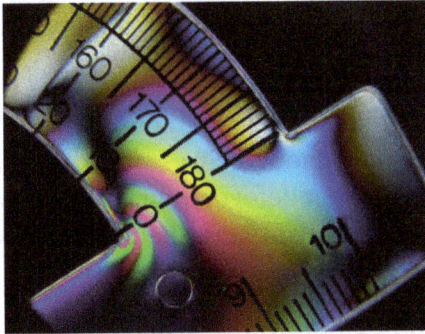

Built-in strain, inside the plastic protractor, developed by the stress of the shape of the protractor, is revealed by the effect of polarized light.

In continuum mechanics, stress is a physical quantity that expresses the internal forces that neighboring particles of a continuous material exert on each other, while strain is the measure of the deformation of the material. For example, when a solid vertical bar is supporting a weight, each particle in the bar pushes on the particles immediately below it. When a liquid is in a closed container under pressure, each particle gets pushed against by all the surrounding particles. The container walls and the pressure-inducing surface (such as a piston) push against them in (Newtonian) reaction. These macroscopic forces are actually the net result of a very large number of intermolecular forces and collisions between the particles in those molecules.

Strain inside a material may arise by various mechanisms, such as stress as applied by external forces to the bulk material (like gravity) or to its surface (like contact forces, external pressure, or friction). Any strain (deformation) of a solid material generates an internal elastic stress, analogous to the reaction force of a spring, that tends to restore the material to its original non-deformed state. In liquids and gases, only deformations that change the volume generate persistent elastic stress. However, if the deformation is gradually changing with time, even in fluids there will usually be some **viscous** stress, opposing that change. Elastic and viscous stresses are usually combined under the name mechanical stress.

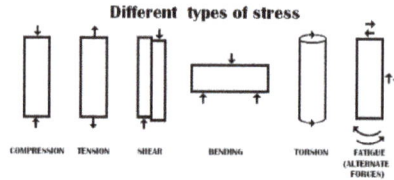

Mechanic stress

Significant stress may exist even when deformation is negligible or non-existent (a common assumption when modeling the flow of water). Stress may exist in the absence of external forces; such built-in stress is important, for example, in prestressed concrete and tempered glass. Stress may also be imposed on a material without the application of net forces, for example by changes in temperature or chemical composition, or by external electromagnetic fields (as in piezoelectric and magnetostrictive materials).

The relation between mechanical stress, deformation, and the rate of change of deformation can be quite complicated, although a linear approximation may be adequate in practice if the quantities are small enough. Stress that exceeds certain strength limits of the material will result in permanent deformation (such as plastic flow, fracture, cavitation) or even change its crystal structure and chemical composition.

In some branches of engineering, the term stress is occasionally used in a looser sense as a synonym of "internal force". For example, in the analysis of trusses, it may refer to the total traction or compression force acting on a beam, rather than the force divided by the area of its cross-section.

History

Roman-era bridge in Switzerland

Since ancient times humans have been consciously aware of stress inside materials. Until the 17th century the understanding of stress was largely intuitive and empirical;

and yet it resulted in some surprisingly sophisticated technology, like the composite bow and glass blowing.

Inca bridge on the Apurimac River

Over several millennia, architects and builders, in particular, learned how to put together carefully shaped wood beams and stone blocks to withstand, transmit, and distribute stress in the most effective manner, with ingenious devices such as the capitals, arches, cupolas, trusses and the flying buttresses of Gothic cathedrals.

Ancient and medieval architects did develop some geometrical methods and simple formulas to compute the proper sizes of pillars and beams, but the scientific understanding of stress became possible only after the necessary tools were invented in the 17th and 18th centuries: Galileo Galilei's rigorous experimental method, René Descartes's coordinates and analytic geometry, and Newton's laws of motion and equilibrium and calculus of infinitesimals. With those tools, Augustin-Louis Cauchy was able to give the first rigorous and general mathematical model for stress in a homogeneous medium. Cauchy observed that the force across an imaginary surface was a linear function of its normal vector; and, moreover, that it must be a symmetric function (with zero total momentum).

The understanding of stress in liquids started with Newton, who provided a differential formula for friction forces (shear stress) in parallel laminar flow.

Overview

Definition

Stress is defined as the force across a "small" boundary per unit area of that boundary, for all orientations of the boundary. Being derived from a fundamental physical quantity (force) and a purely geometrical quantity (area), stress is also a fundamental quantity, like velocity, torque or energy, that can be quantified and analyzed without explicit consideration of the nature of the material or of its physical causes.

Following the basic premises of continuum mechanics, stress is a macroscopic concept. Namely, the particles considered in its definition and analysis should be just small

enough to be treated as homogeneous in composition and state, but still large enough
to ignore quantum effects and the detailed motions of molecules. Thus, the force between two particles is actually the average of a very large number of atomic forces between their molecules; and physical quantities like mass, velocity, and forces that act
through the bulk of three-dimensional bodies, like gravity, are assumed to be smoothly
distributed over them. Depending on the context, one may also assume that the particles are large enough to allow the averaging out of other microscopic features, like the
grains of a metal rod or the fibers of a piece of wood.

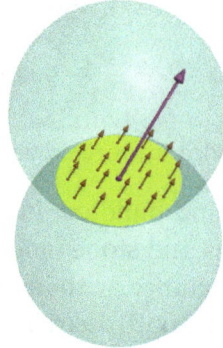

The stress across a surface element (yellow disk) is the force that the material on one side (top ball)
exerts on the material on the other side (bottom ball), divided by the area of the surface.

Quantitatively, the stress is expressed by the Cauchy traction vector T defined as the
traction force F between adjacent parts of the material across an imaginary separating
surface S, divided by the area of S. In a fluid at rest the force is perpendicular to the
surface, and is the familiar pressure. In a solid, or in a flow of viscous liquid, the force
F may not be perpendicular to S; hence the stress across a surface must be regarded a
vector quantity, not a scalar. Moreover, the direction and magnitude generally depend
on the orientation of S. Thus the stress state of the material must be described by a tensor, called the (Cauchy) stress tensor; which is a linear function that relates the normal
vector n of a surface S to the stress T across S. With respect to any chosen coordinate
system, the Cauchy stress tensor can be represented as a symmetric matrix of 3×3 real
numbers. Even within a homogeneous body, the stress tensor may vary from place to
place, and may change over time; therefore, the stress within a material is, in general,
a time-varying tensor field.

Normal and Shear Stress

In general, the stress T that a particle P applies on another particle Q across a surface
S can have any direction relative to S. The vector T may be regarded as the sum of two
components: the normal stress (compression or tension) perpendicular to the surface,
and the shear stress that is parallel to the surface.

If the normal unit vector n of the surface (pointing from Q towards P) is assumed fixed,
the normal component can be expressed by a single number, the dot product $T \cdot n$. This

number will be positive if P is "pulling" on Q (tensile stress), and negative if P is "pushing" against Q (compressive stress) The shear component is then the vector $T - (T \cdot n)n$.

Units

The dimension of stress is that of pressure, and therefore its coordinates are commonly measured in the same units as pressure: namely, pascals (Pa, that is, newtons per square metre) in the International System, or pounds per square inch (psi) in the Imperial system. Because mechanical stresses easily exceed a million Pascals, MPa, which stands for mega pascal, is a common unit of stress. The dimensional formula for stress is $ML^{-1}T^{-2}$

Causes and Effects

Glass vase with the *craquelé* effect. The cracks are the result of brief but intense stress created when the semi-molten piece is briefly dipped in water.

Stress in a material body may be due to multiple physical causes, including external influences and internal physical processes. Some of these agents (like gravity, changes in temperature and phase, and electromagnetic fields) act on the bulk of the material, varying continuously with position and time. Other agents (like external loads and friction, ambient pressure, and contact forces) may create stresses and forces that are concentrated on certain surfaces, lines, or points; and possibly also on very short time intervals (as in the impulses due to collisions). In general, the stress distribution in the body is expressed as a piecewise continuous function of space and time.

Conversely, stress is usually correlated with various effects on the material, possibly including changes in physical properties like birefringence, polarization, and permeability. The imposition of stress by an external agent usually creates some strain (deformation) in the material, even if it is too small to be detected. In a solid material, such strain will in turn generate an internal elastic stress, analogous to the reaction force of a stretched spring, tending to restore the material to its original undeformed state. Fluid materials (liquids, gases and plasmas) by definition can only oppose deformations that would change their volume. However, if the deformation is changing with time, even in fluids there will usually be some viscous stress, opposing that change.

The relation between stress and its effects and causes, including deformation and rate of change of deformation, can be quite complicated (although a linear approximation may be adequate in practice if the quantities are small enough). Stress that exceeds certain strength limits of the material will result in permanent deformation (such as plastic flow, fracture, cavitation) or even change its crystal structure and chemical composition.

Simple Stress

In some situations, the stress within a body may adequately be described by a single number, or by a single vector (a number and a direction). Three such simple stress situations, that are often encountered in engineering design, are the *uniaxial normal stress*, the *simple shear stress*, and the *isotropic normal stress*.

Uniaxial Normal Stress

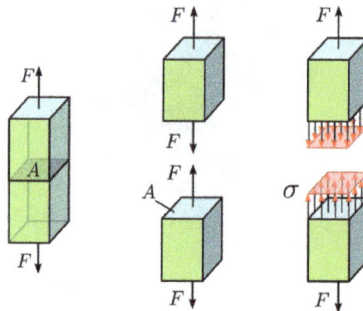

Idealized stress in a straight bar with uniform cross-section.

A common situation with a simple stress pattern is when a straight rod, with uniform material and cross section, is subjected to tension by opposite forces of magnitude along its axis. If the system is in equilibrium and not changing with time, and the weight of the bar can be neglected, then through each transversal section of the bar the top part must pull on the bottom part with the same force F. Therefore, the stress throughout the bar, across any *horizontal* surface, can be described by the number $\sigma = F/A$, where A is the area of the cross-section.

On the other hand, if one imagines the bar being cut along its length, parallel to the axis, there will be no force (hence no stress) between the two halves across the cut.

This type of stress may be called (simple) normal stress or uniaxial stress; specifically, (uniaxial, simple, etc.) tensile stress. If the load is compression on the bar, rather than stretching it, the analysis is the same except that the force F and the stress σ change sign, and the stress is called compressive stress.

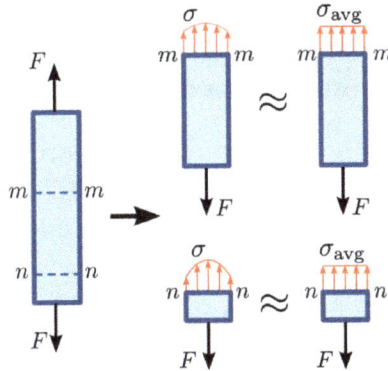

The ratio $\sigma = F / A$ may be only an average stress. The stress may be unevenly distributed over the cross section (m–m), especially near the attachment points (n–n).

This analysis assumes the stress is evenly distributed over the entire cross-section. In practice, depending on how the bar is attached at the ends and how it was manufactured, this assumption may not be valid. In that case, the value $\sigma = F/A$ will be only the average stress, called engineering stress or nominal stress. However, if the bar's length L is many times its diameter D, and it has no gross defects or built-in stress, then the stress can be assumed to be uniformly distributed over any cross-section that is more than a few times D from both ends. (This observation is known as the Saint-Venant's principle).

Normal stress occurs in many other situations besides axial tension and compression. If an elastic bar with uniform and symmetric cross-section is bent in one of its planes of symmetry, the resulting bending stress will still be normal (perpendicular to the cross-section), but will vary over the cross section: the outer part will be under tensile stress, while the inner part will be compressed. Another variant of normal stress is the hoop stress that occurs on the walls of a cylindrical pipe or vessel filled with pressurized fluid.

Simple Shear Stress

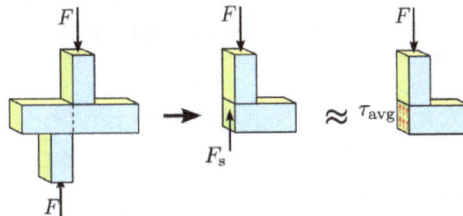

Shear stress in a horizontal bar loaded by two offset blocks.

Another simple type of stress occurs when a uniformly thick layer of elastic material like glue or rubber is firmly attached to two stiff bodies that are pulled in opposite directions by forces parallel to the layer; or a section of a soft metal bar that is being cut by the jaws of a scissors-like tool. Let F be the magnitude of those forces, and M be the midplane of that layer. Just as in the normal stress case, the part of the layer on one side of M must pull the other part with the same force F. Assuming that the direction of the forces is known, the stress across M can be expressed by the single number $\sigma = F/A$, where F is the magnitude of those forces and A is the area of the layer.

However, unlike normal stress, this simple shear stress is directed parallel to the cross-section considered, rather than perpendicular to it. For any plane S that is perpendicular to the layer, the net internal force across S, and hence the stress, will be zero.

As in the case of an axially loaded bar, in practice the shear stress may not be uniformly distributed over the layer; so, as before, the ratio F/A will only be an average ("nominal", "engineering") stress. However, that average is often sufficient for practical purposes. Shear stress is observed also when a cylindrical bar such as a shaft is subjected to opposite torques at its ends. In that case, the shear stress on each cross-section is parallel to the cross-section, but oriented tangentially relative to the axis, and increases with distance from the axis. Significant shear stress occurs in the middle plate (the "web") of I-beams under bending loads, due to the web constraining the end plates ("flanges").

Isotropic Stress

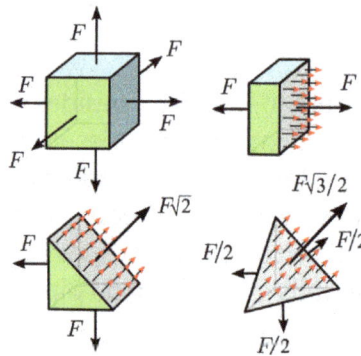

Isotropic tensile stress. Top left: Each face of a cube of homogeneous material is pulled by a force with magnitude F, applied evenly over the entire face whose area is A. The force across any section S of the cube must balance the forces applied below the section. In the three sections shown, the forces are F (top right), $F\sqrt{2}$ (bottom left), and $F\sqrt{3}/2$ (bottom right); and the area of S is A, $A\sqrt{2}$ and $A\sqrt{3}/2$, respectively. So the stress across S is F/A in all three cases.

Another simple type of stress occurs when the material body is under equal compression or tension in all directions. This is the case, for example, in a portion of liquid or gas at rest, whether enclosed in some container or as part of a larger mass of fluid; or inside a cube of elastic material that is being pressed or pulled on all six faces by equal perpendicular forces — provided, in both cases, that the material is homogeneous, without built-in stress, and that the effect of gravity and other external forces can be neglected.

In these situations, the stress across any imaginary internal surface turns out to be equal in magnitude and always directed perpendicularly to the surface independently of the surface's orientation. This type of stress may be called isotropic normal or just isotropic; if it is compressive, it is called hydrostatic pressure or just pressure. Gases by definition cannot withstand tensile stresses, but some liquids may withstand surprisingly large amounts of isotropic tensile stress under some circumstances.

Cylinder Stresses

Parts with rotational symmetry, such as wheels, axles, pipes, and pillars, are very common in engineering. Often the stress patterns that occur in such parts have rotational or even cylindrical symmetry. The analysis of such cylinder stresses can take advantage of the symmetry to reduce the dimension of the domain and/or of the stress tensor.

General Stress

Often, mechanical bodies experience more than one type of stress at the same time; this is called combined stress. In normal and shear stress, the magnitude of the stress is maximum for surfaces that are perpendicular to a certain direction $d,$, and zero across any surfaces that are parallel to $d,$. When the shear stress is zero only across surfaces that are perpendicular to one particular direction, the stress is called biaxial, and can be viewed as the sum of two normal or shear stresses. In the most general case, called triaxial stress, the stress is nonzero across every surface element.

The Cauchy Stress Tensor

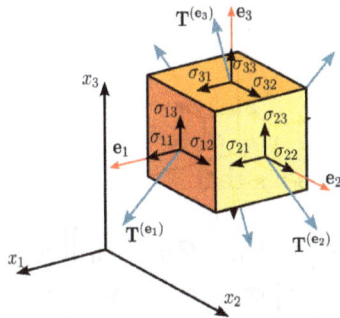

Components of stress in three dimensions

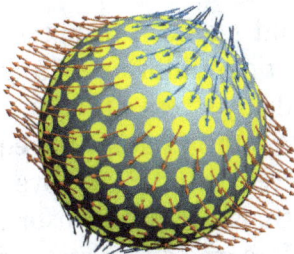

Illustration of typical stresses (arrows) across various surface elements on the boundary of a particle (sphere), in a homogeneous material under uniform (but not isotropic) triaxial stress. The normal stresses on the principal axes are +5, +2, and −3 units.

Combined stresses cannot be described by a single vector. Even if the material is stressed in the same way throughout the volume of the body, the stress across any imaginary surface will depend on the orientation of that surface, in a non-trivial way.

However, Cauchy observed that the stress vector T across a surface will always be a linear function of the surface's normal vector n, the unit-length vector that is perpendicular to it. That is, $T = \sigma(n),$, where the function σ satisfies

$$\sigma(\alpha u + \beta v) = \alpha \sigma(u) + \beta \sigma(v)$$

for any vectors u, v and any real numbers α, β. The function σ, now called the (Cauchy) stress tensor, completely describes the stress state of a uniformly stressed body. (Today, any linear connection between two physical vector quantities is called a tensor, reflecting Cauchy's original use to describe the "tensions" (stresses) in a material.) In tensor calculus, σ is classified as second-order tensor of type (0,2).

Like any linear map between vectors, the stress tensor can be represented in any chosen Cartesian coordinate system by a 3×3 matrix of real numbers. Depending on whether the coordinates are numbered x_1, x_2, x_3 or named $x, y, z,$, the matrix may be written as

$$\begin{bmatrix} \sigma_{11} & \sigma_{12} & \sigma_{13} \\ \sigma_{21} & \sigma_{22} & \sigma_{23} \\ \sigma_{31} & \sigma_{32} & \sigma_{33} \end{bmatrix} \text{ or } \begin{bmatrix} \sigma_{xx} & \sigma_{xy} & \sigma_{xz} \\ \sigma_{yx} & \sigma_{yy} & \sigma_{yz} \\ \sigma_{zx} & \sigma_{zy} & \sigma_{zz} \end{bmatrix}$$

The stress vector $T = 6(n)$ across a surface with normal vector n with coordinates n_1, n_2, n_3 is then a matrix product $T = n \cdot \sigma = \sigma^T \cdot n^T$ (where T in upper index is transposition) (look on Cauchy stress tensor), that is

$$\begin{bmatrix} T_1 \\ T_2 \\ T_3 \end{bmatrix} = \begin{bmatrix} \sigma_{11} & \sigma_{21} & \sigma_{31} \\ \sigma_{12} & \sigma_{22} & \sigma_{32} \\ \sigma_{13} & \sigma_{23} & \sigma_{33} \end{bmatrix} \begin{bmatrix} n_1 \\ n_2 \\ n_3 \end{bmatrix}$$

The linear relation between T and n follows from the fundamental laws of conservation of linear momentum and static equilibrium of forces, and is therefore mathematically exact, for any material and any stress situation. The components of the Cauchy stress tensor at every point in a material satisfy the equilibrium equations (Cauchy's equations of motion for zero acceleration). Moreover, the principle of conservation of angular momentum implies that the stress tensor is symmetric, that is $\sigma_{12} = \sigma_{21}$, $\sigma_{13} = \sigma_{31}$, and $\sigma_{23} = \sigma_{32}$. Therefore, the stress state of the medium at any point and instant can be specified by only six independent parameters, rather than nine. These may be written

$$\begin{bmatrix} \sigma_x & \tau_{xy} & \tau_{xz} \\ \tau_{xy} & \sigma_y & \tau_{yz} \\ \tau_{xz} & \tau_{yz} & \sigma_z \end{bmatrix}$$

where the elements $\sigma_x, \sigma_y, \sigma_z$ are called the orthogonal normal stresses (relative to the chosen coordinate system), and $\tau_{xy}, \tau_{xz}, \tau_{yz}$ the orthogonal shear stresses.

Change of Coordinates

The Cauchy stress tensor obeys the tensor transformation law under a change in the system of coordinates. A graphical representation of this transformation law is the Mohr's circle of stress distribution.

As a symmetric 3×3 real matrix, the stress tensor σ has three mutually orthogonal unit-length eigenvectors e_1, e_2, e_3 and three real eigenvalues $\lambda_1, \lambda_2, \lambda_3$, such that $\sigma e_i = \lambda_i e_i$. Therefore, in a coordinate system with axes e_1, e_2, e_3, the stress tensor is a diagonal matrix, and has only the three normal components $\lambda_1, \lambda_2, \lambda_3$ the principal stresses. If the three eigenvalues are equal, the stress is an isotropic compression or tension, always perpendicular to any surface, there is no shear stress, and the tensor is a diagonal matrix in any coordinate frame.

Stress as a Tensor Field

In general, stress is not uniformly distributed over a material body, and may vary with time. Therefore, the stress tensor must be defined for each point and each moment, by considering an infinitesimal particle of the medium surrounding that point, and taking the average stresses in that particle as being the stresses at the point.

Stress in Thin Plates

A tank car made from bent and welded steel plates.

Man-made objects are often made from stock plates of various materials by operations that do not change their essentially two-dimensional character, like cutting, drilling, gentle bending and welding along the edges. The description of stress in such bodies can be simplified by modeling those parts as two-dimensional surfaces rather than three-dimensional bodies.

In that view, one redefines a "particle" as being an infinitesimal patch of the plate's surface, so that the boundary between adjacent particles becomes an infinitesimal line element; both are implicitly extended in the third dimension, normal to (straight through) the plate. "Stress" is then redefined as being a measure of the internal forces between two adjacent "particles" across their common line element, divided by the length of that line. Some components of the stress tensor can be ignored, but since particles are not infinitesimal in the third dimension one can no longer ignore the torque that a particle applies on its neighbors. That torque is modeled as a bending stress that tends to change the curvature of the plate. However, these simplifications may not hold at welds, at sharp bends and creases (where the radius of curvature is comparable to the thickness of the plate).

Stress in Thin Beams

For stress modeling, a fishing pole may be considered one-dimensional.

The analysis of stress can be considerably simplified also for thin bars, beams or wires of uniform (or smoothly varying) composition and cross-section that are subjected to moderate bending and twisting. For those bodies, one may consider only cross-sections that are perpendicular to the bar's axis, and redefine a "particle" as being a piece of wire with infinitesimal length between two such cross sections. The ordinary stress is then reduced to a scalar (tension or compression of the bar), but one must take into account also a bending stress (that tries to change the bar's curvature, in some direction perpendicular to the axis) and a torsional stress (that tries to twist or un-twist it about its axis).

Other Descriptions of Stress

The Cauchy stress tensor is used for stress analysis of material bodies experiencing small deformations where the differences in stress distribution in most cases can be neglected. For large deformations, also called finite deformations, other measures of stress, such as the first and second Piola–Kirchhoff stress tensors, the Biot stress tensor, and the Kirchhoff stress tensor, are required.

Solids, liquids, and gases have stress fields. Static fluids support normal stress but will flow under shear stress. Moving viscous fluids can support shear stress (dynamic pressure). Solids can support both shear and normal stress, with ductile materials failing under shear and brittle materials failing under normal stress. All materials have temperature dependent variations in stress-related properties, and non-Newtonian materials have rate-dependent variations.

Stress Analysis

Stress analysis is a branch of applied physics that covers the determination of the internal distribution of internal forces in solid objects. It is an essential tool in engineering for the study and design of structures such as tunnels, dams, mechanical parts, and structural frames, under prescribed or expected loads. It is also important in many other disciplines; for example, in geology, to study phenomena like plate tectonics, vulcanism and avalanches; and in biology, to understand the anatomy of living beings.

Goals and Assumptions

Stress analysis is generally concerned with objects and structures that can be assumed to be in macroscopic static equilibrium. By Newton's laws of motion, any external forces are being applied to such a system must be balanced by internal reaction forces, which are almost always surface contact forces between adjacent particles — that is, as stress. Since every particle needs to be in equilibrium, this reaction stress will generally propagate from particle, creating a stress distribution throughout the body.

The typical problem in stress analysis is to determine these internal stresses, given the external forces that are acting on the system. The latter may be body forces (such as gravity or magnetic attraction), that act throughout the volume of a material; or concentrated loads (such as friction between an axle and a bearing, or the weight of a train wheel on a rail), that are imagined to act over a two-dimensional area, or along a line, or at single point.

In stress analysis one normally disregards the physical causes of the forces or the precise nature of the materials. Instead, one assumes that the stresses are related to deformation (and, in non-static problems, to the rate of deformation) of the material by known constitutive equations.

Methods

Stress analysis may be carried out experimentally, by applying loads to the actual artifact or to scale model, and measuring the resulting stresses, by any of several available methods. This approach is often used for safety certification and monitoring. However, most stress analysis is done by mathematical methods, especially during design.

The basic stress analysis problem can be formulated by Euler's equations of motion for continuous bodies (which are consequences of Newton's laws for conservation of linear momentum and angular momentum) and the Euler-Cauchy stress principle, together with the appropriate constitutive equations. Thus one obtains a system of partial differential equations involving the stress tensor field and the strain tensor field, as unknown functions to be determined. The external body forces appear as the independent ("right-hand side") term in the differential equations, while the concentrated forces appear as boundary conditions. The basic stress analysis problem is therefore a boundary-value problem.

Stress analysis for elastic structures is based on the theory of elasticity and infinitesimal strain theory. When the applied loads cause permanent deformation, one must use more complicated constitutive equations, that can account for the physical processes involved (plastic flow, fracture, phase change, etc.).

However, engineered structures are usually designed so that the maximum expected stresses are well within the range of linear elasticity (the generalization of Hooke's law for continuous media); that is, the deformations caused by internal stresses are linearly related to them. In this case the differential equations that define the stress tensor are linear, and the problem becomes much easier. For one thing, the stress at any point will be a linear function of the loads, too. For small enough stresses, even non-linear systems can usually be assumed to be linear.

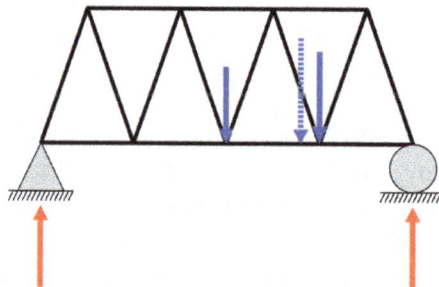

Simplified model of a truss for stress analysis, assuming unidimensional elements under uniform axial tension or compression.

Stress analysis is simplified when the physical dimensions and the distribution of loads allow the structure to be treated as one- or two-dimensional. In the analysis of trusses, for example, the stress field may be assumed to be uniform and uniaxial over each member. Then the differential equations reduce to a finite set of equations (usually linear) with finitely many unknowns. In other contexts one may be able to reduce the three-dimensional problem to a two-dimensional one, and/or replace the general stress and strain tensors by simpler models like uniaxial tension/compression, simple shear, etc.

Still, for two- or three-dimensional cases one must solve a partial differential equation problem. Analytical or closed-form solutions to the differential equations can be obtained when the geometry, constitutive relations, and boundary conditions are simple enough. Otherwise one must generally resort to numerical approximations such as the finite element method, the finite difference method, and the boundary element method.

Alternative Measures of Stress

Other useful stress measures include the first and second Piola–Kirchhoff stress tensors, the Biot stress tensor, and the Kirchhoff stress tensor.

Piola–Kirchhoff Stress Tensor

In the case of finite deformations, the *Piola–Kirchhoff stress tensors* express the stress relative to the reference configuration. This is in contrast to the Cauchy stress tensor which expresses the stress relative to the present configuration. For infinitesimal deformations and rotations, the Cauchy and Piola–Kirchhoff tensors are identical.

Whereas the Cauchy stress tensor $\acute{o}$ relates stresses in the current configuration, the deformation gradient and strain tensors are described by relating the motion to the reference configuration; thus not all tensors describing the state of the material are in either the reference or current configuration. Describing the stress, strain and deformation either in the reference or current configuration would make it easier to define constitutive models (for example, the Cauchy Stress tensor is variant to a pure rotation, while the deformation strain tensor is invariant; thus creating problems in defining a constitutive model that relates a varying tensor, in terms of an invariant one during pure rotation; as by definition constitutive models have to be invariant to pure rotations). The 1st Piola–Kirchhoff stress tensor, **P** is one possible solution to this problem. It defines a family of tensors, which describe the configuration of the body in either the current or the reference state.

The 1st Piola–Kirchhoff stress tensor, **P** relates forces in the *present* configuration with areas in the *reference* ("material") configuration.

$$\mathbf{P} = J\,\sigma\,\mathbf{F}^{-T}$$

where **F** is the deformation gradient and $J = \det \mathbf{F}$ is the Jacobian determinant.

In terms of components with respect to an orthonormal basis, the first Piola–Kirchhoff stress is given by

$$P_{iL} = J\,\sigma_{ik}\,F^{-1}_{Lk} = J\,\sigma_{ik}\frac{\partial X_L}{\partial x_k}$$

Because it relates different coordinate systems, the 1st Piola–Kirchhoff stress is a two-point tensor. In general, it is not symmetric. The 1st Piola–Kirchhoff stress is the 3D generalization of the 1D concept of engineering stress.

If the material rotates without a change in stress state (rigid rotation), the components of the 1st Piola–Kirchhoff stress tensor will vary with material orientation.

The 1st Piola–Kirchhoff stress is energy conjugate to the deformation gradient.

2nd Piola–Kirchhoff Stress Tensor

Whereas the 1st Piola–Kirchhoff stress relates forces in the current configuration to areas in the reference configuration, the 2nd Piola–Kirchhoff stress tensor S relates forces in the reference configuration to areas in the reference configuration. The force in the reference configuration is obtained via a mapping that preserves the relative relationship between the force direction and the area normal in the reference configuration.

$$S = J\, F^{-1}\, \sigma\, F^{-T}$$

In index notation with respect to an orthonormal basis,

$$S_{IL} = J F_{Ik}^{-1} F_{Lm}^{-1} \sigma_{km} = J \frac{\partial X_I}{\partial x_k} \frac{\partial X_L}{\partial x_m} \sigma_{km}$$

This tensor, a one-point tensor, is symmetric.

If the material rotates without a change in stress state (rigid rotation), the components of the 2nd Piola–Kirchhoff stress tensor remain constant, irrespective of material orientation.

The 2nd Piola–Kirchhoff stress tensor is energy conjugate to the Green–Lagrange finite strain tensor.

Deformation (Mechanics)

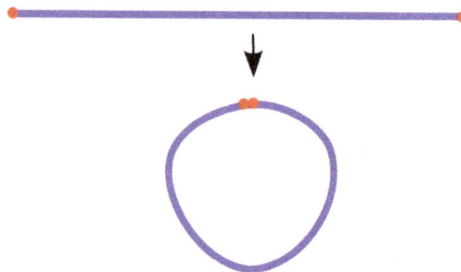

The deformation of a thin straight rod into a closed loop. The length of the rod remains almost unchanged during the deformation, which indicates that the strain is small. In this particular case of bending, displacements associated with rigid translations and rotations of material elements in the rod are much greater than displacements associated with straining.

Deformation in continuum mechanics is the transformation of a body from a *reference* configuration to a *current* configuration. A configuration is a set containing the positions of all particles of the body.

A deformation may be caused by external loads, body forces (such as gravity or electromagnetic forces), or changes in temperature, moisture content, or chemical reactions, etc.

Strain is a description of deformation in terms of *relative* displacement of particles in the body that excludes rigid-body motions. Different equivalent choices may be made for the expression of a strain field depending on whether it is defined with respect to the initial or the final configuration of the body and on whether the metric tensor or its dual is considered.

In a continuous body, a deformation field results from a stress field induced by applied forces or is due to changes in the temperature field inside the body. The relation between stresses and induced strains is expressed by constitutive equations, e.g., Hooke's law for linear elastic materials. Deformations which are recovered after the stress field has been removed are called elastic deformations. In this case, the continuum completely recovers its original configuration. On the other hand, irreversible deformations remain even after stresses have been removed. One type of irreversible deformation is plastic deformation, which occurs in material bodies after stresses have attained a certain threshold value known as the elastic limit or yield stress, and are the result of slip, or dislocation mechanisms at the atomic level. Another type of irreversible deformation is viscous deformation, which is the irreversible part of viscoelastic deformation.

In the case of elastic deformations, the response function linking strain to the deforming stress is the compliance tensor of the material.

Strain

Strain is a measure of deformation representing the displacement between particles in the body relative to a reference length.

A general deformation of a body can be expressed in the form $\mathbf{x} = \boldsymbol{F}(\mathbf{X})$ where $\mathbf{X}$ is the reference position of material points in the body. Such a measure does not distinguish between rigid body motions (translations and rotations) and changes in shape (and size) of the body. A deformation has units of length.

We could, for example, define strain to be

$$\mathring{a} \doteq \frac{\partial}{\partial \mathbf{X}}(\mathbf{x} - \mathbf{X}) = \mathbf{F}' - \mathbf{I},$$

where I is the identity tensor. Hence strains are dimensionless and are usually expressed as a decimal fraction, a percentage or in parts-per notation. Strains measure how much a given deformation differs locally from a rigid-body deformation.

A strain is in general a tensor quantity. Physical insight into strains can be gained by observing that a given strain can be decomposed into normal and shear components. The amount of stretch or compression along material line elements or fibers is the *normal strain*, and the amount of distortion associated with the sliding of plane layers over each other is the *shear strain*, within a deforming body. This could be applied by elongation, shortening, or volume changes, or angular distortion.

The state of strain at a material point of a continuum body is defined as the totality of all the changes in length of material lines or fibers, the *normal strain*, which pass through that point and also the totality of all the changes in the angle between pairs of lines initially perpendicular to each other, the *shear strain*, radiating from this point. However, it is sufficient to know the normal and shear components of strain on a set of three mutually perpendicular directions.

If there is an increase in length of the material line, the normal strain is called *tensile strain*, otherwise, if there is reduction or compression in the length of the material line, it is called *compressive strain*.

Strain Measures

Depending on the amount of strain, or local deformation, the analysis of deformation is subdivided into three deformation theories:

- Finite strain theory, also called *large strain theory*, *large deformation theory*, deals with deformations in which both rotations and strains are arbitrarily large. In this case, the undeformed and deformed configurations of the continuum are significantly different and a clear distinction has to be made between them. This is commonly the case with elastomers, plastically-deforming materials and other fluids and biological soft tissue.

- Infinitesimal strain theory, also called *small strain theory*, *small deformation theory*, *small displacement theory*, or *small displacement-gradient theory* where strains and rotations are both small. In this case, the undeformed and deformed configurations of the body can be assumed identical. The infinitesimal strain theory is used in the analysis of deformations of materials exhibiting elastic behavior, such as materials found in mechanical and civil engineering applications, e.g. concrete and steel.

- *Large-displacement* or *large-rotation theory*, which assumes small strains but large rotations and displacements.

In each of these theories the strain is then defined differently. The *engineering strain* is the most common definition applied to materials used in mechanical and structural engineering, which are subjected to very small deformations. On the other hand, for some materials, e.g. elastomers and polymers, subjected to large deformations, the engineering definition of strain is not applicable, e.g. typical engineering strains greater than 1%, thus other more complex definitions of strain are required, such as *stretch*, *logarithmic strain*, *Green strain*, and *Almansi strain*.

Engineering Strain

The Cauchy strain or engineering strain is expressed as the ratio of total deformation to the initial dimension of the material body in which the forces are being applied. The *engineering normal strain* or *engineering extensional strain* or *nominal strain* e of a material line element or fiber axially loaded is expressed as the change in length ΔL per unit of the original length L of the line element or fibers. The normal strain is positive if the material fibers are stretched and negative if they are compressed. Thus, we have

$$e = \frac{\Delta L}{L} = \frac{l - L}{L}$$

where e is the *engineering normal strain*, L is the original length of the fiber and l is the final length of the fiber. Measures of strain are often expressed in parts per million or microstrains.

The *true shear strain* is defined as the change in the angle (in radians) between two material line elements initially perpendicular to each other in the undeformed or initial configuration. The *engineering shear strain* is defined as the tangent of that angle, and is equal to the length of deformation at its maximum divided by the perpendicular length in the plane of force application which sometimes makes it easier to calculate.

Stretch Ratio

The stretch ratio or extension ratio is a measure of the extensional or normal strain of a differential line element, which can be defined at either the undeformed configuration or the deformed configuration. It is defined as the ratio between the final length l and the initial length L of the material line.

$$\lambda = \frac{l}{L}$$

The extension ratio is approximately related to the engineering strain by

$$e = \frac{l - L}{L} = \lambda - 1$$

This equation implies that the normal strain is zero, so that there is no deformation when the stretch is equal to unity.

The stretch ratio is used in the analysis of materials that exhibit large deformations, such as elastomers, which can sustain stretch ratios of 3 or 4 before they fail. On the other hand, traditional engineering materials, such as concrete or steel, fail at much lower stretch ratios.

True Strain

The logarithmic strain ε, also called, *true strain* or *Hencky strain* (although nothing is particularly "true" about it compared to other valid definitions of strain). Considering an incremental strain (Ludwik)

$$\delta\varepsilon = \frac{\delta l}{l}$$

the logarithmic strain is obtained by integrating this incremental strain:

$$\int \delta\varepsilon = \int_{L}^{l} \frac{\delta l}{l} \varepsilon =$$

$$\ln\left(\frac{l}{L}\right) = \ln(\lambda) = \ln(1+e)$$

$$= e - \frac{e^2}{2} + \frac{e^3}{3} - \cdots$$

where *e* is the engineering strain. The logarithmic strain provides the correct measure of the final strain when deformation takes place in a series of increments, taking into account the influence of the strain path.

Green Strain

The Green strain is defined as:

$$\varepsilon_G = \tfrac{1}{2}\left(\frac{l^2 - L^2}{L^2}\right) = \tfrac{1}{2}(\lambda^2 - 1)$$

Almansi Strain

The Euler-Almansi strain is defined as

$$\varepsilon_E = \tfrac{1}{2}\left(\frac{l^2 - L^2}{l^2}\right) = \tfrac{1}{2}\left(1 - \frac{1}{\lambda^2}\right)$$

Normal and Shear Strain

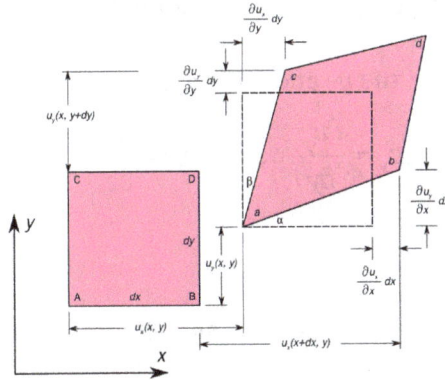

Two-dimensional geometric deformation of an infinitesimal material element.

Strains are classified as either *normal* or *shear*. A *normal strain* is perpendicular to the face of an element, and a *shear strain* is parallel to it. These definitions are consistent with those of normal stress and shear stress.

Normal Strain

For an isotropic material that obeys Hooke's law, a normal stress will cause a normal strain. Normal strains produce *dilations*.

Consider a two-dimensional, infinitesimal, rectangular material element with dimensions $dx \times dy$, which, after deformation, takes the form of a rhombus. From the geometry of the adjacent figure we have

$$\text{length}(AB) = dx$$

and

$$\text{length}(ab) = \sqrt{\left(dx + \frac{\partial u_x}{\partial x} dx \right)^2 + \left(\frac{\partial u_y}{\partial x} dx \right)^2}$$

$$= dx \sqrt{1 + 2\frac{\partial u_x}{\partial x} + \left(\frac{\partial u_x}{\partial x} \right)^2 + \left(\frac{\partial u_y}{\partial x} \right)^2}$$

For very small displacement gradients the squares of the derivatives are negligible and we have

$$\text{length}(ab) \approx dx + \frac{\partial u_x}{\partial x} dx$$

The normal strain in the *x*-direction of the rectangular element is defined by

$$\varepsilon_x = \frac{\text{extension}}{\text{original length}} = \frac{\text{length}(ab) - \text{length}(AB)}{\text{length}(AB)} = \frac{\partial u_x}{\partial x}$$

Similarly, the normal strain in the y- and z-directions becomes

$$\varepsilon_y = \frac{\partial u_y}{\partial y} \quad , \quad \varepsilon_z = \frac{\partial u_z}{\partial z}$$

Shear Strain

The engineering shear strain (γ_{xy}) is defined as the change in angle between lines AC and AB. Therefore,

$$\gamma_{xy} = \alpha + \beta$$

From the geometry of the figure, we have

$$\tan \alpha = \frac{\frac{\partial u_y}{\partial x} dx}{dx + \frac{\partial u_x}{\partial x} dx} = \frac{\frac{\partial u_y}{\partial x}}{1 + \frac{\partial u_x}{\partial x}} \tan \beta$$

$$= \frac{\frac{\partial u_x}{\partial y} dy}{dy + \frac{\partial u_y}{\partial y} dy} = \frac{\frac{\partial u_x}{\partial y}}{1 + \frac{\partial u_y}{\partial y}}$$

For small displacement gradients we have

$$\frac{\partial u_x}{\partial x} \ll 1 \; ; \frac{\partial u_y}{\partial y} \ll 1$$

For small rotations, i.e. α and β are $\ll 1$ we have $\tan \alpha \approx \alpha$, $\tan \beta \approx \beta$. Therefore,

$$\alpha \approx \frac{\partial u_y}{\partial x} \; ; \beta \approx \frac{\partial u_x}{\partial y}$$

thus

$$\gamma_{xy} = \alpha + \beta = \frac{\partial u_y}{\partial x} + \frac{\partial u_x}{\partial y}$$

By interchanging x and y and u_x and u_y, it can be shown that $\gamma_{xy} = \gamma_{yx}$.

Similarly, for the yz- and xz-planes, we have

$$\gamma_{yz} = \gamma_{zy} = \frac{\partial u_y}{\partial z} + \frac{\partial u_z}{\partial y} \quad , \quad \gamma_{zx} = \gamma_{xz} = \frac{\partial u_z}{\partial x} + \frac{\partial u_x}{\partial z}$$

The tensorial shear strain components of the infinitesimal strain tensor can then be expressed using the engineering strain definition, γ, as

$$\underline{\underline{\mathring{a}}} = \begin{bmatrix} \varepsilon_{xx} & \varepsilon_{xy} & \varepsilon_{xz} \\ \varepsilon_{yx} & \varepsilon_{yy} & \varepsilon_{yz} \\ \varepsilon_{zx} & \varepsilon_{zy} & \varepsilon_{zz} \end{bmatrix} = \begin{bmatrix} \varepsilon_{xx} & \frac{1}{2}\gamma_{xy} & \frac{1}{2}\gamma_{xz} \\ \frac{1}{2}\gamma_{yx} & \varepsilon_{yy} & \frac{1}{2}\gamma_{yz} \\ \frac{1}{2}\gamma_{zx} & \frac{1}{2}\gamma_{zy} & \varepsilon_{zz} \end{bmatrix}$$

Metric Tensor

A strain field associated with a displacement is defined, at any point, by the change in length of the tangent vectors representing the speeds of arbitrarily parametrized curves passing through that point. A basic geometric result, due to Fréchet, von Neumann and Jordan, states that, if the lengths of the tangent vectors fulfil the axioms of a norm and the parallelogram law, then the length of a vector is the square root of the value of the quadratic form associated, by the polarization formula, with a positive definite bilinear map called the metric tensor.

Description of Deformation

Deformation is the change in the metric properties of a continuous body, meaning that a curve drawn in the initial body placement changes its length when displaced to a curve in the final placement. If none of the curves changes length, it is said that a rigid body displacement occurred.

It is convenient to identify a reference configuration or initial geometric state of the continuum body which all subsequent configurations are referenced from. The reference configuration need not be one the body actually will ever occupy. Often, the configuration at $t = 0$ is considered the reference configuration, $\kappa_0(B)$. The configuration at the current time t is the *current configuration*.

For deformation analysis, the reference configuration is identified as *undeformed configuration*, and the current configuration as *deformed configuration*. Additionally, time is not considered when analyzing deformation, thus the sequence of configurations between the undeformed and deformed configurations are of no interest.

The components X_i of the position vector X of a particle in the reference configuration, taken with respect to the reference coordinate system, are called the *material or reference coordinates*. On the other hand, the components x_i of the position vector **x** of a particle in the deformed configuration, taken with respect to the spatial coordinate system of reference, are called the *spatial coordinates*

There are two methods for analysing the deformation of a continuum. One description is made in terms of the material or referential coordinates, called material description or Lagrangian description. A second description is of deformation is made in terms of the spatial coordinates it is called the spatial description or Eulerian description.

There is continuity during deformation of a continuum body in the sense that:

- The material points forming a closed curve at any instant will always form a closed curve at any subsequent time.

- The material points forming a closed surface at any instant will always form a closed surface at any subsequent time and the matter within the closed surface will always remain within.

Affine Deformation

A deformation is called an affine deformation if it can be described by an affine transformation. Such a transformation is composed of a linear transformation (such as rotation, shear, extension and compression) and a rigid body translation. Affine deformations are also called homogeneous deformations.

Therefore, an affine deformation has the form

$$\mathbf{x}(\mathbf{X},t) = \mathbf{F}(t) \cdot \mathbf{X} + \mathbf{c}(t)$$

where $\mathbf{x}$ is the position of a point in the deformed configuration, X is the position in a reference configuration, t is a time-like parameter, F is the linear transformer and $\mathbf{c}$ is the translation. In matrix form, where the components are with respect to an orthonormal basis,

$$\begin{bmatrix} x_1(X_1,X_2,X_3,t) \\ x_2(X_1,X_2,X_3,t) \\ x_3(X_1,X_2,X_3,t) \end{bmatrix} = \begin{bmatrix} F_{11}(t) & F_{12}(t) & F_{13}(t) \\ F_{21}(t) & F_{22}(t) & F_{23}(t) \\ F_{31}(t) & F_{32}(t) & F_{33}(t) \end{bmatrix} \begin{bmatrix} X_1 \\ X_2 \\ X_3 \end{bmatrix} + \begin{bmatrix} c_1(t) \\ c_2(t) \\ c_3(t) \end{bmatrix} \quad \begin{bmatrix} x_1(X_1,X_2,X_3,t) \\ x_2(X_1,X_2,X_3,t) \\ x_3(X_1,X_2,X_3,t) \end{bmatrix} = \begin{bmatrix} F_{11}(t) & F_{12}(t) & F_{13}(t) \\ F_{21}(t) & F_{22}(t) & F_{23}(t) \\ F_{31}(t) & F_{32}(t) & F_{33}(t) \end{bmatrix} \begin{bmatrix} X_1 \\ X_2 \\ X_3 \end{bmatrix} + \begin{bmatrix} c_1(t) \\ c_2(t) \\ c_3(t) \end{bmatrix}$$

The above deformation becomes non-affine or inhomogeneous if F = F(X,t) or c = c(X,t).

Rigid Body Motion

A rigid body motion is a special affine deformation that does not involve any shear, extension or compression. The transformation matrix F is proper orthogonal in order to allow rotations but no reflections.

A rigid body motion can be described by

$$\mathbf{x}(\mathbf{X},t) = \mathbf{Q}(t) \cdot \mathbf{X} + \mathbf{c}(t)$$

where

$$\mathbf{Q} \cdot \mathbf{Q}^T = \mathbf{Q}^T \cdot \mathbf{Q} = 1$$

In matrix form,

$$
\begin{bmatrix} x_1(X_1,X_2,X_3,t) \\ x_2(X_1,X_2,X_3,t) \\ x_3(X_1,X_2,X_3,t) \end{bmatrix} = \begin{bmatrix} Q_{11}(t) & Q_{12}(t) & Q_{13}(t) \\ Q_{21}(t) & Q_{22}(t) & Q_{23}(t) \\ Q_{31}(t) & Q_{32}(t) & Q_{33}(t) \end{bmatrix} \begin{bmatrix} X_1 \\ X_2 \\ X_3 \end{bmatrix} + \begin{bmatrix} c_1(t) \\ c_2(t) \\ c_3(t) \end{bmatrix}
$$

Displacement

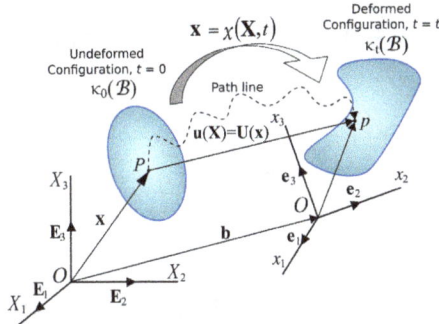

Figure 1. Motion of a continuum body.

A change in the configuration of a continuum body results in a displacement. The displacement of a body has two components: a rigid-body displacement and a deformation. A rigid-body displacement consists of a simultaneous translation and rotation of the body without changing its shape or size. Deformation implies the change in shape and/or size of the body from an initial or undeformed configuration κ_0(B) to a current or deformed configuration κ_t(B) (Figure 1).

If after a displacement of the continuum there is a relative displacement between particles, a deformation has occurred. On the other hand, if after displacement of the continuum the relative displacement between particles in the current configuration is zero, then there is no deformation and a rigid-body displacement is said to have occurred.

The vector joining the positions of a particle P in the undeformed configuration and deformed configuration is called the displacement vector u(X,t) = $u_i e_i$ in the Lagrangian description, or $\mathbf{U}$(x,t) = $U_J \mathbf{E}_J$ in the Eulerian description.

A *displacement field* is a vector field of all displacement vectors for all particles in the body, which relates the deformed configuration with the undeformed configuration. It is convenient to do the analysis of deformation or motion of a continuum body in terms of the displacement field. In general, the displacement field is expressed in terms of the material coordinates as

$$
\mathbf{u}(\mathbf{X},t) = \mathbf{b}(\mathbf{X},t) + \mathbf{x}(\mathbf{X},t) - \mathbf{X} \qquad \text{or} \qquad u_i = \alpha_{iJ}b_J + x_i - \alpha_{iJ}X_J
$$

or in terms of the spatial coordinates as

$$\mathbf{U}(\mathbf{x},t) = \mathbf{b}(\mathbf{x},t) + \mathbf{x} - \mathbf{X}(\mathbf{x},t) \qquad \text{or} \qquad U_J = b_J + \alpha_{Ji}x_i - X_J$$

where α_{Ji} are the direction cosines between the material and spatial coordinate systems with unit vectors $\mathbf{E}_J$ and $\mathbf{e}_i$, respectively. Thus

$$\mathbf{E}_J \cdot \mathbf{e}_i = \alpha_{Ji} = \alpha_{iJ}$$

and the relationship between u_i and U_J is then given by

$$u_i = \alpha_{iJ}U_J \qquad \text{or} \qquad U_J = \alpha_{Ji}u_i$$

Knowing that

$$\mathbf{e}_i = \alpha_{iJ}\mathbf{E}_J$$

then

$$\mathbf{u}(\mathbf{X},t) = u_i\mathbf{e}_i = u_i(\alpha_{iJ}\mathbf{E}_J) = U_J\mathbf{E}_J = \mathbf{U}(\mathbf{x},t)$$

It is common to superimpose the coordinate systems for the undeformed and deformed configurations, which results in $\mathbf{b} = 0$, and the direction cosines become Kronecker deltas:

$$\mathbf{E}_J \cdot \mathbf{e}_i = \delta_{Ji} = \delta_{iJ}$$

Thus, we have

$$\mathbf{u}(\mathbf{X},t) = \mathbf{x}(\mathbf{X},t) - \mathbf{X} \qquad \text{or} \qquad u_i = x_i - \delta_{iJ}X_J = x_i - X_i$$

or in terms of the spatial coordinates as

$$\mathbf{U}(\mathbf{x},t) = \mathbf{x} - \mathbf{X}(\mathbf{x},t) \qquad \text{or} \qquad U_J = \delta_{Ji}x_i - X_J = x_J - X_J$$

Displacement Gradient Tensor

The partial differentiation of the displacement vector with respect to the material coordinates yields the *material displacement gradient tensor* $\nabla_{\mathbf{x}}\mathbf{U}$. Thus we have:

$\mathbf{u}(\mathbf{X},t)$ $= \mathbf{x}(\mathbf{X},t) - \mathbf{X}\nabla_{\mathbf{x}}\mathbf{u}$ $= \nabla_{\mathbf{x}}\mathbf{x} - \mathbf{I}\nabla_{\mathbf{x}}\mathbf{u}$ $= \mathbf{F} - \mathbf{I}$	or	$u_i = x_i - \delta_{iJ}X_J$ $= x_i - X_i \dfrac{\partial u_i}{\partial X_K}$ $= \dfrac{\partial x_i}{\partial X_K} - \delta_{iK}$

where F is the *deformation gradient tensor*.

Similarly, the partial differentiation of the displacement vector with respect to the spatial coordinates yields the *spatial displacement gradient tensor* $\nabla_x U$. Thus we have,

$$
\begin{array}{c|c|c}
\begin{aligned}
(\mathbf{x},t) &= \mathbf{x} - \mathbf{X}(\mathbf{x},t)\nabla_x U \\
&= \mathbf{I} - \nabla_x \mathbf{X}\nabla_x U \\
&= \mathbf{I} - \mathbf{F}^{-1}
\end{aligned}
& \text{or} &
\begin{aligned}
U_J &= \delta_{Ji}x_i - X_J \\
&= x_J - X_J \frac{\partial U_J}{\partial x_k} \\
&= \delta_{Jk} - \frac{\partial X_J}{\partial x_k}
\end{aligned}
\end{array}
$$

Examples of Deformations

Homogeneous (or affine) deformations are useful in elucidating the behavior of materials. Some homogeneous deformations of interest are

- uniform extension
- pure dilation
- simple shear
- pure shear

Plane deformations are also of interest, particularly in the experimental context.

Plane Deformation

A plane deformation, also called *plane strain*, is one where the deformation is restricted to one of the planes in the reference configuration. If the deformation is restricted to the plane described by the basis vectors e_1, e_2, the deformation gradient has the form

$$
\mathbf{F} = F_{11}\mathbf{e}_1 \otimes \mathbf{e}_1 + F_{12}\mathbf{e}_1 \otimes \mathbf{e}_2 + F_{21}\mathbf{e}_2 \otimes \mathbf{e}_1 + F_{22}\mathbf{e}_2 \otimes \mathbf{e}_2 + \mathbf{e}_3 \otimes \mathbf{e}_3
$$

In matrix form,

$$
\mathbf{F} = \begin{bmatrix} F_{11} & F_{12} & 0 \\ F_{21} & F_{22} & 0 \\ 0 & 0 & 1 \end{bmatrix}
$$

From the polar decomposition theorem, the deformation gradient, up to a change of coordinates, can be decomposed into a stretch and a rotation. Since all the deformation is in a plane, we can write

$$\mathbf{F} = \mathbf{R} \cdot \mathbf{U} = \begin{bmatrix} \cos\theta & \sin\theta & 0 \\ -\sin\theta & \cos\theta & 0 \\ 0 & 0 & 1 \end{bmatrix} \begin{bmatrix} \lambda_1 & 0 & 0 \\ 0 & \lambda_2 & 0 \\ 0 & 0 & 1 \end{bmatrix}$$

where θ is the angle of rotation and λ_1, λ_2 are the principal stretches.

Isochoric Plane Deformation

If the deformation is isochoric (volume preserving) then $\det(\mathbf{F}) = 1$ and we have

$$F_{11}F_{22} - F_{12}F_{21} = 1$$

Alternatively,

$$\lambda_1 \lambda_2 = 1$$

Simple Shear

A simple shear deformation is defined as an isochoric plane deformation in which there is a set of line elements with a given reference orientation that do not change length and orientation during the deformation.

If $\mathbf{e}_1$ is the fixed reference orientation in which line elements do not deform during the deformation then $\lambda_1 = 1$ and $\mathbf{F} \cdot \mathbf{e}_1 = \mathbf{e}_1$. Therefore,

$$F_{11}\mathbf{e}_1 + F_{21}\mathbf{e}_2 = \mathbf{e}_1 \quad \Rightarrow \quad F_{11} = 1 \; ; \; F_{21} = 0$$

Since the deformation is isochoric,

$$F_{11}F_{22} - F_{12}F_{21} = 1 \quad \Rightarrow \quad F_{22} = 1$$

Define

$$\gamma := F_{12}$$

Then, the deformation gradient in simple shear can be expressed as

$$\mathbf{F} = \begin{bmatrix} 1 & \gamma & 0 \\ 0 & 1 & 0 \\ 0 & 0 & 1 \end{bmatrix}$$

Now,

$$\mathbf{F} \cdot \mathbf{e}_2 = F_{12}\mathbf{e}_1 + F_{22}\mathbf{e}_2 = \gamma \mathbf{e}_1 + \mathbf{e}_2 \quad \Rightarrow \quad \mathbf{F} \cdot (\mathbf{e}_2 \otimes \mathbf{e}_2) = \gamma \mathbf{e}_1 \otimes \mathbf{e}_2 + \mathbf{e}_2 \otimes \mathbf{e}_2$$

Since

$$\mathbf{e}_i \otimes \mathbf{e}_i = 1$$

we can also write the deformation gradient as

$$\mathbf{F} = 1 + \gamma \mathbf{e}_1 \otimes \mathbf{e}_2$$

Cauchy Stress Tensor

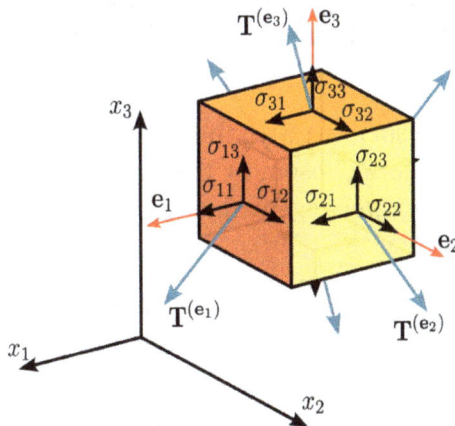

Figure 2.3 Components of stress in three dimensions

In continuum mechanics, the Cauchy stress tensor σ, true stress tensor, or simply called the stress tensor is a second order tensor named after Augustin-Louis Cauchy. The tensor consists of nine components σ_{ij} that completely define the state of stress at a point inside a material in the deformed state, placement, or configuration. The tensor relates a unit-length direction vector n to the stress vector $T^{(n)}$ across an imaginary surface perpendicular to n:

$$T^{(n)} = n\,\sigma \text{ or } T_j^{(n)} = \sigma_{ij} n_i$$

where,

$$\acute{\sigma} = \begin{bmatrix} \sigma_{11} & \sigma_{12} & \sigma_{13} \\ \sigma_{21} & \sigma_{22} & \sigma_{23} \\ \sigma_{31} & \sigma_{32} & \sigma_{33} \end{bmatrix} \equiv \begin{bmatrix} \sigma_{xx} & \sigma_{xy} & \sigma_{xz} \\ \sigma_{yx} & \sigma_{yy} & \sigma_{yz} \\ \sigma_{zx} & \sigma_{zy} & \sigma_{zz} \end{bmatrix} \equiv \begin{bmatrix} \sigma_x & \tau_{xy} & \tau_{xz} \\ \tau_{yx} & \sigma_y & \tau_{yz} \\ \tau_{zx} & \tau_{zy} & \sigma_z \end{bmatrix}$$

The Cauchy stress tensor obeys the tensor transformation law under a change in the system of coordinates. A graphical representation of this transformation law is the Mohr's circle for stress.

The Cauchy stress tensor is used for stress analysis of material bodies experiencing small deformations: It is a central concept in the linear theory of elasticity. For large deformations, also called finite deformations, other measures of stress are required, such as the Piola–Kirchhoff stress tensor, the Biot stress tensor, and the Kirchhoff stress tensor.

According to the principle of conservation of linear momentum, if the continuum body is in static equilibrium it can be demonstrated that the components of the Cauchy stress tensor in every material point in the body satisfy the equilibrium equations (Cauchy's equations of motion for zero acceleration). At the same time, according to the principle of conservation of angular momentum, equilibrium requires that the summation of moments with respect to an arbitrary point is zero, which leads to the conclusion that the stress tensor is symmetric, thus having only six independent stress components, instead of the original nine.

There are certain invariants associated with the stress tensor, whose values do not depend upon the coordinate system chosen, or the area element upon which the stress tensor operates. These are the three eigenvalues of the stress tensor, which are called the principal stresses.

Euler–Cauchy Stress Principle – stress Vector

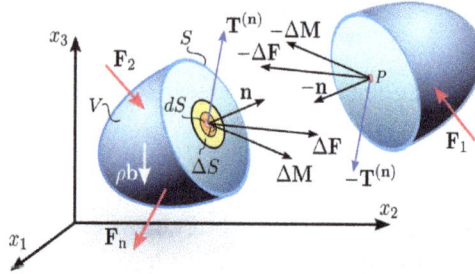

Figure 2.1a Internal distribution of contact forces and couple stresses on a differential $\mathbf{T}^{(n)}$ of the internal surface S in a continuum, as a result of the interaction between the two portions of the continuum separated by the surface

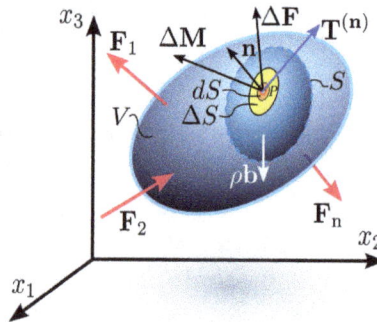

Figure 2.1b Internal distribution of contact forces and couple stresses on a differential dS of the internal surface S in a continuum, as a result of the interaction between the two portions of the continuum separated by the surface

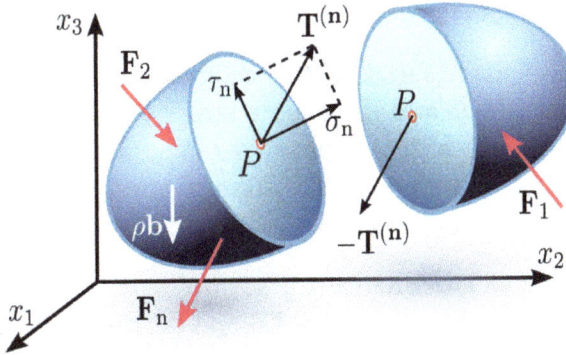

Figure 2.1c Stress vector on an internal surface S with normal vector n. Depending on the orientation of the plane under consideration, the stress vector may not necessarily be perpendicular to that plane, *i.e.* parallel to dS , and can be resolved into two components: one component normal to the plane, called *normal stress* S , and another component parallel to this plane, called the *shearing stress* τ.

The Euler–Cauchy stress principle states that *upon any surface (real or imaginary) that divides the body, the action of one part of the body on the other is equivalent (equipollent) to the system of distributed forces and couples on the surface dividing the body*, and it is represented by a field $\mathbf{T}^{(n)}$, called the stress vector, defined on the surface S and assumed to depend continuously on the surface's unit vector $\mathbf{n}$.

To formulate the Euler–Cauchy stress principle, consider an imaginary surface S passing through an internal material point P dividing the continuous body into two segments, as seen in Figure 2.1a or 2.1b (one may use either the cutting plane diagram or the diagram with the arbitrary volume inside the continuum enclosed by the surface S).

Following the classical dynamics of Newton and Euler, the motion of a material body is produced by the action of externally applied forces which are assumed to be of two kinds: surface forces $\mathbf{F}$ and body forces $\mathbf{b}$. Thus, the total force $\mathcal{F}$ applied to a body or to a portion of the body can be expressed as:

$$\mathcal{F} = \mathbf{b} + \mathbf{F}$$

Only surface forces will be discussed in this article as they are relevant to the Cauchy stress tensor.

When the body is subjected to external surface forces or *contact forces* $\mathbf{F}$, following Euler's equations of motion, internal contact forces and moments are transmitted from point to point in the body, and from one segment to the other through the dividing surface S, due to the mechanical contact of one portion of the continuum onto the other (Figure 2.1a and 2.1b). On an element of area $\ddot{A}S$ containing P, with normal vector $\mathbf{n}$, the force distribution is equipollent to a contact force $\Delta\mathbf{F}$ exerted at point P and surface moment $\Delta\mathbf{M}$. In particular, the contact force is given by

$$\Delta\mathbf{F} = \mathbf{T}^{(\mathbf{n})}\Delta S$$

where $\mathbf{T}^{(\mathbf{n})}$ is the *mean surface traction*.

Cauchy's stress principle asserts that as ΔS becomes very small and tends to zero the ratio $\Delta \mathbf{F} / \Delta S$ becomes $d\mathbf{F} / dS$ and the couple stress vector $\Delta \mathbf{M}$ vanishes. In specific fields of continuum mechanics the couple stress is assumed not to vanish; however, classical branches of continuum mechanics address non-polar materials which do not consider couple stresses and body moments.

The resultant vector $d\mathbf{F} / dS$ is defined as the *surface traction*, also called *stress vector*, *traction*, or *traction vector*. given by $\mathbf{T}^{(\mathbf{n})} = T_i^{(\mathbf{n})} \mathbf{e}_i$ at the point P associated with a plane with a normal vector $\mathbf{n}$:

$$T_i^{(\mathbf{n})} = \lim_{\Delta S \to 0} \frac{\Delta F_i}{\Delta S} = \frac{dF_i}{dS}.$$

This equation means that the stress vector depends on its location in the body and the orientation of the plane on which it is acting.

This implies that the balancing action of internal contact forces generates a *contact force density* or *Cauchy traction field* $\mathbf{T}(\mathbf{n}, \mathbf{x}, t)$ that represents a distribution of internal contact forces throughout the volume of the body in a particular configuration of the body at a given time t. It is not a vector field because it depends not only on the position $\mathbf{x}$ of a particular material point, but also on the local orientation of the surface element as defined by its normal vector $\mathbf{n}$.

Depending on the orientation of the plane under consideration, the stress vector may not necessarily be perpendicular to that plane, *i.e.* parallel to $\mathbf{n}$, and can be resolved into two components (Figure 2.1c):

- one normal to the plane, called *normal stress*

$$\sigma_n = \lim_{\Delta S \to 0} \frac{\Delta F_n}{\Delta S} = \frac{dF_n}{dS},$$

  where dF_n is the normal component of the force $d\mathbf{F}$ to the differential area dS

- and the other parallel to this plane, called the *shear stress*

$$\tau = \lim_{\Delta S \to 0} \frac{\Delta F_s}{\Delta S} = \frac{dF_s}{dS},$$

  where dF_s is the tangential component of the force $d\mathbf{F}$ to the differential surface area dS. The shear stress can be further decomposed into two mutually perpendicular vectors.

Cauchy's Postulate

According to the *Cauchy Postulate*, the stress vector $\mathbf{T}^{(\mathbf{n})}$ remains unchanged for all surfaces passing through the point P and having the same normal vector $\mathbf{n}$ at P, i.e., having a common tangent at P. This means that the stress vector is a function of the normal vector $\mathbf{n}$ only, and is not influenced by the curvature of the internal surfaces.

Cauchy's Fundamental Lemma

A consequence of Cauchy's postulate is *Cauchy's Fundamental Lemma*, also called the *Cauchy reciprocal theorem*, which states that the stress vectors acting on opposite sides of the same surface are equal in magnitude and opposite in direction. Cauchy's fundamental lemma is equivalent to Newton's third law of motion of action and reaction, and is expressed as

$$-\mathbf{T}^{(\mathbf{n})} = \mathbf{T}^{(-\mathbf{n})}.$$

Cauchy's Stress Theorem—stress Tensor

The state of stress at a point in the body is then defined by all the stress vectors $\mathbf{T}^{(\mathbf{n})}$ associated with all planes (infinite in number) that pass through that point. However, according to *Cauchy's fundamental theorem*, also called *Cauchy's stress theorem*, merely by knowing the stress vectors on three mutually perpendicular planes, the stress vector on any other plane passing through that point can be found through coordinate transformation equations.

Cauchy's stress theorem states that there exists a second-order tensor field $\boldsymbol{\sigma}(\mathbf{x}, t)$, called the Cauchy stress tensor, independent of $\mathbf{n}$, such that T is a linear function of n:

$$T^{(n)} = n\,\sigma \text{ or } T_j^{(n)} = \sigma_{ij}n_i$$

This equation implies that the stress vector $\mathbf{T}^{(\mathbf{n})}$ at any point P in a continuum associated with a plane with normal unit vector $\mathbf{n}$ can be expressed as a function of the stress vectors on the planes perpendicular to the coordinate axes, *i.e.* in terms of the components σ_{ij} of the stress tensor $\boldsymbol{\sigma}$.

To prove this expression, consider a tetrahedron with three faces oriented in the coordinate planes, and with an infinitesimal area dA oriented in an arbitrary direction specified by a normal unit vector n (Figure 2.2). The tetrahedron is formed by slicing the infinitesimal element along an arbitrary plane n. The stress vector on this plane is denoted by $T^{(n)}$. The stress vectors acting on the faces of the tetrahedron are denoted as $T^{(e_1)}$, $T^{(e_2)}$, and $T^{(e_3)}$, and are by definition the components σ_{ij} of the stress tensor $\boldsymbol{\sigma}$. This tetrahedron is sometimes called the *Cauchy tetrahedron*. The equilibrium of forces, *i.e.* Euler's first law of motion (Newton's second law of motion), gives:

$$\mathbf{T}^{(\mathbf{n})}\,dA - \mathbf{T}^{(e_1)}\,dA_1 - \mathbf{T}^{(e_2)}\,dA_2 - \mathbf{T}^{(e_3)}\,dA_3 = \rho\left(\frac{h}{3}\,dA\right)\mathbf{a},$$

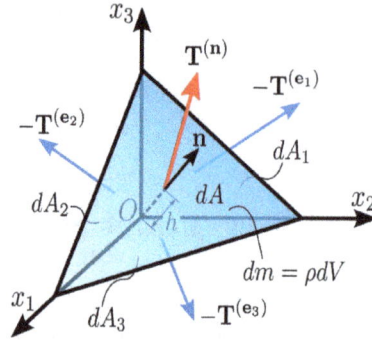

Figure 2.2. Stress vector acting on a plane with normal unit vector n.A note on the sign convention: The tetrahedron is formed by slicing a parallelepiped along an arbitrary plane **n**. So, the force acting on the plane **n** is the reaction exerted by the other half of the parallelepiped and has an opposite sign.

where the right-hand-side represents the product of the mass enclosed by the tetrahedron and its acceleration: ρ is the density, **a** is the acceleration, and h is the height of the tetrahedron, considering the plane **n** as the base. The area of the faces of the tetrahedron perpendicular to the axes can be found by projecting dA into each face (using the dot product):

$$dA_1 = \left(\mathbf{n}\cdot\mathbf{e}_1\right)dA = n_1\ dA,$$

$$dA_2 = \left(\mathbf{n}\cdot\mathbf{e}_2\right)dA = n_2\ dA,$$

$$dA_3 = \left(\mathbf{n}\cdot\mathbf{e}_3\right)dA = n_3\ dA,$$

and then substituting into the equation to cancel out dA:

$$\mathbf{T}^{(\mathbf{n})} - \mathbf{T}^{(\mathbf{e}_1)}n_1 - \mathbf{T}^{(\mathbf{e}_2)}n_2 - \mathbf{T}^{(\mathbf{e}_3)}n_3 = \rho\left(\frac{h}{3}\right)\mathbf{a}.$$

To consider the limiting case as the tetrahedron shrinks to a point, h must go to 0 (intuitively, the plane **n** is translated along n toward O). As a result, the right-hand-side of the equation approaches 0, so

$$\mathbf{T}^{(\mathbf{n})} = \mathbf{T}^{(\mathbf{e}_1)}n_1 + \mathbf{T}^{(\mathbf{e}_2)}n_2 + \mathbf{T}^{(\mathbf{e}_3)}n_3.$$

Assuming a material element (Figure 2.3) with planes perpendicular to the coordinate axes of a Cartesian coordinate system, the stress vectors associated with each of the element planes, i.e. $\mathbf{T}^{(\mathbf{e}_1)}$, $\mathbf{T}^{(\mathbf{e}_2)}$, and $\mathbf{T}^{(\mathbf{e}_3)}$ can be decomposed into a normal component and two shear components, *i.e.* components in the direction of the three coordinate axes. For the particular case of a surface with normal unit vector oriented in the direction of the x_1-axis, denote the normal stress by σ_{11}, and the two shear stresses as σ_{12} and σ_{13}:

$$\mathbf{T}^{(\mathbf{e}_1)} = T_1^{(\mathbf{e}_1)}\mathbf{e}_1 + T_2^{(\mathbf{e}_1)}\mathbf{e}_2 + T_3^{(\mathbf{e}_1)}\mathbf{e}_3 = \sigma_{11}\mathbf{e}_1 + \sigma_{12}\mathbf{e}_2 + \sigma_{13}\mathbf{e}_3,$$

$$\mathbf{T}^{(\mathbf{e}_2)} = T_1^{(\mathbf{e}_2)}\mathbf{e}_1 + T_2^{(\mathbf{e}_2)}\mathbf{e}_2 + T_3^{(\mathbf{e}_2)}\mathbf{e}_3 = \sigma_{21}\mathbf{e}_1 + \sigma_{22}\mathbf{e}_2 + \sigma_{23}\mathbf{e}_3,$$

$$\mathbf{T}^{(\mathbf{e}_3)} = T_1^{(\mathbf{e}_3)}\mathbf{e}_1 + T_2^{(\mathbf{e}_3)}\mathbf{e}_2 + T_3^{(\mathbf{e}_3)}\mathbf{e}_3 = \sigma_{31}\mathbf{e}_1 + \sigma_{32}\mathbf{e}_2 + \sigma_{33}\mathbf{e}_3,$$

In index notation this is

$$\mathbf{T}^{(\mathbf{e}_i)} = T_j^{(\mathbf{e}_i)}\mathbf{e}_j = \sigma_{ij}\mathbf{e}_j.$$

The nine components σ_{ij} of the stress vectors are the components of a second-order Cartesian tensor called the *Cauchy stress tensor*, which completely defines the state of stress at a point and is given by

$$\mathbf{\acute{o}} = \sigma_{ij} = \begin{bmatrix} \mathbf{T}^{(\mathbf{e}_1)} \\ \mathbf{T}^{(\mathbf{e}_2)} \\ \mathbf{T}^{(\mathbf{e}_3)} \end{bmatrix} = \begin{bmatrix} \sigma_{11} & \sigma_{12} & \sigma_{13} \\ \sigma_{21} & \sigma_{22} & \sigma_{23} \\ \sigma_{31} & \sigma_{32} & \sigma_{33} \end{bmatrix} \equiv \begin{bmatrix} \sigma_{xx} & \sigma_{xy} & \sigma_{xz} \\ \sigma_{yx} & \sigma_{yy} & \sigma_{yz} \\ \sigma_{zx} & \sigma_{zy} & \sigma_{zz} \end{bmatrix} \equiv \begin{bmatrix} \sigma_x & \tau_{xy} & \tau_{xz} \\ \tau_{yx} & \sigma_y & \tau_{yz} \\ \tau_{zx} & \tau_{zy} & \sigma_z \end{bmatrix},$$

where σ_{11}, σ_{22}, and σ_{33} are normal stresses, and σ_{12}, σ_{13}, σ_{21}, σ_{23}, σ_{31}, and σ_{32} are shear stresses. The first index i indicates that the stress acts on a plane normal to the x_i-axis, and the second index j denotes the direction in which the stress acts. A stress component is positive if it acts in the positive direction of the coordinate axes, and if the plane where it acts has an outward normal vector pointing in the positive coordinate direction.

Thus, using the components of the stress tensor

$$\begin{aligned} \mathbf{T}^{(\mathbf{n})} &= \mathbf{T}^{(\mathbf{e}_1)}n_1 + \mathbf{T}^{(\mathbf{e}_2)}n_2 + \mathbf{T}^{(\mathbf{e}_3)}n_3 \\ &= \sum_{i=1}^{3}\mathbf{T}^{(\mathbf{e}_i)}n_i \\ &= \left(\sigma_{ij}\mathbf{e}_j\right)n_i \\ &= \sigma_{ij}n_i\mathbf{e}_j \end{aligned}$$

or, equivalently,

$$T_j^{(\mathbf{n})} = \sigma_{ij}n_i.$$

Alternatively, in matrix form we have

$$\begin{bmatrix} T_1^{(\mathbf{n})} & T_2^{(\mathbf{n})} & T_3^{(\mathbf{n})} \end{bmatrix} = \begin{bmatrix} n_1 & n_2 & n_3 \end{bmatrix} \cdot \begin{bmatrix} \sigma_{11} & \sigma_{12} & \sigma_{13} \\ \sigma_{21} & \sigma_{22} & \sigma_{23} \\ \sigma_{31} & \sigma_{32} & \sigma_{33} \end{bmatrix}.$$

The Voigt notation representation of the Cauchy stress tensor takes advantage of the symmetry of the stress tensor to express the stress as a six-dimensional vector of the form:

$$\sigma = \begin{bmatrix} \sigma_1 & \sigma_2 & \sigma_3 & \sigma_4 & \sigma_5 & \sigma_6 \end{bmatrix}^T \equiv \begin{bmatrix} \sigma_{11} & \sigma_{22} & \sigma_{33} & \sigma_{23} & \sigma_{13} & \sigma_{12} \end{bmatrix}^T.$$

The Voigt notation is used extensively in representing stress–strain relations in solid mechanics and for computational efficiency in numerical structural mechanics software.

Transformation Rule of the Stress Tensor

It can be shown that the stress tensor is a contravariant second order tensor, which is a statement of how it transforms under a change of the coordinate system. From an x_i-system to an x_i'-system, the components σ_{ij} in the initial system are transformed into the components σ_{ij}' in the new system according to the tensor transformation rule (Figure 2.4):

$$\sigma_{ij}' = a_{im} a_{jn} \sigma_{mn} \quad \text{or} \quad \sigma' = \mathbf{A}\sigma\mathbf{A}^T,$$

where $\mathbf{A}$ is a rotation matrix with components a_{ij}. In matrix form this is

$$\begin{bmatrix} \sigma_{11}' & \sigma_{12}' & \sigma_{13}' \\ \sigma_{21}' & \sigma_{22}' & \sigma_{23}' \\ \sigma_{31}' & \sigma_{32}' & \sigma_{33}' \end{bmatrix} \begin{bmatrix} a_{11} & a_{12} & a_{13} \\ a_{21} & a_{22} & a_{23} \\ a_{31} & a_{32} & a_{33} \end{bmatrix} \begin{bmatrix} \sigma_{11} & \sigma_{12} & \sigma_{13} \\ \sigma_{21} & \sigma_{22} & \sigma_{23} \\ \sigma_{31} & \sigma_{32} & \sigma_{33} \end{bmatrix} \begin{bmatrix} a_{11} & a_{21} & a_{31} \\ a_{12} & a_{22} & a_{32} \\ a_{13} & a_{23} & a_{33} \end{bmatrix}$$

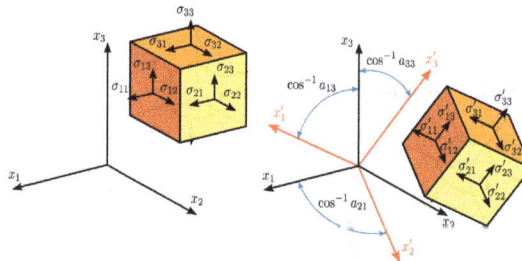

Figure 2.4 Transformation of the stress tensor

Expanding the matrix operation, and simplifying terms using the symmetry of the stress tensor, gives

$$\sigma_{11'} = a_{11}^2 \sigma_{11} + a_{12}^2 \sigma_{22} + a_{13}^2 \sigma_{33} + 2a_{11}a_{12}\sigma_{12} + 2a_{11}a_{13}\sigma_{13} + 2a_{12}a_{13}\sigma_{23},$$

$$\sigma_{22'} = a_{21}^2 \sigma_{11} + a_{22}^2 \sigma_{22} + a_{23}^2 \sigma_{33} + 2a_{21}a_{22}\sigma_{12} + 2a_{21}a_{23}\sigma_{13} + 2a_{22}a_{23}\sigma_{23},$$

$$\sigma_{33'} = a_{31}^2 \sigma_{11} + a_{32}^2 \sigma_{22} + a_{33}^2 \sigma_{33} + 2a_{31}a_{32}\sigma_{12} + 2a_{31}a_{33}\sigma_{13} + 2a_{32}a_{33}\sigma_{23},$$

$$\sigma_{12'} = a_{11}a_{21}\sigma_{11} + a_{12}a_{22}\sigma_{22} + a_{13}a_{23}\sigma_{33}$$
$$+ (a_{11}a_{22} + a_{12}a_{21})\sigma_{12} + (a_{12}a_{23} + a_{13}a_{22})\sigma_{23} + (a_{11}a_{23} + a_{13}a_{21})\sigma_{13},$$

$$\begin{aligned}
\sigma_{23'} = {} & a_{21}a_{31}\sigma_{11} + a_{22}a_{32}\sigma_{22} + a_{23}a_{33}\sigma_{33} \\
& + (a_{21}a_{32} + a_{22}a_{31})\sigma_{12} + (a_{22}a_{33} + a_{23}a_{32})\sigma_{23} + (a_{21}a_{33} + a_{23}a_{31})\sigma_{13},
\end{aligned}$$

$$\begin{aligned}
\sigma_{13'} = {} & a_{11}a_{31}\sigma_{11} + a_{12}a_{32}\sigma_{22} + a_{13}a_{33}\sigma_{33} \\
& + (a_{11}a_{32} + a_{12}a_{31})\sigma_{12} + (a_{12}a_{33} + a_{13}a_{32})\sigma_{23} + (a_{11}a_{33} + a_{13}a_{31})\sigma_{13}.
\end{aligned}$$

The Mohr circle for stress is a graphical representation of this transformation of stresses.

Normal and Shear Stresses

The magnitude of the normal stress component σ_n of any stress vector $\mathbf{T}^{(n)}$ acting on an arbitrary plane with normal unit vector $\mathbf{n}$ at a given point, in terms of the components σ_{ij} of the stress tensor $\boldsymbol{\sigma}$, is the dot product of the stress vector and the normal unit vector:

$$\begin{aligned}
\sigma_n &= \mathbf{T}^{(n)} \cdot \mathbf{n} \\
&= T_i^{(n)} n_i \\
&= \sigma_{ij} n_i n_j.
\end{aligned}$$

The magnitude of the shear stress component τ_n, acting orthogonal to the vector $\mathbf{n}$, can then be found using the Pythagorean theorem:

$$\begin{aligned}
\tau_n &= \sqrt{\left(T^{(n)}\right)^2 - \sigma_n^2} \\
&= \sqrt{T_i^{(n)} T_i^{(n)} - \sigma_n^2},
\end{aligned}$$

where

$$\left(T^{(n)}\right)^2 = T_i^{(n)} T_i^{(n)} = \left(\sigma_{ij} n_j\right)\left(\sigma_{ik} n_k\right) = \sigma_{ij}\sigma_{ik} n_j n_k.$$

Balance Laws – Cauchy's Equations of Motion

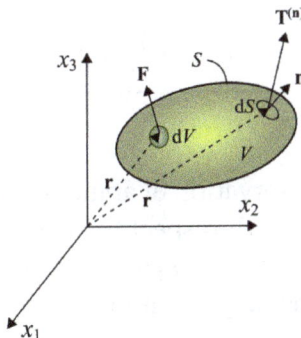

Figure 4. Continuum body in equilibrium

Cauchy's First Law of Motion

According to the principle of conservation of linear momentum, if the continuum body is in static equilibrium it can be demonstrated that the components of the Cauchy stress tensor in every material point in the body satisfy the equilibrium equations.

$$\sigma_{ji,j} + F_i = 0$$

For example, for a hydrostatic fluid in equilibrium conditions, the stress tensor takes on the form:

$$\sigma_{ij} = -p\delta_{ij},$$

where p is the hydrostatic pressure, and δ_{ij} is the kronecker delta.

Derivation of equilibrium equations

Consider a continuum body occupying a volume V, having a surface area S, with defined traction or surface forces $T_i^{(n)}$ per unit area acting on every point of the body surface, and body forces F_i per unit of volume on every point within the volume V. Thus, if the body is in equilibrium the resultant force acting on the volume is zero, thus:

$$\int_S T_i^{(n)} dS + \int_V F_i dV = 0$$

By definition the stress vector is $T_i^{(n)} = \sigma_{ji} n_j$, then

$$\int_S \sigma_{ji} n_j dS + \int_V F_i dV = 0$$

Using the Gauss's divergence theorem to convert a surface integral to a volume integral gives

$$\int_V \sigma_{ji,j} dV + \int_V F_i dV = 0$$

$$\int_V (\sigma_{ji,j} + F_i) dV = 0$$

For an arbitrary volume the integral vanishes, and we have the *equilibrium equations*

$$\sigma_{ji,j} + F_i = 0$$

Cauchy's Second Law of Motion

According to the principle of conservation of angular momentum, equilibrium requires that the summation of moments with respect to an arbitrary point is zero, which leads to the conclusion that the stress tensor is symmetric, thus having only six independent stress components, instead of the original nine:

$$\sigma_{ij} = \sigma_{ji}$$

Derivation of symmetry of the stress tensor

Summing moments about point O (Figure 4) the resultant moment is zero as the body is in equilibrium. Thus,

$$M_O = \int_S (\mathbf{r} \times \mathbf{T})dS + \int_V (\mathbf{r} \times \mathbf{F})dV = 00$$

$$= \int_S \varepsilon_{ijk} x_j T_k^{(n)} dS + \int_V \varepsilon_{ijk} x_j F_k dV$$

where $\mathbf{r}$ is the position vector and is expressed as

$$\mathbf{r} = x_j \mathbf{e}_j$$

Knowing that $T_k^{(n)} = \sigma_{mk} n_m$ and using Gauss's divergence theorem to change from a surface integral to a volume integral, we have

$$0 = \int_S \varepsilon_{ijk} x_j \sigma_{mk} n_m \, dS + \int_V \varepsilon_{ijk} x_j F_k \, dV$$

$$= \int_V (\varepsilon_{ijk} x_j \sigma_{mk})_{,m} dV + \int_V \varepsilon_{ijk} x_j F_k \, dV$$

$$= \int_V (\varepsilon_{ijk} x_{j,m} \sigma_{mk} + \varepsilon_{ijk} x_j \sigma_{mk,m}) dV + \int_V \varepsilon_{ijk} x_j F_k \, dV$$

$$= \int_V (\varepsilon_{ijk} x_{j,m} \sigma_{mk}) dV + \int_V \varepsilon_{ijk} x_j (\sigma_{mk,m} + F_k) dV$$

The second integral is zero as it contains the equilibrium equations. This leaves the first integral, where $x_{j,m} = \delta_{jm}$, therefore

$$\int_V (\varepsilon_{ijk} \sigma_{jk}) dV = 0$$

For an arbitrary volume V, we then have

$$\varepsilon_{ijk} \sigma_{jk} = 0$$

which is satisfied at every point within the body. Expanding this equation we have

$$\sigma_{12} = \sigma_{21}, \ \sigma_{23} = \sigma_{32}, \text{ and } \sigma_{13} = \sigma_{31}$$

or in general

$$\sigma_{ij} = \sigma_{ji}$$

This proves that the stress tensor is symmetric

However, in the presence of couple-stresses, i.e. moments per unit volume, the stress tensor is non-symmetric. This also is the case when the Knudsen number is close to one, $K_n \to 1$, , or the continuum is a non-Newtonian fluid, which can lead to rotationally non-invariant fluids, such as polymers.

Principal Stresses and Stress Invariants

At every point in a stressed body there are at least three planes, called *principal planes*, with normal vectors , called *principal directions*, where the corresponding stress vector is perpendicular to the plane, i.e., parallel or in the same direction as the normal vector , and where there are no normal shear stresses . The three stresses normal to these principal planes are called *principal stresses*.

The components of the stress tensor depend on the orientation of the coordinate system at the point under consideration. However, the stress tensor itself is a physical quantity and as such, it is independent of the coordinate system chosen to represent it. There are certain invariants associated with every tensor which are also independent of the coordinate system. For example, a vector is a simple tensor of rank one. In three dimensions, it has three components. The value of these components will depend on the coordinate system chosen to represent the vector, but the magnitude of the vector is a physical quantity (a scalar) and is independent of the Cartesian coordinate system chosen to represent the vector. Similarly, every second rank tensor (such as the stress and the strain tensors) has three independent invariant quantities associated with it. One set of such invariants are the principal stresses of the stress tensor, which are just the eigenvalues of the stress tensor. Their direction vectors are the principal directions or eigenvectors.

A stress vector parallel to the normal unit vector **n** is given by:

$$T^{(n)} = \lambda n = \sigma_n n$$

where λ is a constant of proportionality, and in this particular case corresponds to the magnitudes of the normal stress vectors or principal stresses.

Knowing that $T_i^{(n)} = \sigma_{ij} n_j$ and $n_i = \delta_{ij} n_j$, we have

$$T_i^{(\ )} = \lambda n_i \sigma_{ij} n_j = \lambda n_i \sigma_{ij} n_j - \lambda n_i$$
$$= \left(\sigma_{ij} - \lambda \delta_{ij} \right)_{j}$$

This is a homogeneous system, i.e. equal to zero, of three linear equations where are the unknowns. To obtain a nontrivial (non-zero) solution for , the determinant matrix of the coefficients must be equal to zero, i.e. the system is singular. Thus,

$$\left|\sigma_{ij} - \lambda\delta_{ij}\right| = \begin{vmatrix} \sigma_{11} - \lambda & \sigma_{12} & \sigma_{13} \\ \sigma_{21} & \sigma_{22} - \lambda & \sigma_{23} \\ \sigma_{31} & \sigma_{32} & \sigma_{33} - \lambda \end{vmatrix} = 0$$

Expanding the determinant leads to the *characteristic equation*

$$\left|\sigma_{ij} - \lambda\delta_{ij}\right| = -\lambda^3 + I_1\lambda^2 - I_2\lambda + I_3 = 0$$

where

$$I_1 = \sigma_{11} + \sigma_{22} + \sigma_{33}$$
$$= \sigma_{kk} = \mathrm{tr}(\sigma)$$

$$I_2 = \begin{vmatrix} \sigma_{22} & \sigma_{23} \\ \sigma_{32} & \sigma_{33} \end{vmatrix} + \begin{vmatrix} \sigma_{11} & \sigma_{13} \\ \sigma_{31} & \sigma_{33} \end{vmatrix} + \begin{vmatrix} \sigma_{11} & \sigma_{12} \\ \sigma_{21} & \sigma_{22} \end{vmatrix}$$

$$= \sigma_{11}\sigma_{22} + \sigma_{22}\sigma_{33} + \sigma_{11}\sigma_{33} - \sigma_{12}^2 - \sigma_{23}^2 - \sigma_{31}^2$$

$$= \frac{1}{2}\left(\sigma_{ii}\sigma_{jj} - \sigma_{ij}\sigma_{ji}\right)$$

$$I_3 = \det(\sigma_{ij}) = \det(\sigma)$$

$$= \sigma_{11}\sigma_{22}\sigma_{33} + 2\sigma_{12}\sigma_{23}\sigma_{31} - \sigma_{12}^2\sigma_{33} - \sigma_{23}^2\sigma_{11} - \sigma_{31}^2\sigma_{22}$$

The characteristic equation has three real roots λ_i, i.e. not imaginary due to the symmetry of the stress tensor. The $\sigma_1 = \max\left(\lambda_1, \lambda_2, \lambda_3\right)$, , $\sigma_3 = \min\left(\lambda_1, \lambda_2, \lambda_3\right)$ and $\sigma_2 = I_1 - \sigma_1 - \sigma_3$, , are the principal stresses, functions of the eigenvalues λ_i.. The eigenvalues are the roots of the characteristic polynomial. The principal stresses are unique for a given stress tensor. Therefore, from the characteristic equation, the coefficients I_1, I_2 and I_3, called the first, second, and third *stress invariants*, respectively, always have the same value regardless of the coordinate system's orientation.

For each eigenvalue, there is a non-trivial solution for n_j in the equation $\left(\sigma_{ij} - \lambda\delta_{ij}\right)n_j = 0$. These solutions are the principal directions or eigenvectors defining the plane where the principal stresses act. The principal stresses and principal directions characterize the stress at a point and are independent of the orientation.

A coordinate system with axes oriented to the principal directions implies that the normal stresses are the principal stresses and the stress tensor is represented by a diagonal matrix:

$$\sigma_{ij} = \begin{bmatrix} \sigma_1 & 0 & 0 \\ 0 & \sigma_2 & 0 \\ 0 & 0 & \sigma_3 \end{bmatrix}$$

The principal stresses can be combined to form the stress invariants, I_1, I_2, and I_3, The first and third invariant are the trace and determinant respectively, of the stress tensor. Thus,

$$I_1 = \sigma_1 + \sigma_2 + \sigma_3$$
$$I_2 = \sigma_1\sigma_2 + \sigma_2\sigma_3 + \sigma_3\sigma_1$$
$$I_3 = \sigma_1\sigma_2\sigma_3$$

Because of its simplicity, the principal coordinate system is often useful when considering the state of the elastic medium at a particular point. Principal stresses are often expressed in the following equation for evaluating stresses in the x and y directions or axial and bending stresses on a part. The principal normal stresses can then be used to calculate the von Mises stress and ultimately the safety factor and margin of safety.

$$\sigma_1, \sigma_2 = \frac{\sigma_x + \sigma_y}{2} \pm \sqrt{\left(\frac{\sigma_x - \sigma_y}{2}\right)^2 + \tau_{xy}^2}$$

Using just the part of the equation under the square root is equal to the maximum and minimum shear stress for plus and minus. This is shown as:

$$\tau_{max}, \tau_{min} = \pm\sqrt{\left(\frac{\sigma_x - \sigma_y}{2}\right)^2 + \tau_{xy}^2}$$

Maximum and Minimum Shear Stresses

The maximum shear stress or maximum principal shear stress is equal to one-half the difference between the largest and smallest principal stresses, and acts on the plane that bisects the angle between the directions of the largest and smallest principal stresses, i.e. the plane of the maximum shear stress is oriented $45°$ from the principal stress planes. The maximum shear stress is expressed as

$$\tau_{max} = \frac{1}{2}|\sigma_{max} - \sigma_{min}|$$

Assuming Assuming $\sigma_1 \geq \sigma_2 \geq \sigma_3$ then

$$\tau_{max} = \frac{1}{2}|\sigma_1 - \sigma_3|$$

When the stress tensor is non zero the normal stress component acting on the plane for the maximum shear stress is non-zero and it is equal to

$$\sigma_n = \frac{1}{2}(\sigma_1 + \sigma_3)$$

Stress Deviator Tensor

The stress tensor σ_{ij} can be expressed as the sum of two other stress tensors:

1. a *mean hydrostatic stress tensor* or *volumetric stress tensor* or *mean normal stress tensor*, $\pi\delta_{ij}$, which tends to change the volume of the stressed body; and

2. a deviatoric component called the *stress deviator tensor*, s_{ij}, which tends to distort it.

So:

$$\sigma_{ij} = s_{ij} + \pi\delta_{ij},$$

where π is the mean stress given by

$$\pi = \frac{\sigma_{kk}}{3} = \frac{\sigma_{11} + \sigma_{22} + \sigma_{33}}{3} = \tfrac{1}{3}I_1.$$

Pressure (p) is generally defined as negative one-third the trace of the stress tensor minus any stress the divergence of the velocity contributes with, i.e.

$$p = \lambda\nabla\cdot\vec{u} - \pi = \lambda\frac{\partial u_k}{\partial x_k} - \pi = \sum_k \lambda\frac{\partial u_k}{\partial x_k} - \pi,$$

where is a proportionality constant, is the divergence operator, is the k:th Cartesian coordinate, is the velocity and is the k:th Cartesian component of .

The deviatoric stress tensor can be obtained by subtracting the hydrostatic stress tensor from the Cauchy stress tensor:

$$s_{ij} = \sigma_{ij} - \frac{\sigma_{kk}}{3}\delta_{ij},$$

$$
\begin{bmatrix} s_{11} & s_{12} & s_{13} \\ s_{21} & s_{22} & s_{23} \\ s_{31} & s_{32} & s_{33} \end{bmatrix}
=
\begin{bmatrix} \sigma_{11} & \sigma_{12} & \sigma_{13} \\ \sigma_{21} & \sigma_{22} & \sigma_{23} \\ \sigma_{31} & \sigma_{32} & \sigma_{33} \end{bmatrix}
-
\begin{bmatrix} \pi & 0 & 0 \\ 0 & \pi & 0 \\ 0 & 0 & \pi \end{bmatrix}
$$

$$
=
\begin{bmatrix} \sigma_{11} - \pi & \sigma_{12} & \sigma_{13} \\ \sigma_{21} & \sigma_{22} - \pi & \sigma_{23} \\ \sigma_{31} & \sigma_{32} & \sigma_{33} - \pi \end{bmatrix}.
$$

Invariants of the Stress Deviator Tensor

As it is a second order tensor, the stress deviator tensor also has a set of invariants, which can be obtained using the same procedure used to calculate the invariants of the stress tensor. It can be shown that the principal directions of the stress deviator tensor s_{ij} are the same as the principal directions of the stress tensor σ_{ij}. Thus, the characteristic equation is

$$\left| s_{ij} - \lambda \delta_{ij} \right| = -\lambda^3 + J_1 \lambda^2 - J_2 \lambda + J_3 = 0,$$

where , and are the first, second, and third *deviatoric stress invariants*, respectively. Their values are the same (invariant) regardless of the orientation of the coordinate system chosen. These deviatoric stress invariants can be expressed as a function of the components of or its principal values and or alternatively, as a function of or its principal values and Thus,

$$J_1 = s_{kk} = 0,$$

$$J_2 = \frac{1}{2} s_{ij} s_{ji} = \tfrac{1}{2} \operatorname{tr}(\mathbf{s}^2)$$

$$= \tfrac{1}{2} (s_1^2 + s_2^2 + s_3^2)$$

$$= \tfrac{1}{6} \left[(\sigma_{11} - \sigma_{22})^2 + (\sigma_{22} - \sigma_{33})^2 + (\sigma_{33} - \sigma_{11})^2 \right] + \sigma_{12}^2 + \sigma_{23}^2 + \sigma_{31}^2$$

$$= \tfrac{1}{6} \left[(\sigma_1 - \sigma_2)^2 + (\sigma_2 - \sigma_3)^2 + (\sigma_3 - \sigma_1)^2 \right] = \tfrac{1}{3} I_1^2 - I_2 = \frac{1}{2} \left[\operatorname{tr}(\mathbf{\acute{\sigma}}^2) - \frac{1}{3} \operatorname{tr}(\mathbf{\acute{\sigma}})^2 \right],$$

$$J_3 = \det(s_{ij})$$

$$= \tfrac{1}{3} s_{ij} s_{jk} s_{ki} = \tfrac{1}{3} \operatorname{tr}(\mathbf{s}^3)$$

$$= s_1 s_2 s_3$$

$$= \tfrac{2}{27} I_1^3 - \tfrac{1}{3} I_1 I_2 + I_3 = \tfrac{1}{3} \left[\operatorname{tr}(\mathbf{\acute{\sigma}}^3) - \operatorname{tr}(\mathbf{\acute{\sigma}}^2) \operatorname{tr}(\mathbf{\acute{\sigma}}) + \tfrac{2}{9} \operatorname{tr}(\mathbf{\acute{\sigma}})^3 \right]$$

Because $s_{kk} = 0$, , the stress deviator tensor is in a state of pure shear.

A quantity called the equivalent stress or von Mises stress is commonly used in solid mechanics. The equivalent stress is defined as

$$\sigma_e = \sqrt{3 J_2} = \sqrt{\tfrac{1}{2} \left[(\sigma_1 - \sigma_2)^2 + (\sigma_2 - \sigma_3)^2 + (\sigma_3 - \sigma_1)^2 \right]}.$$

Octahedral Stresses

Considering the principal directions as the coordinate axes, a plane whose normal vector makes equal angles with each of the principal axes (i.e. having direction cosines equal to) is called an *octahedral plane*. There are a total of eight octahedral planes (Figure 6). The normal and shear components of the stress tensor on these planes are called *octahedral normal stress* and *octahedral shear stress* , respectively.

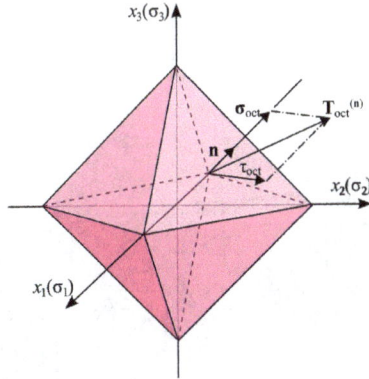

Figure 6. Octahedral stress planes

Knowing that the stress tensor of point O (Figure 6) in the principal axes is

$$\sigma_{ij} = \begin{bmatrix} \sigma_1 & 0 & 0 \\ 0 & \sigma_2 & 0 \\ 0 & 0 & \sigma_3 \end{bmatrix}$$

the stress vector on an octahedral plane is then given by:

$$\mathbf{T}_{oct}^{(\mathbf{n})} = \sigma_{ij} n_i \mathbf{e}_j$$
$$= \sigma_1 n_1 \mathbf{e}_1 + \sigma_2 n_2 \mathbf{e}_2 + \sigma_3 n_3 \mathbf{e}_3$$
$$= \tfrac{1}{\sqrt{3}} (\sigma_1 \mathbf{e}_1 + \sigma_2 \mathbf{e}_2 + \sigma_3 \mathbf{e}_3)$$

The normal component of the stress vector at point O associated with the octahedral plane is

$$\sigma_{oct} = T_i^{(n)} n_i = \sigma_{ij} n_i n_j$$
$$= \sigma_1 n_1 n_1 + \sigma_2 n_2 n_2 + \sigma_3 n_3 n_3$$
$$= \tfrac{1}{3}(\sigma_1 + \sigma_2 + \sigma_3) = \tfrac{1}{3} I_1$$

which is the mean normal stress or hydrostatic stress. This value is the same in all eight octahedral planes. The shear stress on the octahedral plane is then

$$\tau_{oct} = \sqrt{T_i^{(n)} T_i^{(n)} - \sigma_n^2}$$
$$= \left[\tfrac{1}{3}(\sigma_1^2 + \sigma_2^2 + \sigma_3^2) - \tfrac{1}{9}(\sigma_1 + \sigma_2 + \sigma_3)^2 \right]^{1/2}$$
$$= \tfrac{1}{3}\left[(\sigma_1 - \sigma_2)^2 + (\sigma_2 - \sigma_3)^2 + (\sigma_3 - \sigma_1)^2 \right]^{1/2}$$
$$= \tfrac{1}{3}\sqrt{2 I_1^2 - 6 I_2} = \sqrt{\tfrac{2}{3} J_2}$$

Residual Stress

Residual stress in a roll formed HSS tubing visible during band-saw slitting

Residual stresses are stresses that remain in a solid material after the original cause of the stresses has been removed. Residual stress may be desirable or undesirable. For example, laser peening imparts deep beneficial compressive residual stresses into metal components such as turbine engine fan blades, and it is used in toughened glass to allow for large, thin, crack- and scratch-resistant glass displays on smartphones. However, unintended residual stress in a designed structure may cause it to fail prematurely.

Residual stresses can occur through a variety of mechanisms including inelastic (plastic) deformations, temperature gradients (during thermal cycle) or structural changes (phase transformation). Heat from welding may cause localized expansion, which is taken up during welding by either the molten metal or the placement of parts being welded. When the finished weldment cools, some areas cool and contract more than others, leaving residual stresses. Another example occurs during semiconductor fabrication and microsystem fabrication when thin film materials with different thermal and crystalline properties are deposited sequentially under different process conditions. The stress variation through a stack of thin film materials can be very complex and can vary between compressive and tensile stresses from layer to layer.

Applications

While uncontrolled residual stresses are undesirable, some designs rely on them. In particular, brittle materials can be toughened by including compressive residual stress, as in the case for toughened glass and pre-stressed concrete. The predominant mechanism for failure in brittle materials is brittle fracture, which begins with initial crack formation. When an external tensile stress is applied to the material, the crack tips concentrate stress, increasing the local tensile stresses experienced at the crack tips to a greater extent than the average stress on the bulk material. This causes the initial crack to enlarge quickly (propagate) as the surrounding material is overwhelmed by the stress concentration, leading to fracture.

A material having compressive residual stress helps to prevent brittle fracture because the initial crack is formed under compressive (negative tensile) stress. To cause brittle fracture by crack propagation of the initial crack, the external tensile stress must overcome the compressive residual stress before the crack tips experience sufficient tensile stress to propagate.

The manufacture of some swords utilises a gradient in martensite formation to produce particularly hard edges (notably the katana). The difference in residual stress between the harder cutting edge and the softer back of the sword gives such swords their characteristic curve.

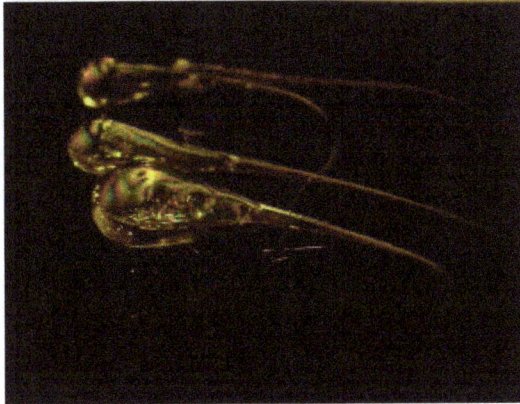

Prince Rupert's Drops

In toughened glass, compressive stresses are induced on the surface of the glass, balanced by tensile stresses in the body of the glass. Due to the residual compressive stress on the surface, toughened glass is more resistant to cracks, but shatter into small shards when the outer surface is broken. A demonstration of the effect is shown by Prince Rupert's Drop, a material-science novelty in which a molten glass globule is quenched in water: Because the outer surface cools and solidifies first, when the volume cools and solidifies, it "wants" to take up a smaller volume than the outer "skin" has already defined; this puts much of the volume in tension, pulling the "skin" in, putting the "skin" in compression. As a result, the solid globule is extremely tough, able to be hit with a hammer, but if its long tail is broken, the balance of forces is upset, causing the entire piece to shatter violently.

In certain types of gun barrels made with two tubes forced together, the inner tube is compressed while the outer tube stretches, preventing cracks from opening in the rifling when the gun is fired.

Premature Failure

Castings may also have large residual stresses due to uneven cooling. Residual stress is often a cause of premature failure of critical components, and was probably a factor in the collapse of the Silver Bridge in West Virginia, United States in December 1967.

The eyebar links were castings which showed high levels of residual stress, which in one eyebar, encouraged crack growth. When the crack reached a critical size, it grew catastrophically, and from that moment, the whole structure started to fail in a chain reaction. Because the structure failed in less than a minute, 46 drivers and passengers in cars on the bridge at the time were killed as the suspended roadway fell into the river below.

The collapsed Silver Bridge, as seen from the Ohio side

Compressive Residual Stress

Common methods to induce compressive residual stress are shot peening for surfaces and High frequency impact treatment for weld toes. Depth of compressive residual stress varies depending of the method. Both methods can increase lifetime of contructions significiantly.

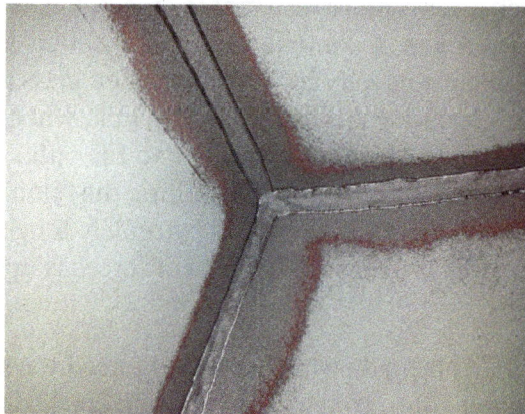

Example of a HiFIT treated assembly

Creation of Residual Stress

There are some techniques which are used to create uniform residual stress in a beam. For example, the four point bend allows inserting residual stress by applying a load on a beam using two cylinders.

Measurement Techniques

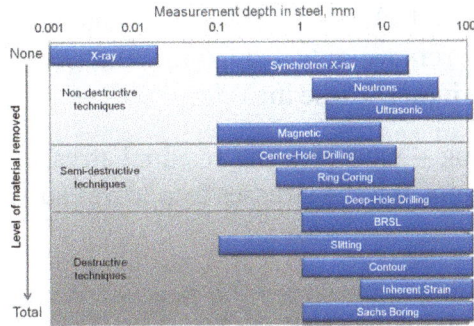

A diagram comparing residual stress measurement techniques, showing the measurement length scale, penetration, and level of destruction to the measured component.

Overview

There are many techniques used to measure residual stresses, which are broadly categorised into destructive, semi-destructive and non-destructive techniques. The selection of the technique depends on the information required and the nature of the measurement specimen. Factors include the depth/penetration of the measurement (surface or through-thickness), the length scale to be measured over (macroscopic, mesoscopic or microscopic), the resolution of the information required, and also the composition geometry and location of the specimen. Additionally, some of the techniques need to be performed in specialised laboratory facilities, meaning that "on-site" measurements are not possible for all of the techniques.

Destructive Techniques

The destructive techniques is such that they result in a large and irreparable structural change to the specimen, meaning that either the specimen cannot not returned to service or a mock-up or spare must be used. These techniques function using a "strain release" principle; cutting the measurement specimen to relax the residual stresses and then measuring the deformed shape. As these deformations are usually elastic, there is an exploitable linear relationship between the magnitude of the deformation and magnitude of the released residual stress. Destructive techniques include:

- Contour Method – measures the residual stress on a 2D plane section through a specimen, in a uniaxial direction normal to a surface cut through the specimen with wire EDM.

- Slitting (Crack Compliance) – measures residual stress through the thickness of a specimen, at a normal to a cut "slit".

- Block Removal/Splitting/Layering

- Sach's Boring

Semi-destructive Techniques

Similarly to the destructive techniques, these also function using the "strain release" principle. However, they remove only a small amount of material, leaving the overall integrity of the structure intact. These include:

- Deep Hole Drilling – measures the residual stresses through the thickness of a component by relaxing the stresses in a "core" surrounding a small diameter drilled hole.

- Centre Hole Drilling – measures the near surface residual stresses by strain release corresponding to a small shallow drilled hole with a strain gauge rosette.

- Ring Core – similar to Centre Hole Drilling, but with greater penetration, and with the cutting taking place around the strain gauge rosette rather than through its centre.

Non-destructive Techniques

The non-destructive techniques measure the effects of relationships between the residual stresses and their action of crystallographic properties of the measured material. Some of these work by measuring the diffraction of high frequency electromagnetic radiation through the atomic lattice spacing (which has been deformed due to the stress) relative to a stress-free sample. The Ultrasonic and Magnetic techniques exploit the acoustic and ferromagnetic properties of materials to perform relative measurements of residual stress. Non-destructive techniques include:

- Neutron Diffraction

- Synchrotron Diffraction

- X-Ray Diffraction

- Ultrasonic

- Magnetic

Relief of Residual Stress

When undesired residual stress is present from prior metalworking operations, the amount of residual stress may be reduced using several methods. These methods may be classified into thermal and mechanical (or nonthermal) methods. All the methods involve processing the part to be stress relieved as a whole.

Thermal Method

The thermal method involves changing the temperature of the entire part uniformly, either through heating or cooling. When parts are heated for stress relief, the process may also be known as stress relief bake. Cooling parts for stress relief is known as cryogenic stress relief and is relatively uncommon.

Stress Relief Bake

Most metals, when heated, experience a reduction in yield strength. If the material's yield strength is sufficiently lowered by heating, locations within the material that experienced residual stresses greater than the yield strength (in the heated state) would yield or deform. This leaves the material with residual stresses that are at most as high as the yield strength of the material in its heated state.

Stress relief bake should not be confused with annealing or tempering, which are heat treatments to increase ductility of a metal. Although those processes also involve heating the material to high temperatures and reduce residual stresses, they also involve a change in metallurgical properties, which may be undesired.

For certain materials such as low alloy steel, care must be taken during stress relief bake so as not to exceed the temperature at which the material achieves maximum hardness.

Cryogenic Stress Relief

Cryogenic stress relief involves placing the material (usually steel) into a cryogenic environment such as liquid nitrogen. In this process, the material to be stress relieved will be cooled to a cryogenic temperature for a long period, then slowly brought back to room temperature.

Nonthermal Methods

Mechanical methods to relieve undesirable surface tensile stresses and replace them with beneficial compressive residual stresses include shot peening and laser peening. Each works the surface of the material with a media: shot peening typically uses a metal or glass material; laser peening uses high intensity beams of light to induce a shock wave that propagates deep into the material.

References

- Gordon, J.E. (2003). Structures, or, Why things don't fall down (2. Da Capo Press ed.). Cambridge, MA: Da Capo Press. ISBN 0306812835.

- Wai-Fah Chen and Da-Jian Han (2007), "Plasticity for Structural Engineers". J. Ross Publishing ISBN 1-932159-75-4

- Peter Chadwick (1999), "Continuum Mechanics: Concise Theory and Problems". Dover Publications, series "Books on Physics". ISBN 0-486-40180-4. pages

- Ronald L. Huston and Harold Josephs (2009), "Practical Stress Analysis in Engineering Design".

3rd edition, CRC Press, 634 pages. ISBN 9781574447132

- Donald Ray Smith and Clifford Truesdell (1993) "An Introduction to Continuum Mechanics after Truesdell and Noll". Springer. ISBN 0-7923-2454-4

- Rees, David (2006). Basic Engineering Plasticity: An Introduction with Engineering and Manufacturing Applications. Butterworth-Heinemann. ISBN 0-7506-8025-3.

- Rees, David (2006). Basic Engineering Plasticity: An Introduction with Engineering and Manufacturing Applications. Butterworth-Heinemann. p. 41. ISBN 0-7506-8025-3.

- Peter Chadwick (1999), "Continuum Mechanics: Concise Theory and Problems". Dover Publications, series "Books on Physics". ISBN 0-486-40180-4. pages

- Yuan-cheng Fung and Pin Tong (2001) "Classical and Computational Solid Mechanics". World Scientific. ISBN 981-02-4124-0

- G. Thomas Mase and George E. Mase (1999), "Continuum Mechanics for Engineers" (2nd edition). CRC Press. ISBN 0-8493-1855-6

- Teodor M. Atanackovic and Ardéshir Guran (2000), "Theory of Elasticity for Scientists and Engineers". Springer. ISBN 0-8176-4072-X

- Keith D. Hjelmstad (2005), "Fundamentals of Structural Mechanics" (2nd edition). Prentice-Hall. ISBN 0-387-23330-X

- Wai-Fah Chen and Da-Jian Han (2007), "Plasticity for Structural Engineers". J. Ross Publishing ISBN 1-932159-75-4

- Rabindranath Chatterjee (1999), "Mathematical Theory of Continuum Mechanics". Alpha Science. ISBN 81-7319-244-8

- John Conrad Jaeger, N. G. W. Cook, and R. W. Zimmerman (2007), "Fundamentals of Rock Mechanics" (4th edition). Wiley-Blackwell. ISBN 0-632-05759-9

- Mohammed Ameen (2005), "Computational Elasticity: Theory of Elasticity and Finite and Boundary Element Methods" (book). Alpha Science, ISBN 1-84265-201-X

Permissions

We would like to thank the editorial team for lending their expertise to make the book truly unique. They have played a crucial role in the development of this book. Without their invaluable contributions this book wouldn't have been possible. They have made vital efforts to compile up to date information on the varied aspects of this subject to make this book a valuable addition to the collection of many professionals and students.

This book was conceptualized with the vision of imparting up-to-date and integrated information in this field. To ensure the same, a matchless editorial board was set up. Every individual on the board went through rigorous rounds of assessment to prove their worth. After which they invested a large part of their time researching and compiling the most relevant data for our readers.

The editorial board has been involved in producing this book since its inception. They have spent rigorous hours researching and exploring the diverse topics which have resulted in the successful publishing of this book. They have passed on their knowledge of decades through this book. To expedite this challenging task, the publisher supported the team at every step. A small team of assistant editors was also appointed to further simplify the editing procedure and attain best results for the readers.

Apart from the editorial board, the designing team has also invested a significant amount of their time in understanding the subject and creating the most relevant covers. They scrutinized every image to scout for the most suitable representation of the subject and create an appropriate cover for the book.

The publishing team has been an ardent support to the editorial, designing and production team. Their endless efforts to recruit the best for this project, has resulted in the accomplishment of this book. They are a veteran in the field of academics and their pool of knowledge is as vast as their experience in printing. Their expertise and guidance has proved useful at every step. Their uncompromising quality standards have made this book an exceptional effort. Their encouragement from time to time has been an inspiration for everyone.

The publisher and the editorial board hope that this book will prove to be a valuable piece of knowledge for students, practitioners and scholars across the globe.

Index

www.ingramcontent.com/pod-product-compliance
Lightning Source LLC
Chambersburg PA
CBHW061929190326
41458CB00009B/2701